S0-CAH-402

Schreier / Bernreuther / Huffer
Analysis of Chiral Organic Molecules

Peter Schreier
Alexander Bernreuther
Manfred Huffer

Analysis of Chiral Organic Molecules

Methodology and Applications

Walter de Gruyter · Berlin · New York 1995

Professor Dr. Peter Schreier
Dr. Alexander Bernreuther
Dr. Manfred Huffer
Institut für Pharmazie und Lebensmittelchemie
Universität Würzburg
Am Hubland
97074 Würzburg

The book contains 72 figures and formulas and 42 tables.

♾ Printed on acid-free paper which falls within the guidelines of the ANSI to ensure permanence and durability.

Library of Congress Cataloging-in-Publication Data

Schreier, Peter, 1942–
Analysis of chiral organic molecules : methodology and applications / Peter Schreier, Alexander Bernreuther, Manfred Huffer.
p. cm.
Includes bibliographical references (p. –) and index.
ISBN 3-11-013659-7
1. Optical isomers–Analysis. 2. Chirality. I. Bernreuther, Alexander, 1961– . II. Huffer, Manfred, 1960– . III. Title.
QD471.S38 1996
547′.3–dc20 95-32104
CIP

Die Deutsche Bibliothek – Cataloging-in-Publication Data

Schreier, Peter:
Analysis of chiral organic molecules : methodology and applications / Peter Schreier ; Alexander Bernreuther ; Manfred Huffer. – Berlin ; New York : de Gruyter, 1995
ISBN 3-11-013659-7
NE: Bernreuther, Alexander:; Huffer, Manfred:

Printed in Germany.
Printing: Gerike GmbH, Berlin. – Binding: Lüderitz & Bauer GmbH, Berlin.
Cover Design: Hansbernd Lindemann, Berlin.

PREFACE

In the course of many years' work in the applied research field of the analysis of volatile aroma compounds and their non-volatile precursors, we were (and still are) continuously confronted with the problem of selecting the most appropriate method for the analysis of chiral molecules. In all areas of 'chiral analysis', i.e. from classical optical rotation to the modern chromatographic and electrophoretic methods, excellent comprehensive reviews and monographs can be found and for applications in liquid chromatography and gas chromatography even databases are available. However, a concise introduction to and guide through this rapidly developing field covering all facets of methodologies in the analysis of chiral organic molecules is lacking to date. This book is an attempt to fill this gap, its primary objective being to introduce the practical considerations involved in 'chiral analysis', including chiroptical methods (polarimetry, optical rotation dispersion, circular dichroism), nuclear magnetic resonance, chromatographic (liquid chromatography, gas chromatography, supercritical fluid chromatography, planar chromatography, counter-current chromatography) and electrophoretic techniques. Some knowledge of theory is essential to attain this goal, but neither a comprehensive nor a rigorous treatment of the theories is presented here. In order to extend the utility of this book to the largest possible number of users, we have stressed simplicity, particularly in explanations. We sincerely hope that this book succeeds in facilitating the approach to 'chiral analysis' and enabling analysts to pinpoint the most appropriate analytical methods quickly and easily.

We are grateful to Dr. M. Herderich, Dr. H. U. Humpf, and Dr. W. Schwab for their helpful discussions und contributions. Our particular thanks are due to M. Kleinschnitz for his intensive work in preparing the 'camera-ready' manuscript. Finally, the kind support provided by the publisher is gratefully acknowledged.

Würzburg
April 1995

P. Schreier
A. Bernreuther
M. Huffer

CONTENTS

List of Abbreviations

α	Relative retention
A	Absorbance
Å	Symbol for Angström unit
ACC	N-Acetylcysteine
ACE	Affinity capillary electrophoresis
AcOH	Acetic acid
AEC	Affinity electrokinetic chromatography
AGE	Affinity gel chromatography
AGP	α_1-Acid glycoprotein
AHNS	4-Amino-5-hydroxy-2,7-naphtalene disulphonate
ala	Alanine
all	Allose
ANA	5-Amino-2-naphtalene sulphonate
AP	2-Aminopyridine
ara	Arabinose
arg	Arginine
asn	Asparagine
asp	Aspartic acid
ASTM	American society for testing and materials
BE	Butyl ester
BGE	Background electrolyte
BOC	tert-Butyloxycarbonyl
BSA	Bovine serum albumine
CAE	Capillary affinity electrophoresis
CAGE	Capillary affinity gel electrophoresis
CAZE	Capillary affinity zone electrophoresis
CCC	Countercurrent chromatography
CCGLC	Continuous countercurrent gas liquid chromatography
CCI	Chiral counter-ion
CD	Circular dichroism
	Cyclodextrin
CDA	Chiral derivatizing agent
CE	Capillary electrophoresis
CEC	Capillary electrochromatography
CES	Capillary electroseparation
CGE	Capillary gel electrophoresis
CGS	Centimeter, gram, second-system

chromat.	chromatographic; chromatography
CIEF	Capillary isoelectric focusing
CITP	Capillary isotachophoresis
CLC	Centrifugal layer chromatography
CLEC	Chiral liquid exchange chromatography
CLSR	Chiral lanthanide shift reagent
CM	Carboxy methyl
CMA	Chiral mobile phase additive
CME	Carboxymethylethyl
CMEC	Capillary micellar electrokinetik chromatography
cov.	covalent
CSA	Chiral solvating agent
CSP	Chiral stationary phase
CSR	Chiral shift reagent
CT	Charge transfer
CTA	Cellulose triacetate
cys	Cysteine
CZE	Capillary zone electrophoresis
d	day
Da	Dalton
Dansyl	5-Dimethylamino-1-naphthalenesulphonyl
DBT	Dibutyl tartrate
DC	Direct current
DCC	Dicyclohexyl carbodiimide
DCC(C)	Droplet counter-current chromatography
DE	Displacement electrophoresis
der	derivative
DIOP	2,3-Isopropylidene-2,3-dihydroxy-1,4-bis(diphenylphosphino)butane
DIPTA	Diisopropyl tartraric diamide
diss	dissolved in
DMA	Dimethylamino
DMAP	4-(Dimethylamino)pyridine
DNBC	3,5-Dinitrobenzoyl chloride
DNA	3,5-Dinitroaniline
DNB	3,5-Dinitrobenzoyl
DNP	Dinitrophenyl
DNS	Dansylated, cf. Dansyl
DOPA	3,4-Dihydroxyphenylalanine
EC	Electrochromatography
ECC	Electrokinetik capillary chromatography
ee	Enantiomeric excess
	$\%\ ee = (R - S)/(R + S) \cdot 100$; for $R > S$
EEOQ	N-Ethoxycarbonyl-2-ethoxy-1,2-dihydroquinoline
EKC	Electrokineticchromatography
ELISA	Enzyme linked immunosorbent assay

$Eu(dcm)_3$	Tris-(d,d-dicampholylmethanato)-europium(III)
$Eu(fod)_3$	[Tris-(6,6,7,78,8,8-heptafluoro-2,2-dimethyl-3,5-octanedionato)-europium]
$Eu(hfc)_3$	Tris-[3-(heptafluoropropyl-hydroxymethylene)-d-camphorato]-europium(III)
$Eu(pvc)_3$	Tris-(3-tertbutyl-hydroxymethylene-1*R*-camphorato)-europium(III)
$Eu(tfc)_3$	Tris-[3-(trifluoromethyl-hydroxymethylene)-d-camphorato]-europium(III)
f	Femto (= 10^{-15})
FFPLC	Forced-flow planar liquid chromatography
FFTLC	Forced-flow thin-layer chromatography
FID	Flame ionization detector
FMOC	9-Fluorenylmethoxycarbonyl
frc	Fructose
FS	Fused silica
FSCE	Free solution capillary electrophoresis
FTIR	Fourier transform infrared spectroscopy
fuc	Fucose
FZCE	Free zone capillary electrophoresis
FZE	Free zone electrophoresis
gal	Galactose
GC	Gas chromatography
GC-CIMS	Gas chromatography-chemical ionization mass spectrometry
GC-MS	Gas chromatography-mass spectrometry
GE	Gel electrophoresis
GITC	2,3,4,6-Tetra-O-acetyl-ß-glucopyranosyl isothiocyanate
GLC	Gas liquid chromatography
glc	Glucose
gln	Glutamine
glu	Glutamic acid
gly	Glycine
GPC	Gel permeation chromatography
GSC	Gas solid chromatography
GZE	Gel zone electrophoresis
h	Hour
H	Length of a column equivalent to one theoretical plate: h = L/n (cf. HETP)
HEC	Hydroxyethylcellulose
HETP	Height equivalent to a theoretical plate
HFB	Heptafluorobutanoyl
his	Histidine
HPCE	High performance capillary electrophoresis
HPCGE	High performance capillary gel electrophoresis
HPCZE	High performance capillary zone electrophoresis

HPE	High performance electrophoresis
HPLC	High performance (pressure) liquid chromatography
HPPE	High performance paper electrophoresis
HPPLC	High performance planar liquid chromatography
HPTLC	High performance thin-layer chromatography
HPZE	High performance zone electrophoresis
HRGC	High resolution (capillary) gas chromatography
HRP	Horseradish peroxidase
HTAB	Hexadecyltrimethylammonium bromide
HVE	High voltage electrophoresis
HVPE	High voltage paper electrophoresis
Hz	Hertz
ICDNA	N-Imidazole-N'-carbonic acid-3,5-dinitroanilide
i.d.	Inner diameter
IEF	Isoelectric focusing
ile	Isoleucine
iPA	Isopropylamide
iPC	Isopropylcarbamate
iPE	Isopropylester
IPG	Immobilized pH gradient
iPU	Isopropylureido
ITP	Isotachophoresis
IUPAC	International union of pure and applied chemistry
k	Partition ratio
K_m	Michaelis constant
LC	Liquid chromatography
LCP	Left circularly polarized light
LE	Leading electrolyte
LEE	Ligand exchange chromatography
leu	Leucine
LIF	Laser induced fluorescence
LSR	Lanthanide shift reagent
lys	Lysine
lyx	Lyxose
man	Mannose
MCD	Magnetic circular dichroism
MCE	Microemulsion capillary electrophoresis
MDGC	Multidimensional gas chromatography
MEC	Micellar electrokinetik chromatography
MECC	Micellar electrokinetik capillary chromatography
MEEKC	Microemulsion electrokinetik chromatography
ME(K)C	Micellar electrokinetik chromatography
MES	Morpholine ethansulphonic acid monohydrate

met	Methionine
Methyldopa	3,4-Dihydroxy-α-methylphenylalanine
MHEC	Methylhydroxyethylcellulose
min	Minute
MMA	Monomethylamino
MORD	Magnetic optical rotatory dispersion
M_r	Molecular mass
M-RPC	Microchamber rotation planar chromatography
MS	Mass spectrometry
MTH	Methylthiohydantoin
MTPA	α-Methoxy-α-(trifluoromethyl)phenylacetic acid
n	Theoretical plate number ($n = t_R/\sigma)^2$
N	Effective theoretical plate number ($N = t'_R/s)^2$
NC-1	1-Naphthoyl chloride
NEA	1,1'-Naphtylethylamino
NIC-1	1-Naphthylisothiocyanate
NMA-1	1-Naphthalenemethylamine
NMR	Nuclear magnetic resonance
o.d.	Outer diameter
OD	Optical density
OPA	o-Phthaldialdehyde
OPLC	Over-pressured layer chromatography
OPPC	Over-pressured planar chromatography
OPTLC	Over-pressured thin-layer chromatography
ORD	Optical rotation dispersion
orn	Ornithine
OV	Ovomucoid
P	Optical purity
PAA	Polyacrylamide
PAGE	Polyacrylamide gel electrophoresis
PE	Paper electrophoresis
PEA	Phenylethylamine
PEG	Polyethyleneglycol
PFP	Pentafluoropropanoyl
phe	Phenylalanine
pI	Isoelectric point
PLC	Planar liquid chromatography
PPAA	Poly(ethylenephenylalanineamide)
PPL	Porcine pancreatic lipase
$Pr(hfc)_3$	Tris-[3-(heptafluoropropyl-hydroxymethylene)-d-camphorato]-praseodym(III)
$Pr(tfc)_3$	Tris-[3-(trifluoromethyl-hydroxymethylene)-d-camphorato]-praseodym(III)

pro	Proline
PSD	Phase sensitive detector
PTFE	Poly(tetrafluoroethylene)
PTH	Phenylthiohydantoin
PVA	Polyvinylalcohol
PVC	Polyvinylchloride
PVPP	Polyvinylpolypyrrolidone
R_S	Peak resolution
rac.	racemic
RCP	Right circularly polarized light
rha	Rhamnose
RI	Refraction index
R_i	Retention index
RIA	Radioimmunoassay
rib	Ribose
RLCC(C)	Rotation locular countercurrent chromatography
ROA	Raman optical activity
RPC	Rotation planar chromatography
σ	Standard deviation in a Gaussian peak
s	Second
SA	Serum albumine
SBE	Sulphobutyl ether
SDS	Sodium dodecyl sulphate
ser	Serine
SFC	Supercritical fluid chromatography
SIM	Single ion monitoring
sor	Sorbose
SubFC	Subcritical fluid chromatography
t_M	Gas holdup time
t_R	Retention time
t'_R	Adjusted retention time
TAC	Triacetylcellulose
tal	Talose
tBA	tert-Butylamide
tBC	tert-Butylcarbamate
tBE	tert-Butylester
TBS-HSO_4	Tetra-n-butyl ammonium hydrogen sulphate
TE	Terminating electrolyte
TEAA	Triethylammonium acetate
TFA	Trifluoroacetyl
TFAE	1-(9-Anthryl)-2,2,2-trifluoroethanol
TFM	Trifluoromethyl
thr	Threonine

TLC	Thin-layer chromatography
TLE	Thin-layer electrophoresis
TMA	Tetramethylammonium
TMS	Trimethylsilyl
trp	Tryptophan
tyr	Tyrosine
U-RPC	Ultra-microchamber rotation planar chromatography
UV	Ultraviolet
val	Valine
VCD	Vibrational circular dichroism
VOA	Vibrational optical density
w_b	Peak width at base
w_h	Peak width at half height
xyl	Xylose
$Yb(hfc)_3$	Tris-[3-(heptafluoropropyl-hydroxymethylene)-d-camphorato]-ytterbium(III)
$Yb(tfc)_3$	Tris-[3-(trifluoromethyl-hydroxymethylene)-d-camphorato]-ytterbium(III)
ZE	Zone electrophoresis

1 Introduction

An understanding of the current methods of analysis of chiral organic molecules requires a fundamental knowledge of the most important advances that have been made in analytical stereochemistry and separation techniques. The next chapter, therefore, will provide a brief summary of stereochemical concepts. The comprehensive third chapter is devoted to the various techniques used in the analysis of optically active organic compounds. In two main parts the methods not using separations and those employing separations are treated.

There are many reasons for the increasing interest in the analysis of chiral organic molecules in recent years. The phenomena associated with the optical rotation features of asymmetric molecules have long been studied by molecular spectroscopists. The role of chiral compounds has been decisive in the elucidation of reaction mechanisms and their dynamic behaviour in organic chemistry. The reaction mechanisms in organic chemistry would not be understandable without studies of optically active molecules. The knowledge issuing from these investigations, which have mostly been based on classical polarimetry, has tremendously stimulated organic chemistry. In addition, the increasing interest in the analysis of chiral organic molecules is also related to the often high biological activity devoted to diastereoselective and enantioselective reactions.

Both in diastereoselective synthesis (cf. [1-6]) and in the different areas of enantioselective synthesis, i.e. (i) kinetic resolution of racemates [7], (ii) biotransformation of prochiral substrates [8], (iii) diastereoselective reactions with optically pure reagents [9], and (iv) enantioselective reactions (cf. examples in Figures 1.1-1.3 and monographs [10-13]), it is essential to select the most appropriate analytical method for stereocontrol. High-sophisticated syntheses in natural product chemistry, as recently performed, e.g., for calicheamicin [14] and taxol [15,16], imply the knowledge of all the facets of the analysis of chiral organic molecules.

R1 —OH R2 R3 —— $Ti(OiPr)_4$, $tBuO_2H$, CH_2Cl_2 / L-(+)-diethyl tartrate ——→ R1 O —OH R2 R3

R1 = H, alkyl
R2 = H, alkyl, phenyl; R3 = H, alkyl

> 90 % ee

Figure 1.1 Sharpless epoxidation of allylic alcohols [17].

In addition to synthetic organic chemical approaches to enantioselective syntheses, biotransformations have become key technologies. Representative examples are the chemistry/biotransformation approach to both D- and L-amino acids, the use of

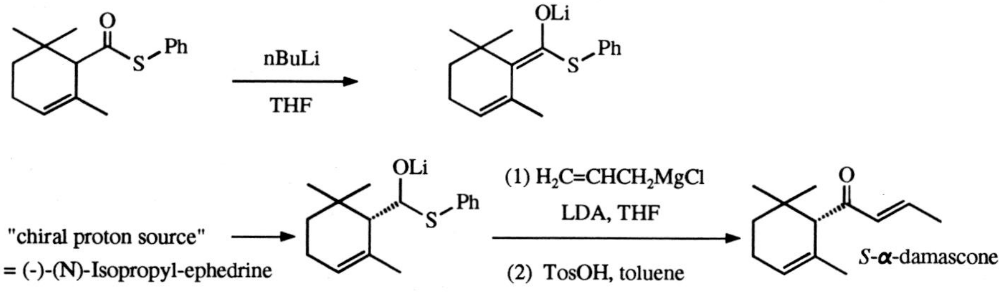

O
S
Ph
nBuLi
THF
OLi
S
Ph

"chiral proton source"
= (-)-(N)-Isopropyl-ephedrine
OLi
S
Ph
(1) $H_2C=CHCH_2MgCl$
LDA, THF
(2) TosOH, toluene
O
S-α-damascone

Figure 1.2 Enantioselective synthesis of α-damascone [18].

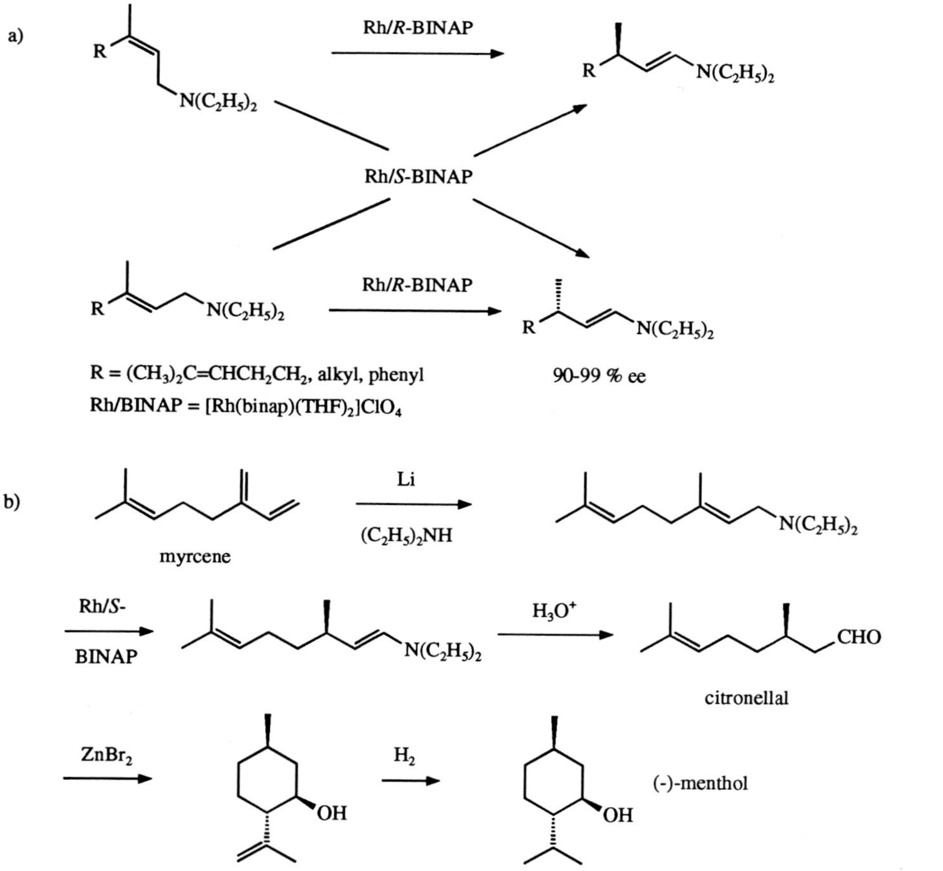

a)
R
$N(C_2H_5)_2$
Rh/*R*-BINAP
R
$N(C_2H_5)_2$
Rh/*S*-BINAP
R
$N(C_2H_5)_2$
Rh/*R*-BINAP
R
$N(C_2H_5)_2$

R = $(CH_3)_2C=CHCH_2CH_2$, alkyl, phenyl
Rh/BINAP = $[Rh(binap)(THF)_2]ClO_4$
90-99 % ee

b)
myrcene
Li
$(C_2H_5)_2NH$
$N(C_2H_5)_2$
Rh/*S*-
BINAP
$N(C_2H_5)_2$
H_3O^+
CHO
citronellal
$ZnBr_2$
OH
H_2
OH
(-)-menthol

Figure 1.3 a) Isomerization of allylic amines by Rh/*R*- or *S*-BINAP; b) application of (a) for the synthesis of (-)-menthol [19].

aldolases in the production of phenylserines, and the nitril hydratase/amidase catalyzed process to synthesize the antibiotic potentiator cilastatin, an N-substituted *S*-2,2-dimethylcyclopropanecarboxamide. Moreover, chiral recognition of substrates is one of the most important facts of biological activity. The interaction between biologically active compounds and receptor proteins often shows a high or complete stereoselectivity. Classical examples of different physiological behaviour are found in the taste and smell of enantiomers. Whereas D-amino acids generally exhibit a sweet taste, the L-enantiomers are tasteless or bitter (cf. asparagine in Table 1.1) [20].

Table 1.1 Chirospecific physiological effects of selected compounds.

No.	Compound	Characteristics	Reference
1	*R*-(-)-Sodium glutamate	no synergistic action	[23]
	S-(+)-	flavour enhancing	
2	D-(-)-Ascorbic acid	no activity	[24]
	L-(+)-	antiscorbutic	
3	*R*-(+)-Penicillamine	extremely toxic	[25]
	S-(-)-	antiarthritic	
4	*R*-(+)-Thalidomide	soporific (no secondary effect)	[25]
	S-(-)-	teratogenic (one strain of mice and Natal rats)[a]	
5	*R*-(-)-Asparagine	sweet taste	[25]
	S-(+)-	bitter taste	
6	P-(+)-Gossypol	no activity	[26]
	M-(-)-	antispermatogenic	
7	(+)-Androst-4,16-dien-3-one	strong sweaty, urine-like (sexual hormone of boars)	[27]
	(-)-	odourless	
8	*R*-(-)-Carvone	spearmint	[28]
	S-(+)-	caraway	
9	*R*-(+)-Limonene	orange-like	[29]
	S-(-)-	terpene-like	
10	5*R*,6*S*,8*R*-(+)-Nootkatone	grapefruit	[30]
	5*S*,6*R*,8*S*-(-)-	woody-spicy	
11	*R*-(-)-4-Methyl-3-heptanone	blocks the effect of *S* at *Popillia japonica*	[31]
	S-(+)-	alarm pheromone	[32]
12	1*R*,3*R*,4*S*-(-)-Menthol	sweet, fresh, mint, strong cooling effect	[33]
	1*S*,3*S*,4*R*-(+)-	herbaceous, less mint-like, weak cooling effect	

[a] Differences given according to a 1979 study of the University of Bonn, Germany, but investigations as early as 1967 at Mary's Hospital Medical School in London showed that both enantiomers are teratogenic in rabbits. In 1984, a study at the University of Münster, Germany, revealed that each enantiomer racemizes at physiological pH *in vitro* or after injection into rabbits. Today, clinical tests provide results to indicate that thalidomide can reduce high levels of the immune system messenger substance tumor necrosis factor-alpha (TNF-α) in AIDS.

In the flavour and pheromone field, nature's ability to produce and convert chiral compounds with remarkable stereoselectivity is well-known. In Table 1.1 a few representative examples are given, in particular, for the enantioselective action of flavour and pheromone substances. The different physiological effect of the enantiomers is well documented in the literature [21].

A few additional examples of chiral chemicals with enantioselective physiological or pharmacological effects are listed in Table 1.1. They comprise the enantiomers of sodium glutamate, ascorbic acid, penicillamine, and gossypol. Their structures are outlined in Figure 1.4.

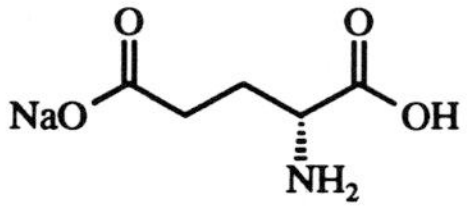

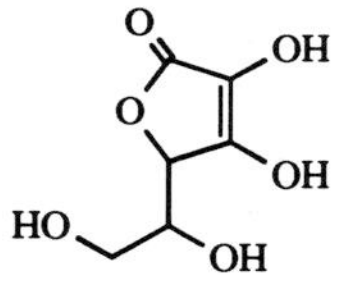

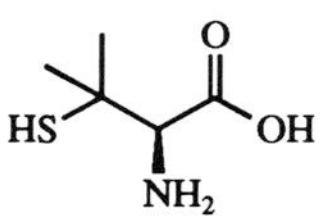

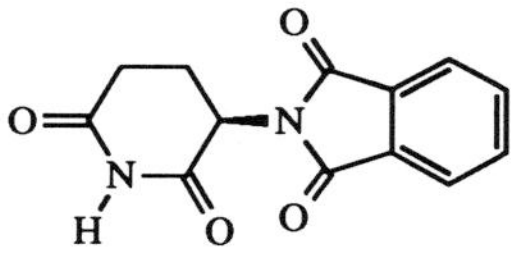

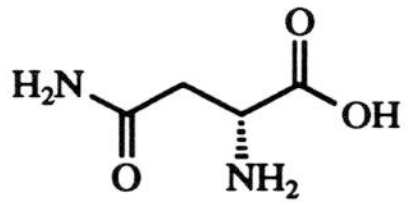

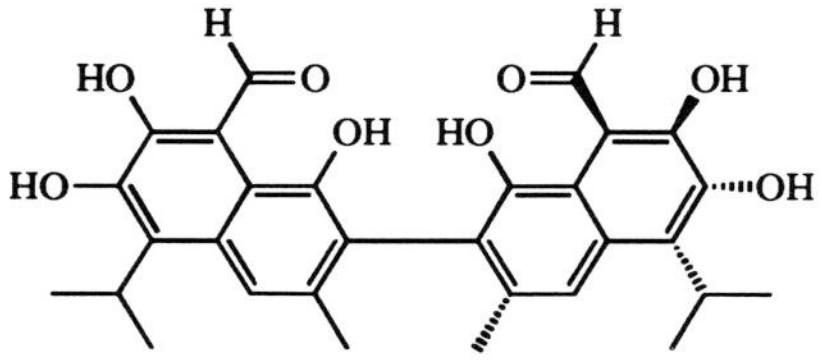

Figure 1.4 Structures of chiral compounds **1-12**. The enantiomer first mentioned in Table 1.1 is represented.

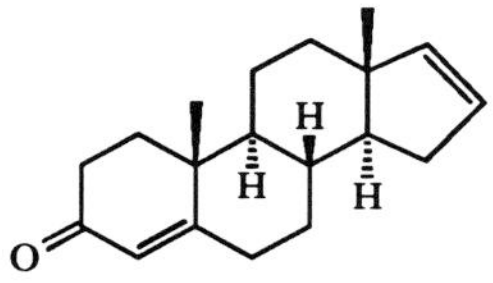

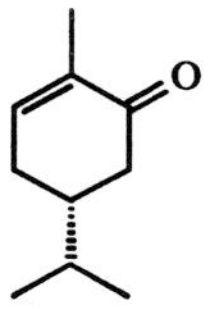

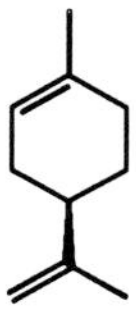

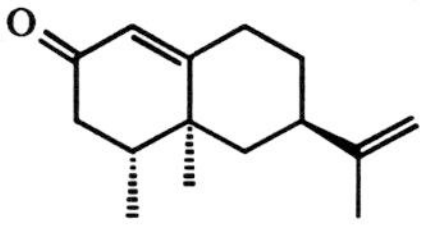

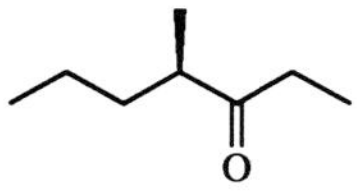

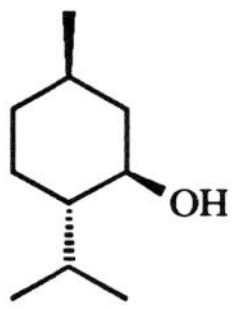

Numerous drugs are synthetic racemic compounds that have been used as pharmaceuticals for a long time. Similar to flavour substances and pheromones, however, often distinct enantioselective physiological effects can be observed. One of the enantiomers may exhibit no or even undesirable effects. Thus, today, the drug industry has to be interested in methods for resolving racemates into their enantiomers in order to be able to subject them individually to pharmacological tests [22].

In fact, perhaps the most important trend in the chiral drug industry is that a steady, rising flow of single-isomer forms of chiral drugs onto world markets is creating an increasing demand for enantiomeric intermediates and bulk active compounds as well as enantioselective technology and analytical services. The activity in developing new chiral drugs has also resulted in soaring demand for certain highly specialized intermediates, such as the enantiomers of glycidol (2,3-epoxypropanol), glycidyl triphenyl methyl ether, 2,3-isopropylideneglycerol (2,2-dimethyl-3-hydroxymethyldioxolane), and 3-aminoquinuclidine (3-amino-1-azabicyclo[2.2.2]octane) in particular. This activity has further led to progress in manufacturing chiral drugs and their enantiomeric precursors. A recent development is the emergence of liquid chromatographic techniques from relative disfavour to become highly promising ton-scale processes. In addition to its new promise as a production process ('process chromatography'), liquid chromatography has become the drug industry's method of choice to determine enantiomeric purity (cf. Section 3.4).

Whether chromatographic or other, drug companies will usually have to develop enantioselective analyses for both single isomers and racemates of chiral drugs. Such analyses may have to be done on both the bulk active substances, the dosage form of the drug, and any key intermediates. Finally, such analyses should include tests of identity and quantitative determinations of the amount of a single isomer in a sample.

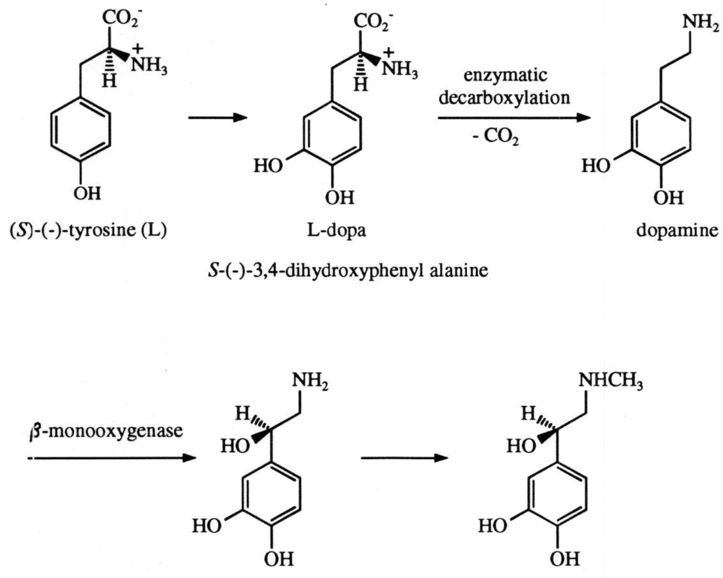

(*S*)-(-)-tyrosine (L)

L-dopa

S-(-)-3,4-dihydroxyphenyl alanine

enzymatic decarboxylation

$- CO_2$

dopamine

β-monooxygenase

(*R*)-(-)-norepinephrine (L)

(*R*)-(-)-epinephrine (L)

Figure 1.5 Stereochemical biosynthesis of catecholamines.

Further classic examples of chiral discrimination in physiological reactions are found among the hormones. For instance, the human endogenous, tyrosine-derived catecholamines are potent regulatory hormones, affecting blood pressure and other essential body functions. Their immediate progenitor is 3-hydroxy-L-tyrosine (L-dopa), which yields dopamine after enzymatic decarboxylation (Figure 1.5). Dopamine is then converted in a completely stereospecific hydroxylation step to norepinephrine (noradrenaline) which is finally transformed into epinephrine (adrenaline) by N-methylation. Only the L-enantiomer is accepted as substrate by dopa decarboxylase.

In addition to L-dopa, methyldopa has to be mentioned which is used to treat high blood pressure. As shown in Figure 1.6, its synthesis begins with vanillin, which is converted to a racemic acetaminonitrile corresponding to the final amino acid product. It is this acetaminonitrile that is the racemic conglomerate, resolvable into its enantiomers by spontaneous crystallisation. The L-isomer is converted to methyldopa. The D-form, when heated with sodium cyanide in dimethylsulphoxide, racemizes via a symmetrical acetimino compound. The recovered racemate is combined with virgin racemate entering the process. Thus, 100 % of acetaminonitrile is theoretically available for the conversion to methyldopa.

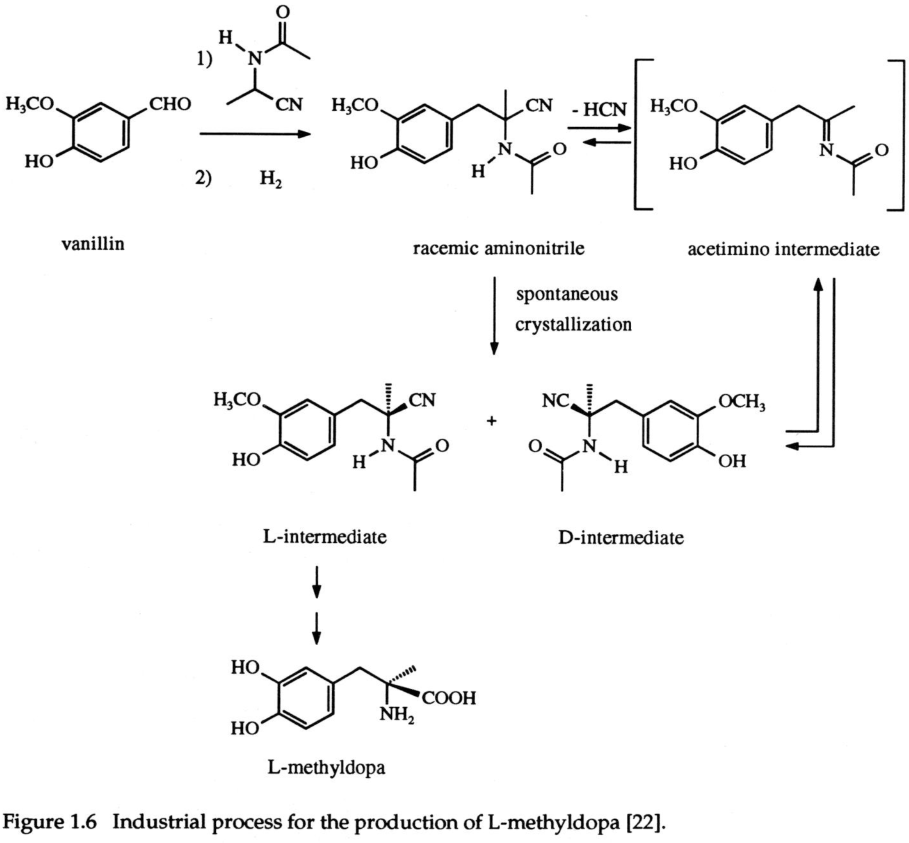

Figure 1.6 Industrial process for the production of L-methyldopa [22].

In recent years, the high selectivity of enzymatic reactions is increasingly being used in organic synthesis [34-36]. A recent example is the selective reduction of organic hydroperoxides by peroxidases. For example, chloroperoxidase from *Caldariomyces fumago* has been found to catalyze the oxidation of sulphides to prepare *R*-sulphoxides with excellent enantiomeric excess (ee) (97-100 %). The enzyme was also used in the regioselective bromohydration of certain saccharide glycals with KBr and H_2O_2 to give the corresponding 2-deoxy-2-bromosaccharides [37]. In addition, the more easily available horseradish peroxidase (HRP) has recently been found to catalyze the selective reduction of chiral hydroperoxides, allowing the preparation of optically pure alcohols and hydroperoxides (Figure 1.7) [38]. In constrast, the preparation of enantiomerically pure hydroperoxides by nonenzymatic means is quite cumbersone.

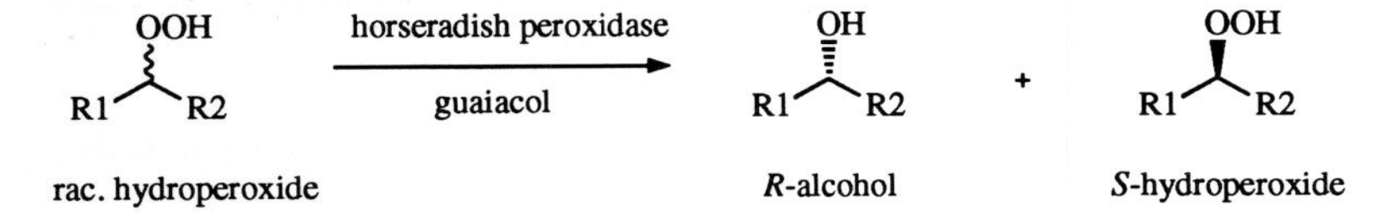

Figure 1.7 HRP- catalyzed kinetic resolution of chiral hydroperoxides [38].

These few examples of enantioselective effects in biological, pharmaceutical and organic chemistry demonstrate that the analysis of chiral organic molecules is of considerable importance in diverse fields of application.

References

[1] Meyers, A.I. *Pure Appl. Chem.* (1979), *51*, 1255

[2] Evans, D.A.; Takacs, J.M.; McGee, L.R.; Ennis, M.D.; Mathre, D.J.; Bartroli, *J. Pure Appl. Chem.* (1981), *53*, 1109

[3] Schöllkopf, U. *Pure Appl. Chem.* (1983), *55*, 1791

[4] Oppolzer, W. *Angew. Chem.* (1984), *96*, 840; *Angew. Chem. Intern. Ed. Engl.* (1984), 23, 876

[5] Helmchen, G.; Leikauf, U.; Taufer-Knöpfel, I. *Angew. Chem.* (1985), *97*, 874; *Angew. Chem. Intern. Ed. Engl.* (1985), 24, 874

[6] Seebach, D.; Imwinkelried, R.; Weber, T. In *Modern Synthetic Methods;* Scheffold, R., Ed.; Springer: Heidelberg, 1986; p. 128

[7] Chen, C.S.; Fujimoto, Y.; Girdaukas, G.; Sih, C.J. *J. Amer. Chem. Soc.* (1982), *104*, 7294

[8] Crout, D.H.G.; Christen, M. In *Modern Synthetic Methods,* Scheffold, R., Ed.; Springer: Heidelberg: 1989; p. 1

[9] Nogrady, M. *Stereoselective Synthesis,* VCH Verlagsges..: Weinheim, 1987

[10] Koskinen, A.M.P. *Asymmetric Synthesis of Natural Products,* Wiley, New York, 1993

[11] Ojima, I. Catalytic *Asymmetric Synthesis,* VCH Verlagsges.: New York, 1993

[12] Sheldon, R. *Chirotechnology,* Dekker, New York, 1993

[13] Brunner, H., Zettlmeier, W. *Handbook of Enantioselective Catalysis with Transition Metal Compounds*, VCH Verlagsges.: Weinheim, 1993

[14] Nicolaou, K.C.; Hummel, C.W.; Pitsinos, E.N.; Nakada, M.; Smith, A.L.; Shibayama, K.; Saimoto, H. *J. Amer. Chem. Soc.* (1992), *114*, 10082. Nicolaou et al., *ibid.* (1993), *115*, 7593.

[15] Holton, R.A., Somoza, C., Kim, H.B., Liang, F., Biedinger, R.J., Boatman, P.D., Shindo, M., Smith C.C., Kim, S., Nadizeh, H., Suzuki, Y., Tao, C., Vu, P., Tang, S., Zhang, P., Murthi, K.K., Gentile, L.N., Liu, J.H. *J. Amer. Chem. Soc.* (1994), *116*, 1597; *J. Amer. Chem. Soc.* (1994), *116*, 1599

[16] Nicolaou, K.C., Yang, Z., Liu, J.J., Ueno, H., Nantermet, P.G., Guy, R.K., Clalborne, C.F., Renaud, J., Couladouros, E.A., Paulvannan, K., Sorensen, E.J. *Nature* (1994), *367*, 630

[17] Rossiter, B.E. In *Asymmetric Synthesis*; Morrison, J.D., Ed.; Vol. 5, Academic Press: New York, 1985; p. 194. Finn, M.G.; Sharpless, K.B., *ibid.*, p. 247

[18] Fehr, C.; Galindo, J. *J. Amer. Chem. Soc.* (1988), *110*, 6909

[19] Noyori, R.; Kitamura, M. In *Modern Synthetic Methods*; Scheffold, R., Ed.; Springer: Heidelberg, 1989, p. 115

[20] Belitz, H.D.; Wieser, H. *Food Rev. Intern.* (1985), *1*, 271

[21] Ohloff, G. *Riechstoffe und Geruchssinn. Die molekulare Welt der Düfte*, Springer: Berlin, Heidelberg, New York, 1990

[22] Stinson, S.C. *Chem. Eng. News* (1993), *39*, 38

[23] Solms, J. *Chimia* (1967), *21*, 169

[24] Daniels, T.C.; Jorgensen, E.C. In: *Wilson and Grisvold's Textbook of Organic Medicinal and Pharmaceutical Chemistry*, Doerge, R.F., Ed.; 8th ed.; Lipincott: Philadelphia, 1982

[25] Enders, D.; Hoffmann, W. *Chem. Uns. Zeit* (1985), *19*, 177

[26] Meyer, V.R. *Pharm. Uns. Zeit* (1989), *18*, 140

[27] Prelog, V.; Ruzicka, L.; Meister, P.; Wieland, P. *Helv. Chim. Acta* (1945), *28*, 618

[28] Leitereg, T.J.; Guadagni, D.G.; Harris, J.; Mon, T.R.; Teranishi, R. *Nature* (1971), *230*, 455

[29] Friedman, L.; Miller, J.G. *Science* (1971), *172*, 1044

[30] Haring, H.G.; Rijkens, F.; Boelens, H.; Van der Gen, A. *J. Agric. Food Chem.* (1972), *20*, 1018

[31] Riley, R.G.; Silverstein, R.M. *Tetrahedron* (1974), *30*, 1171

[32] Tumlinson, J.H. In: *Chemical Ecology: Odour Communication in Animals*, Ritter, F.J., Ed.; Elsevier: Amsterdam, 1979, p. 1301

[33] Emberger, R.; Hopp, R. In: *Topics in Flavour Research*, Berger, R.G.; Nitz, S.; Schreier, P., Eds.; Eichner: Marzling, 1985

[34] Poppe, L.; Novák, L. *Selective Biocatalysis. A Synthetic Approach*, VCH Publ.: Weinheim, 1992

[35] Holland, H.L. *Organic Synthesis with Oxidative Enzymes*, VCH Verlagsges.: Weinheim, 1992

[36] Wong, C.H., Whitesides, G.M. *Enzymes in Synthetic Organic Chemistry*, Pergamon: Oxford, 1994

[37] Fu, H.; Kondo, H.; Ichikawa, Y.; Look, G.C.; Wong, C.H. *J. Org. Chem.* (1992), *57*, 7265

[38] Adam, W.; Hoch, U.; Saha-Möller, C.R.; Schreier, P. *Angew. Chem.* (1993), *105*, 1800; *Angew. Chem. Int. Ed. Engl.* (1993), *32*, 1737

2 The development of stereochemical concepts

In 1908 the French physicist Malus [1] discovered that the light reflected under certain angles from opaque or transparent material exhibits particular properties. These properties of so-called linearly polarized light can be explained by light vectors that vibrate in only one plane. Already in 1801 the French mineralogist Hauy had noticed that certain quartz crystals show hemihedral characteristics. Hemihedral crystals cannot be made coincident with their mirror-images. In addition, the British astronomer Sir John Herschel [2] discovered that all quartz crystals whose hemihedral faces are inclined in the same direction towards the crystallographic axial system rotate the vibrational plane of light in the same sense. In contrast, the mirror-image crystals with hemihedral faces of opposite inclination exhibited the inversel rotatory power.

Subsequent to an observation made by Arago [3], in 1813, the French physicist Biot discovered that a quartz plate cut vertically to the main axis of the crystal rotates the vibrational plane of the linearly polarized light; the inclination of the angle was proportional to the strength of the plate. Some quartz crystals showed dextrorotatory, others laevorotatory power. Two years later, Biot [4] founded organic stereochemistry by discovering the rotation power of certain liquid or dissolved organic compounds such as turpentine, camphor, sugar, and tartaric acid. Only quartz crystals exhibit this rotatory power, which depends, in addition, on the direction at which the crystal is passed by the light beam. As for the above-mentioned organic compounds, the reason for their rotatory power has to be sought in the individual molecule. This means that the effect does not disappear when the compound is melted, dissolved or evaporated.

The relation between symmetry and rotatory power, as it exists in crystals, was first recognized for single molecules by Louis Pasteur. From a number of observations made on optically active compounds, in 1860, Pasteur concluded that the reason for the permanent rotatory power of an optically active molecule lay in an asymmetric grouping of atoms. According to Pasteur molecules of the same structure that rotate the vibrational plane of linearly polarized light by the same amount in one case to the right, in another case to the left, bear the same relation to each other as an image does to its mirror-image [5].

At the same time the development of organic structural chemistry was decisively stimulated by Kekulés recognition of the general tetravalence of carbon atoms [6]. Subsequently, the idea ripened - in van't Hoff in Utrecht [7] and in Le Bel in Paris [8] - that the four valencies of carbon are tetrahedrally arranged. A sufficient but, as shown below, not necessary condition for the phenomenon of optical activity is the non-uniformity of the four atoms (or atom groups) at one of the carbon atoms of a molecule. In such a case two arrangements are possible which cannot be made coincident (Figure 2.1). According to van't Hoff and Le Bel one of these arrangements corresponds to the isomer that rotates the vibrational plane of the light in one direction, whereas the isomer causing the opposite rotation shows the other arrangement. In cases where two of the atoms (or atom groups) are the same, the two forms can be made coincident, i.e. they are no longer geometrically different; such molecules do not cause optical rotation.

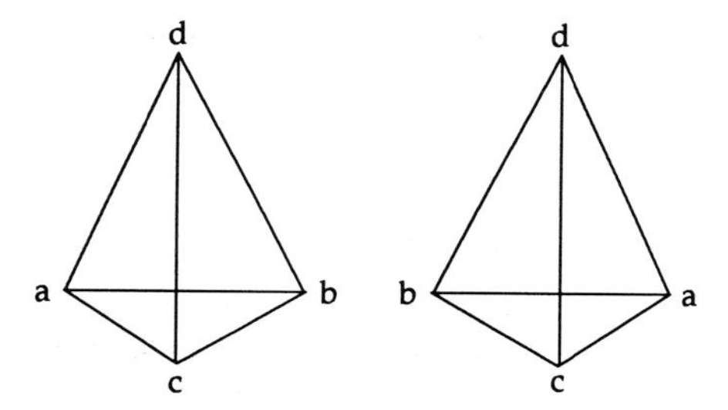

Figure 2.1 Scheme of molecules (type Cabcd), which are mirror-images.

This earlier explanation of optical activity based on the tetrahedral arrangement of valencies of the carbon atom has been confirmed by physical measurement, e.g., by electron diffraction analysis, and by theoretical analyses based upon quantum mechanics.

2.1 Chirality and molecular structure

Chiral molecules are of different nature and shape, but they can all be classified on the basis of symmetry elements, into three categories, i.e. exhibiting central, axial, or planar chirality. Three-dimensional space can be occupied asymmetrically about a chiral centre, a chiral axis or a chiral plane. In addition, steric crowding in a molecule may lead to molecular distortion and hence chirality, as found, e.g., in helicenes. In Table 2.1.1 the elements of chirality together with several classes of compounds are summarized (formulas, cf. Figure 2.1.1).

Table 2.1.1 The elements of chirality with several classes of compounds.

Element	Class	Example	No.
Central	*Aliphatics*	2-Butanol	**1**
	Alicycles	2-Methylcyclohexanone	**2**
	Heterocycles	Tetrahydro-2-methylfuran	**3**
Axial	*Allenes*	Ethyl 3,4-hexadienoate	**4**
	Alkylidene cycloalkanes	1-Ethylidene-4-methylcyclohexane	**5**
	Spiranes	2,6-Dimethyl-spiro[3.3]heptane	**6**
	1,1'-Biaryl derivatives	Gossypol	**7**

Element	Class	Example	No.
Planar	*E-Cycloalkanes*	E-1,2-Cyclooctene	**8**
	Cyclophanes	[2.2]Paracyclophane carboxylic acid	**9**
Helical	*Arylderivatives*	Hexahelicene	**10**
	Polypeptides		

2.1.1 *Molecules with asymmetric atoms*

The above-mentioned tetrahedral structures formed by four different groups around atoms of elements such as carbon, silicon, nitrogen, phosphorus or sulphur are well-known; numerous examples of optically active compounds of this type have been described in the literature. The formulas **1-3** in Figure 2.1.1 represent a few structures.

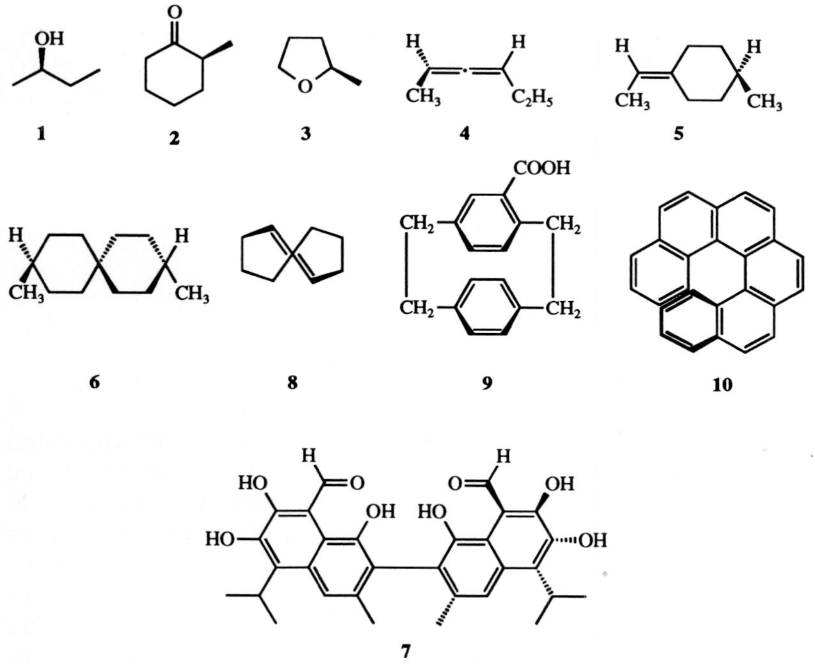

Figure 2.1.1 Examples of chiral molecules (cf. Table 2.1.1).

Elements of groups V or VI in the periodic table can also form non-planar structures with three ligands and the lone-pair electrons of the central atom being regarded as the fourth ligand. Thus, a planar configuration is often possible, the formation of which will permit interconversion within the chiral pyramidal structure. Some examples of chiral molecular structures with central atoms other than carbon are given in Figure 2.1.2.

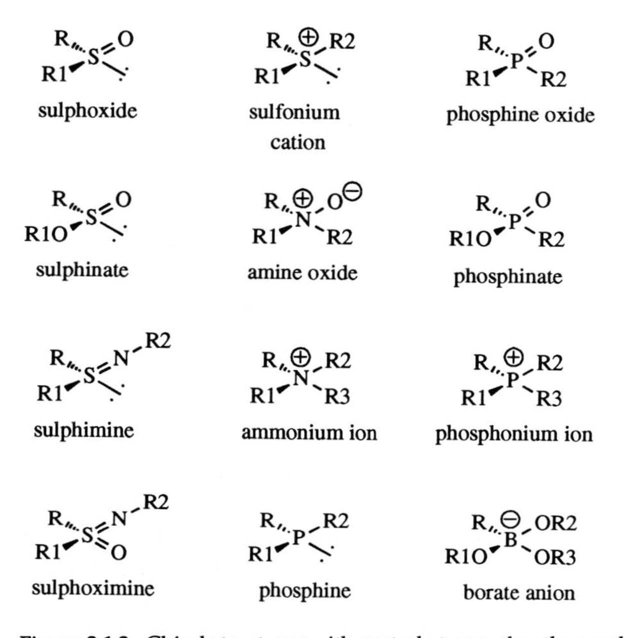

Figure 2.1.2 Chiral structures with central atoms other than carbon.

2.1.2 *Other types of chiral molecular structures*

One of the simplest cases of a chiral structure not involving an asymmetric central atom is derived from allene, $H_2C=C=CH_2$. It is readily seen that substitution of one of the hydrogen atoms at each carbon atom by a substituent R is sufficient to generate chirality, i.e. a non-superimposable mirror image will result (cf. **4** in Figure 2.1.1).

As to axial chirality a few examples are given: alkyliden cycloalkanes (**5**), spiranes (**6**), and 1,1'-biaryl derivatives (**7**). In the latter case, the so-called atropisomerism, restricted rotation around the central bond is caused by the steric effect of the substituents. The groups are simply too large to pass each other, resulting in configurational stability.

As two representative examples for planar chirality the structures of *E*-1,2-cyclooctene (**8**) and [2,2]paracyclophane carboxylic acid (**9**) are given.

In the condensed aromatic hydrocarbons called helicenes (cf. hexahelicene **10**) the phenomenon of steric crowding is obvious. Steric constraints give rise to a helical form with an energy barrier to interconversion between the right- and left-handed enantiomers which is high enough to permit their resolution.

2.2 Definitions and nomenclature

In 1951, the configuration of D-glucose (**11**), which had been arbitrarily assigned by Fischer, who had related it configurationally to D-glyceraldehyde (**12**), was confirmed [9]. Later, this result has been verified on the basis of new stereochemical investigations [10].

CHO
H–|–OH
HO–|–H
H–|–OH
H–|–OH
CH_2OH

11

CHO
H–|–OH
CH_2OH

12

The D,L-nomenclature system, which relates a variety of optically active compounds to each other, in particular carbohydrates, hydroxy acids, and amino acids, is still much used and has not been replaced in the case of carbohydrates. The system is only applicable, however, to compounds having asymmetric carbon atoms. Therefore, Cahn, Ingold and Prelog [11,12] presented a new, more generally useful system, called the *R,S*-nomenclature. The system is briefly outlined below; for details the reader is referred to the original literature. The principle is based on three steps: (i) arranging the ligands associated with an element of chirality into a sequence; (ii) using this sequence to trace a chiral path; and (iii) using the chiral sense of this path to classify the element of chirality.

(i) The ligands around a centre of chirality are ordered in a sequence according to the following basic rules:

(a) higher atomic number is given priority;
(b) higher atomic mass is given priority;
(c) cis is prior to trans;
(d) like pairs, i.e. *R,R* or *S,S*, are prior to unlike pairs, i.e. *R,S* or *S,R*;
(e) lone-pair electrons are assigned the atomic number 0.

(ii) Using these rules, the ligands, ordered in the sequence A>B>C>D, are viewed in such a way that D (of lowest priority) points away from the viewer.

(iii) The remaining ligands are then counted, starting from the one of highest priority (i.e. first A, second B, third C). If this operation is clockwise for the viewer, the designation will be *R* (rectus), otherwise it will be *S* (sinister). Thus, the example shown is an *R*-configuration:

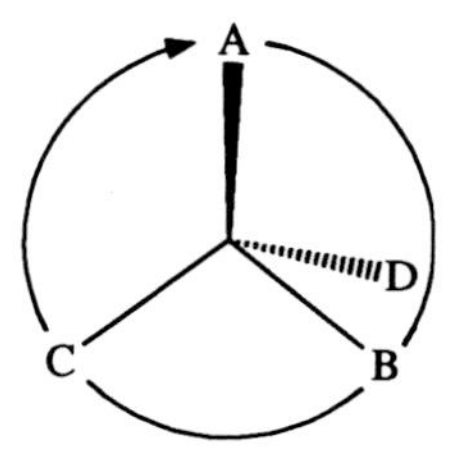

The selection for axial chirality implies that the atoms closest to the axis are considered in a priority sequence, e.g., the ortho-carbon atoms in a biaryl compound.

With regard to a molecule exhibiting planar chirality, a plane of chirality has first to be selected. The second step involves the determination of a pilot atom P which should be bound directly to an atom of the plane and located at the preferred ("nearer") side. P is selected according to the sequence rules. In the next step one passes from P to the in-plane atom to which it is directly bound (a). This atom is then the atom of highest priority of the in-plane sequence. The second atom of this sequence is the in-plane atom (b), bound directly to (a), which is most preferred by the standard subrules. After completion of the sequence the chirality rule can be applied (c). The paracyclophane **9** illustrates the principle.

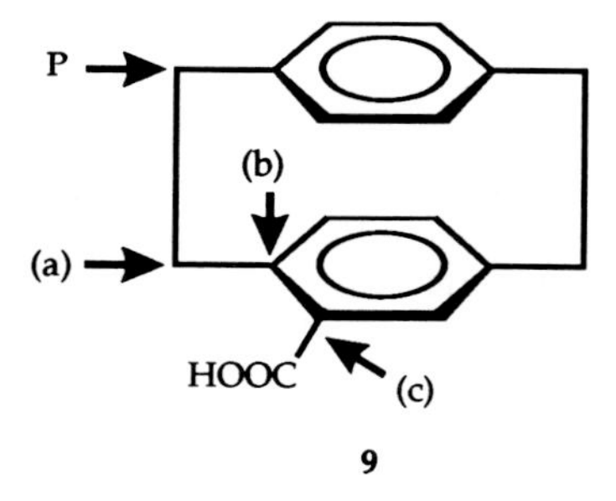

9

The helicenes can be treated as axially chiral molecules, but they are preferentially regarded as secondary structures. Thus, for hexahelicene **10** the (-)-form represented below forms a lefthanded helix [M (= minus) helicity], which is designated M-(-). The opposite enantiomer is called P (plus).

This M,P-nomenclature is also often employed for chiral biaryls. Firstly, an axis is drawn through the single bond around which conformation is defined and the smallest torsion angle formed between the carbon atoms bearing the groups of highest priority is used to define the helix.

The following definitions are used throughout this book: If structural isomers with the same constitution differ in the spatial arrangement of their substituents, they are called stereoisomers. To classify stereoisomers according to their symmetry one can differentiate between *enantiomers* and *diastereomers* (Figure 2.2.1). Each stereoisomer can be regarded as a chiral object (from the Greek word "*cheir*" = hand), which means that the object is not superimposable on its mirror image.

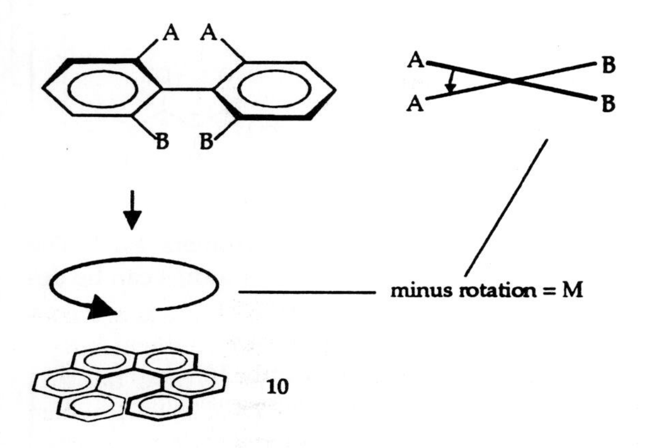

If between both stereoisomers an enantiomeric relation exists, they are called *enantiomers*. Enantiomers (or optical antipodes) are chiral molecules related to each other by reflection symmetry exhibiting identical internal energies. Stereoisomers that are not enantiomers, are called *diastereomers*. Diastereomers are not related to each other by reflection symmetry and they have non-identical internal energies. A molecule lacking any symmetry element is called *asymmetric*. In the case of *cis-trans* isomerism the exact terminology is *π-diastereomerism*.

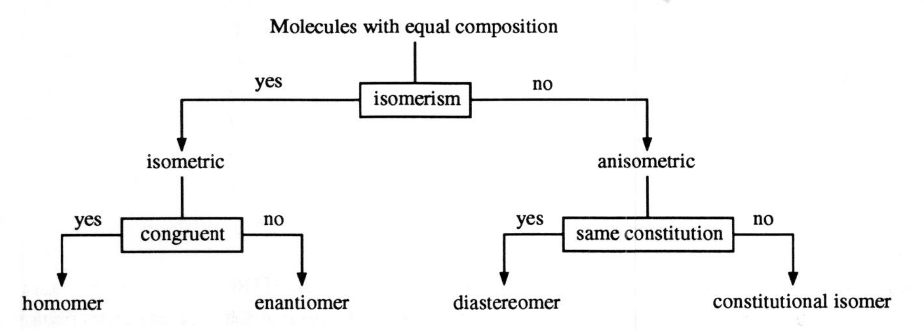

Figure 2.2.1 Classification of the relations between isomeric molecules [13].

Two diastereomers from the achiral compound 1,2-dichloroethene (**13** and **14**) exist, which are diastereomeric to each other, but not enantiomeric. At the same time, two stereoisomers cannot have the relation enantiomeric and diastereomeric to each other, i.e. enantiomerism and diastereomerism exclude each other [13]. If a mixture contains the two enantiomers in the ratio of R:S = 50:50, the mixture is called *racemic* or *racemate*, and it shows no optical activity. If the ratio differs from 50:50, the mixture is *non-racemic*; it may show optical activity. Sometimes in the literature enantiomerically (optically) pure compounds (R:S = 100:0 or 0:100) are called *homochiral* or *monochiral* and racemic mixtures are called *heterochiral*, but these terms do not satisfy IUPAC rules.

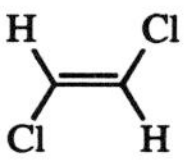

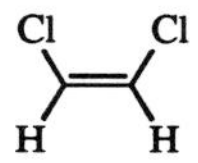

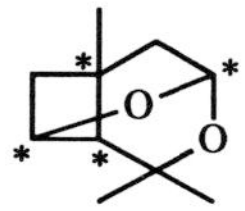

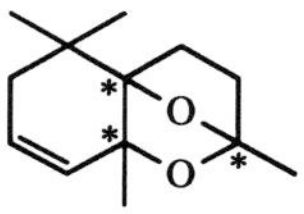

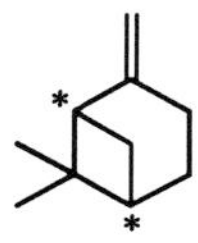

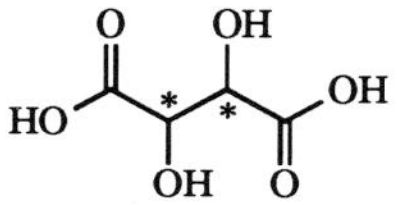

13 **14** **15** **16** **17** **18**

For molecules with more independent chiral centres, 2^n stereoisomers exist. The number of stereoisomers of molecules with dependent asymmetric atoms can be determined only empirically; e.g., theoretically there exist for

lineatin (**15**)	$2^4 = 16$ stereoisomers, but in reality only 2;
"Riesling acetal" (**16**)	$2^3 = 8$ stereoisomers, but in reality only 4;
ß-pinene (**17**)	$2^2 = 4$ stereoisomers, but in reality only 2.

As to open-chain molecules exhibiting a minimum of two asymmetric centres, but a symmetric constitution, $2^{(n-1)} + 2^{(n-2)/2}$ stereoisomers exist, if n is an even number, and $2^{(n-1)}$, if n is an odd number. A classic example of a molecule with an even number of chiral centres is tartaric acid (**18**). It is easy to recognize that both asymmetric carbon atoms carry the same substituents. According to the Cahn-Ingold-Prelog system, the right-rotating form shows the *R,R* configuration and the left-rotating form exhibits the *S,S* configuration. The second expected enantiomeric pair does not exist, since both forms, *R,S* and *S,R*, are congruent to each other and are, therefore, identical and achiral. Such an optically inactive stereoisomer is called the meso form; it has a diastereomeric relation to the two other stereoisomers.

References

[1] Malus, E.L. *Mém. Soc. Arcueil* (1809), *2*, 143

[2] Herschel, J.F.W. *Trans Cambridge Phil. Soc.* (1821), *1*, 43

[3] Arago, D.F. *Mém. Classe Sci. Math. Phys. Int. Imp. France* (1811), *12*, 115

[4] Biot, J.B. *Bull. Soc. Philomath. Paris* (1815), *190*; (1816), *125*; *Mém. Acad. Roy. Sci. Inst. France* (1817), *2*, 41

[5] Pasteur, L.; *Lectures from 20.1. and 3.2.1860 at the Soc. Chim. Paris* (cf. Richardson, G.M. The Foundations of Stereochemistry, Amer. Book Comp.: New York, 1921)

[6] Kekulé, A. *Liebigs Ann. Chem.* (1858), *196*, 154

[7] Van't Hoff, J.H. *Bull. Soc. Chim. France* (1875), *23*, 295 (cf. Richardson, G.M. *The Foundations of Stereochemistry*, Amer. Book Comp.: New York, 1921)

[8] Le Bel, J.A. *Bull. Soc. Chim. France* (1874), *22*, 337

[9] Bijvoet, J.M. Peerdeman, A.F.; Van Bommel, A.J. *Nature* (1951), *168*, 271

[10] Buding, H.; Deppisch, B.; Musso, H.; Snatzke, G. *Angew. Chem.* (1985), *97*, 503

[11] Cahn, R.S.; Ingold, C.K.; Prelog, V. *Experientia* (1956), *12*, 81

[12] Cahn, R.S.; Ingold, C.K.; Prelog, V. *Angew. Chem. Intern. Ed.* (1966), *5*, 511

[13] Mislow, K. *Introduction to Stereochemistry*, Benjamin: Menlo Park, 1965

3 Techniques used in the analysis of optically active compounds

3.1 Chiroptical methods

Chiroptical methods comprise polarimetry, optical rotatory dispersion (ORD), and circular dichroism (CD). Detection is based on the interaction between a chiral center in the analyte and the incident polarized electromagnetic radiation. Previous applications focused primarily on the elucidation of molecular structures, particularly of natural products for which a technique capable of confirming or determining the absolute stereochemistry was critical. In recent years the application of these techniques has become more and more significant to analytical chemistry.

Among the various requirements of analytical methodologies the properties of analytical selectivity and breadth of application are of prime importance. Analytical selectivity depends on the structural properties of the analyte and the ability of the selected detector to differentiate between the analyte and a potentially high number of interfering compounds. The optimum number of molecular properties necessary to achieve an acceptable level of selectivity appears to be two. If only one property is necessary, separation is essential unless a more sophisticated procedure, which is either time- or phase-sensitive, is used. If three or more properties are necessary, the number of potential analytes is greatly diminished. The most widely used chiroptical method is CD, which measures both rotation and absorbance simultaneously.

Several comprehensive articles on the physical phenomena of chirality and the manifestation of its interaction with polarized light are available [1-5]. For chemical analysis, an elementary understanding of the nature of the interactions and their relationships to each other, as well as the dependence of the experimentally measured parameters on the concentration of the optically active species, is sufficient [6].

3.1.1 Theoretical background of optical activity

A molecule will absorb light strongly only if the transition from ground state to excited state involves a translation of charge. This is the basis of linear dichroism. Thus, the lowest energy singlet valence electronic transition in bicyclohexylidene can be shown to be polarized along the double bond, with light linearly polarized along the double bond being preferentially absorbed to light with a perpendicular linear polarization (Figure 3.1.1). For a known transition polarization, linear dichroism measurements can supply information about the orientation of the absorbing group with respect to the axes of the linearly polarized light. Linear dichroism is concerned with the relationship between electronic movements in a molecule (chromophore) and the oscillating electric vector of electromagnetic radiation.

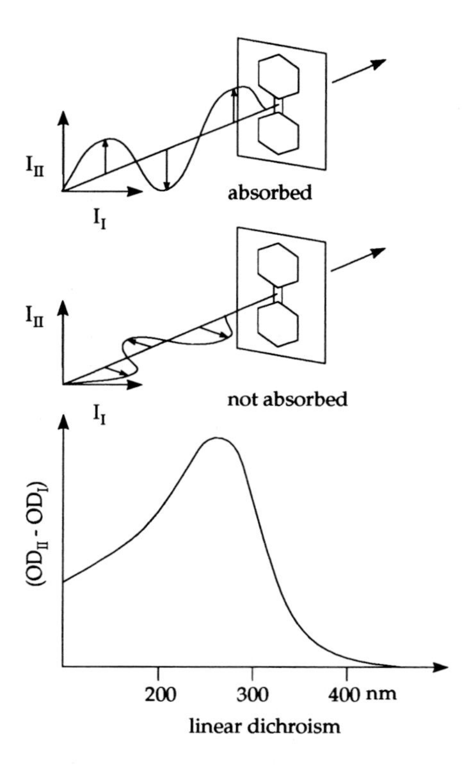

Figure 3.1.1 The origin of linear dichroism for the example of bicyclohexylidene [7].

Optical activity results from differences in the ability of a chromophore in a chiral molecule to absorb the two hands of circularly polarized light. In case of $A_L > A_R$ a positive CD will be obtained (where $A_{L,R}$ are the absorbances for left and right circularly polarized light). The molecule has interacted preferentially with left circularly polarized light (the converse would be true for the enantiomer with $A_R > A_L$). All theories of optical activity are concerned with matching electronic movements in a chromophore (or infrared vibration) with the oscillating electric field of one of the hands of circularly polarized light. This electronic chirality originates from interactions between the chromophore and its associated molecular structure (absolute configuration, conformation). To exhibit optical activity a transition must have collinear electric and magnetic transition dipole moments. Figure 3.1.2 shows the two possible electric chiralities. Parallel electric (μ) and magnetic (m) dipole moments give a right-handed electronic chirality (positive CD); antiparallel moments give a left-handed electronic chirality (negative CD).

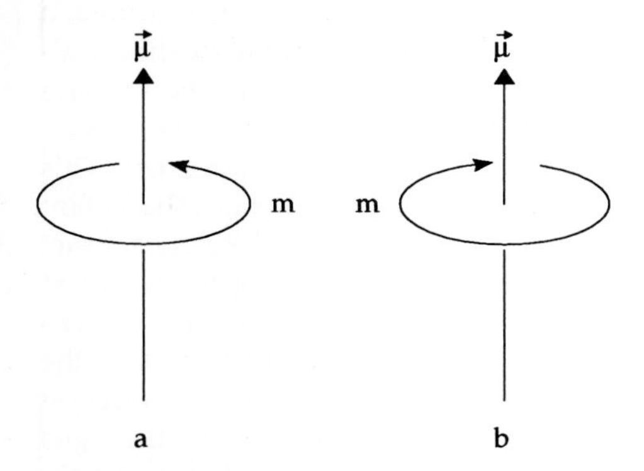

Figure 3.1.2 Electronic chiralities of a spectroscopic transition. (a) Collinear parallel and (b) antiparallel electric (μ) and magnetic (m) dipole moments [7].

In electronic spectroscopy, the mechanisms for generating these chiralities can be grouped into four classes (Figure 3.1.3).

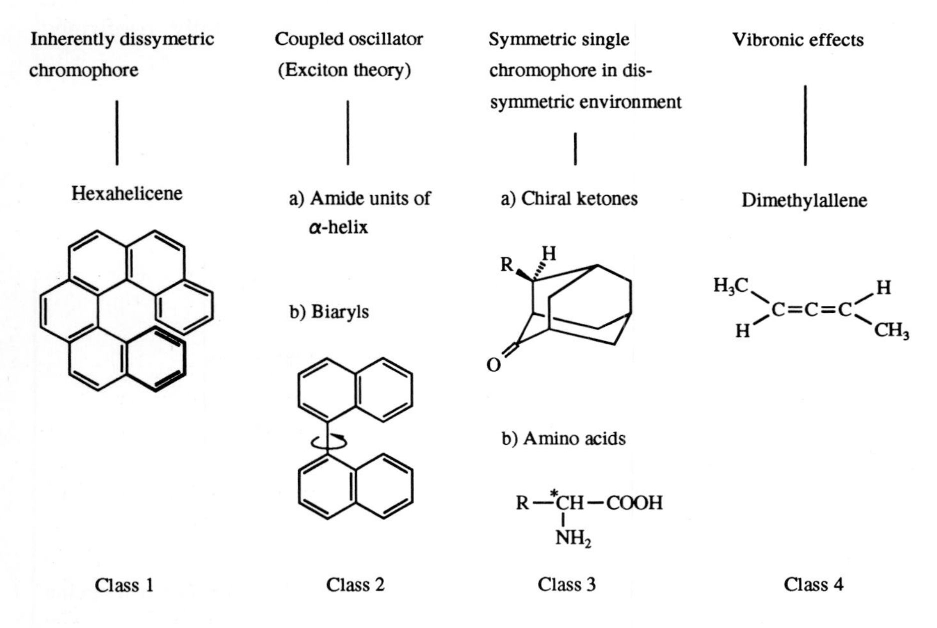

Figure 3.1.3 Four classes of chromophores capable of generating electronic chirality [7].

Class 1 includes molecules where the chromophore itself is chiral and the associated electronic transitions inherently possess transition electric and magnetic dipole moments. Class 3 comprises molecules with isolated chromophores whose transitions have only one moment or neither of the required moments. For example, a ketone has an n-π^* transition at approximately 300 nm which is magnetic dipole allowed (rotation of charge) with a low ordinary extinction coefficient; the missing collinear electric dipole moment can be generated by electrostatic interactions with the polarizability of the various bonds in the surrounding molecular structure. This has been taken as the theoretical basis for different rules [5]. In particular, the 'octant rule' has to be mentioned, which often allows, especially for steroid ketones, exact prediction of the sign of the cotton effect. By the three nodal planes of the n and π^* orbitals the space around the C=O group is divided up into the octants of Cartesian coordinate system (Figure 3.1.4a). Looking in the direction of the O=C axis of the carbonyl group, the four rear, more important octants are represented in projection (Figure 3.1.4b). Atoms arranged in a nodal plane do not contribute to CD; the signs of the contributions of atoms within the various octants are outlined in Figure 3.1.4b. It may be assumed that the disymmetric disturbance of the basically symmetric C=O chromophor is caused by Coulomb interactions of the other nuclei of the molecule which are insufficiently shielded by their electron shells. Theoretically, two substituents in neighboring octants contribute opposite signs to the cotton effect. The octant distribution of the signs was determined semi-empirically by means of various calculations and a large number of collected data. The 'octant rule' cannot be applied to 2,3-unsaturated ketones. In this case, the experimentally confirmed helicity rule is valid.

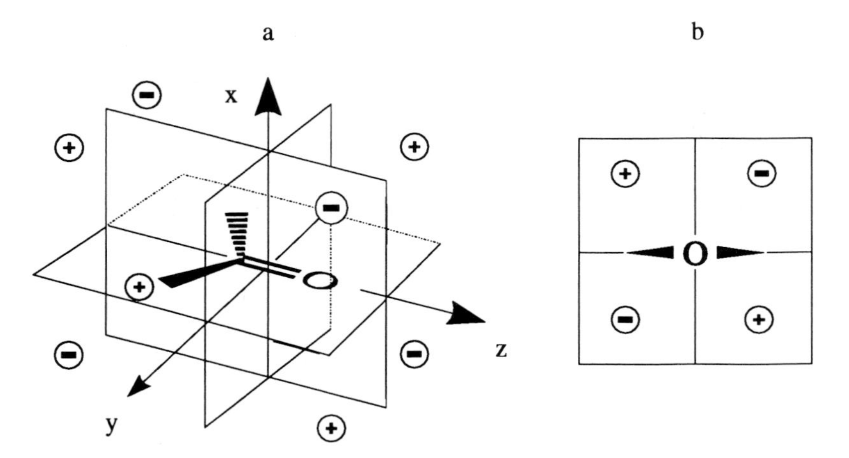

Figure 3.1.4 The 'octant rule' of saturated ketones. (a) All eight octants, (b) the four rear octants with the sign of the CD effect.

Nonetheless, the number and relative importance of chromophore-bond interactions makes the assessment of electronic chirality uncertain. Therefore, an absolute determination of configuration is not generally attempted for this class. It is normally undertaken relatively by comparison of the unknown with a 'library' of reference standards.

The most important class in electronic optical activity (class 2) includes molecules in which the required collinear moments are derived from the interaction of two electric dipole allowed transitions ('Exciton Coupling'). For example, the absolute configuration of the alcohol **1** belongs to class 3; it has an optical activity that happens to be exceedingly small. Its benzoylated derivative **2** contains two chromophores with electric dipole allowed transitions around 239 nm, one on the phenanthrene and the other on the benzoate which is a charge transfer transition polarized along the long axis (determined from linear dichroism measurements). The two transition moments can be 'in-phase' or 'out-of-phase', giving rise to the characteristic 'Exciton Coupling' with a CD sign pattern deriving from the electronic chirality of the two transition electronic configurations, which therefore determines the absolute stereochemistry of the alcohol **1**. This aspect of optical activity has been discussed extensively in Nakanishi's books [8].

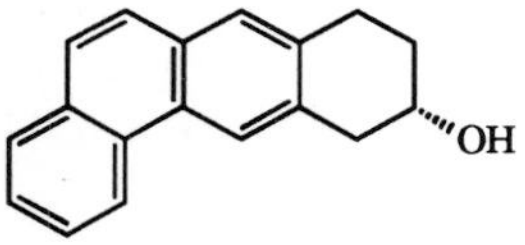

OH

1

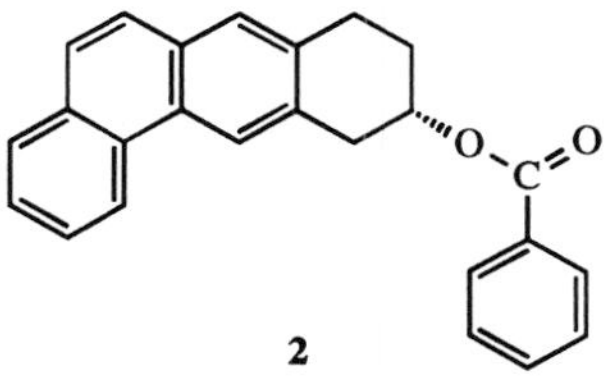

O C O

2

Polarimetry and ORD both determine the extent to which a beam of linearly polarized light is rotated on transmission through the medium containing the chiral sample. The two techniques are entirely equivalent for nonabsorbing chiral species and differ only in that ORD yields a spectral response, whereas polarimetric measurements usually are restricted to a limited number of preselected wavelengths.

3.1.2 *Polarimetry*

The comparison of the chiroptical properties of an optically active compound of given enantiomeric composition with that of the pure enantiomer (of either sign of optical rotation) represents a direct quantitative measure for optical purity. Since polarimetric equipment is available at most research facilities and the measurement of optical rotations goes back almost two centuries, the determination of the optical purity P by polarimetry is the most popular method for evaluating enantiomeric compositions. Optical rotation is the angle by which the polarization plane is rotated as plane-polarized light passes through a sample of optically active molecules. Figure 3.1.5 explains the expression 'plane-polarized light'.

Ordinary visible light is radiant energy, of a certain range of frequencies or wavelengths, which is transmitted as a result of vibrations of an electromagnetic character. According to physical theory these vibrations occur in all directions at right angles to the direction of propagation of the light. By passing a beam of light from a mono- or polychromatic source through certain optical devices, e.g., a Nicol prism (the so-called polarizer), all the vibrations except those in one particular plane are absorbed. This emergent beam is then said to be plane-polarized.

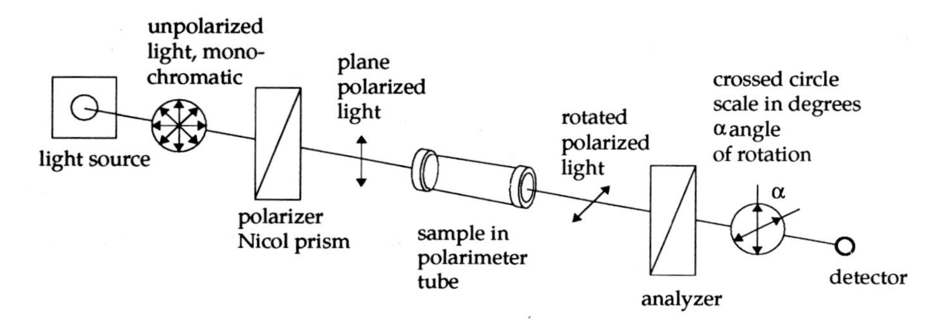

Figure 3.1.5 Basic elements of a polarimeter.

This light will pass through a second Nicol prism (the so-called analyzer), if it is held at exactly the same orientation to the polarizer because they both transmit light in the same plane. If this second prism is rotated through 90° about an axis in the direction of the beam of light, it will now absorb the vibrations transmitted by the first prism. If the analyzer of a polarimeter containing water or some other achiral solvent is rotated until no light passes through, this is the zero point for the instrument. If an optically active compound is now placed in the flow cell, a certain rotation of the light will take place. To measure this rotation the analyzer prism can be rotated again until zero is found. This gives the optical rotation α of the solution.

For a solution of the optically active sample the well-known expression formulated by Biot is used:

$$[\alpha]^{T}_{\lambda} = 100\,\alpha / l\,c$$

$[\alpha]^{T}_{\lambda}$	= *specific* rotation at a given temperature and wavelength
α	= *optical* rotation in degrees (the observed angle by which the polarization plane is rotated as plane-polarized light passes through a sample of optically active molecules)
T	= temperature in degrees Centigrade
λ	= wavelength (for historical reasons the sodium D-line, 589 nm)
l	= cell path length in decimeters (= 10 cm)
c	= concentration in grams per 100 ml solution at the temperature T.

The magnitude and sign of [α] are functions of these variables. Under defined conditions the specific rotation of one enantiomer has the same magnitude, but opposite sign as that of its antipode. When the specific rotation of a pure liquid is cited in literature its density d must also be cited. An error is introduced into [α] when the density measurement is inaccurate. An alternative way to report the rotation of a pure liquid would be by the observed rotation α. Since the observed rotation is dependent on cell length, the path length must also be given [10].

The specific rotation can be converted from percent composition to the molecular rotation [Φ] by relating it to the molecular mass, M. This expression is used frequently in ORD, but not in polarimetry.

$$[\Phi]^{T}_{\lambda} = M\,[\alpha] / 100$$

The terms 'optically active' and 'chiral' are often used synonymously, although a chiral molecule shows optical activity only when exposed to plane-polarized light and occasionally a chiral molecule may possess no measurable optical activity. The optical purity P is defined as follows:

$$P = [\alpha] / [\alpha]_{max}$$

[α] = *specific* rotation in degrees without CGS dimensions
$[\alpha]_{max}$ = *specific* rotation in degrees of the pure enantiomer (absolute rotation).

It has to be stressed that literature data must be carefully evaluated before the degree of enantiomeric purity is accepted, since the potential for error in obtaining an optical rotation is significant [11]. The sources of errors in the determination of enantiomeric compositions with regard to the optical purity will now be discussed in detail.

(i) *Concentration dependence:* Errors may arise from the concentration-dependence of the specific rotation. For example, the specific rotation of malic acid in water changes its sign with increasing concentration [12]. Dependence of the specific rotation on dilution has been observed for 2-phenylpropanal (hydratropaldehyde) in benzene [13]. Non-linear rotations may be observed in highly dilute solutions [14].

In some extreme cases, a change in sign of the rotation may occur on dilution; these changes become more pronounced at wavelengths close to the optically active absorption bands [11]. It has, therefore, been recommended to measure at different wavelength when the specific rotation of a new compound is reported [11]. For the purpose of determining the optical purity P, it is essential to use the same concentration of the given solvent when comparing solutions on an absolute basis.

(ii) *Nature of the solvent:* The nature of the solvent decisively influences the magnitude and sign of [α], owing to the interactions between the rotatory power and molecular actions between the solute and solvent, formation of solvates, conformational changes and variations in ionic species. For example, the sign of the optical rotation of tartaric acid is positive when it is dissolved in water and negative when dissolved in benzene/ethanol.

In practice, the solvent will often be selected according to literature data; such properties as the pH, the solubility of the solute, the magnitude of optical rotation, the absence of molecular association and chemical reaction should be considered. Another parameter is the purity of the solvent:

(iii) *Purity of the solvent:* A high purity grade of the solute is also important in polarimetry. Small impurities in a chiral compound exhibiting a very high rotatory power may strongly influence the accuracy of polarimetric measurements. Achiral impurities as well can change, sometimes even increase, the value of specific rotations through chemical interaction with the solute. For instance, traces of water may alter the specific rotation of solutes prone to hydrogen-bonding such as amines, alcohols and carboxylic acids. Thus, impurities both in the solute and in the solvent may impair the accuracy of specific rotations. Careful purification of an isolated sample of unknown enantiomeric purity is therefore of prime importance. Care has to be taken, in particular, not to involve any crystallization steps for solids, since accidental optical fractionation can alter (in most cases increase) the enantiomeric composition as compared to the original ratio. Distillation or chromatographic methods in an achiral environment are recommended purification steps.

(iv) *Temperature:* The specific rotation is temperature-dependent. The effect of temperature on $[\alpha]$ arises from at least three main sources [11,15]: the density and concentration, the equilibrium constants for molecular association and dissociation, and the relative population of the chiral conformations change with temperature. Thus, for accurate measurements of certain compounds, e.g., tartaric acid derivatives, precise control of the temperature is very important [11]. The specific rotations $[\alpha]$ and $[\alpha]_{max}$ must be measured at the same temperature when the optical purity P is determined.

(v) *Molecular self-association:* The optical purity (which describes the ratio of the specific rotation of a mixture of enantiomers to that of the pure enantiomer) is linearly related to the enantiomeric purity (which describes the actual composition) only when the enantiomers do not interact with each other. It has been shown that the optical purity markedly deviates from the true enantiomeric composition if the enantiomers undergo molecular self-association [16]. The oligomers formed in solution display their own individual rotatory power and, depending on their concentration, contribute to the overall specific rotation.

The prerequisite for determining enantiomeric compositions via optical purity measurements is a medium to high rotatory power of the sample, permitting the correct determination of small differences in enantiomeric excess, 'ee' [defintion, % ee = $(R\text{-}S)/(R\text{+}S)\cdot 100$; for R>S]. Specific rotations may range from very high values (e.g., helicenes) to very low values (e.g., molecules owing their chirality to isotopic substitution). Some chiral hydrocarbons are even optically inactive under conventional conditions, for example 5-ethyl-5-propyl-undecane [17].

The determination of the optical purity of a chiral sample requires that the specific rotation of the pure enantiomer, $[\alpha]_{max}$ (absolute rotation), be known. The absolute rotation may be established by calculation or determined directly. Whereas the semi-empirical calculation of optical rotations presents difficulties, Horeau [18,19] and Schoofs and Guetté [20] have described methods for calculating absolute rotations based on the principle of kinetic resolution. The maximum specific rotation of an enantiomer can be calculated by the method of using asymmetric destruction of the corresponding racemic mixture or by the method of using two reciprocal kinetic resolutions.

A direct estimation of the absolute rotation can be achieved by enzymatic destruction of one enantiomer. The kinetic resolution methods using enzymes require that one enantiomer reacts quantitatively in the presence of the other enantiomer which is completely inert. Natural products originating from enzymatic reactions) are usually believed to be enantiomerically pure and may, therefore, serve as reference standards for the determination of the absolute rotation. There is increasing evidence, however, that natural products (e.g., pheromones [21,22]) are not always enantiomerically pure.

The absolute rotation of a sample may be ascertained by crystallization to constant rotation. In rare cases, however, the constant rotation of a sample may also conform to a composition below 100 % ee. The classical resolution by crystallization has been reviewed in detail [23,24].

Since many direct non-chiroptical methods are available for determining enantiomeric purities, absolute rotations can be extrapolated from the specific rotation of a sample of known ee value. This procedure requires that the optical purity - based on the rotation - and the enantiomeric purity - based on independent physical methods - be identical. The uncritical use of literature data of absolute rotations may lead to errors in predicting the optical yield of enantioselective syntheses.

For accurate determination of the optical purity P the specific rotation α and the absolute rotation α_{max} have to be determined under the same experimental conditions [25,26], such as the pH of the solvent, the purity grade of sample and solvent, instrumentation and cell parameters. The analyst should not rely on literature data for $[\alpha]_{max}$ but determine the standard on his own and check the enantiomeric purity by an independent method. It has been recommended that the optical rotation be preferably measured at several wavelengths and in at least two solvents [11].

Even when these precautions are observed, the determination of the optical purity P will be affected by systematic error. Errors in measurements of the optical rotation resulting from temperature and concentration effects have been reported to be at least +/- 4%. The main systematic error arises from instrument reading, in particular, when low specific rotations are recorded, which may be due to low rotatory power or low enantiomeric purity of a sample. Thus, polarimetric measurements are not recommended for compounds of low optical rotatory power or for near-racemic mixtures. But optical yields of greater than 97% may be questionable as well, unless experimental conditions are clearly stated.

Polarimetry represents a convenient and popular routine method for obtaining optical purity data, but its use for the correct determination of enantiomeric composition is limited by the conditions summarized in the following [28]:

(i) The accurate knowledge of the maximum optical rotation $[\alpha]_{max}$ of the pure enantiomer (absolute optical rotation) is essential.
(ii) Relatively large sample sizes are required.
(iii) The substance must exhibit medium to high optical rotatory power, permitting the correct determination of small differences in ee.
(iv) Isolation and purification of the chiral substance have to be performed without accidental enantiomer enrichment.
(v) The accuracy of optical rotation depends on temperature, solvents and traces of optically active (or inactive) impurities.
(vi) 'Optical purity' may not, a priori, conform to enantiomeric composition.

3.1.3 *Optical rotation dispersion (ORD)*

The degree of rotation depends on the rotatory strength of the chiral center, the concentration of the chirophore, and the path length. Previously, unusual terms and concentration units have been formulated [2-4] when dealing with solution media, such as

$$[\Phi] = 10^{-2}\, M\, [\alpha] \text{ and}$$
$$[\alpha] = 100\, \alpha / (c\, d)$$

where α, $[\alpha]$ and $[\Phi]$ are the rotation, specific rotation, and molecular rotation, respectively; M is the molecular mass; d is the sample path length; and c is the solute concentration, expressed either as a percent or as g/dL (IUPAC recommends retaining this concentration unit because of the high number of references employing it in the literature). Combining these equations yields an equation that is analogous to the Beer-Lambert law, namely

$$\alpha = [\Phi]\, d\, c.$$

Experimental values for α are usually on the order of millidegrees (m°) unless laser sources are used, in which case microdegrees can be measured. In the absence of absorption, the plain ORD spectrum changes monotonically with the wavelength. This change can be either positive or negative (see Figure 3.1.6a). For chiral media that absorb the polarized light beam, anomalous rotations in the ORD spectrum are produced if the chiral center and the chromophore are structurally adjacent to each other in an arrangement called a chirophore. This anomalous behaviour is referred to as the Cotton effect [29] and is limited to the wavelength range of the absorption band, where it is seen superimposed on the monotonically changing plain curve. The anomaly takes the form of a sigmoidal curve with peak and through extrema whose wavelength values are bisected at an intermediate crossover point at which the rotation is zero (Figure 3.1.6b). In the simplest case, where a single Cotton effect exists, the height between the extrema can be used for quantitative measurements.

ORD has not been extensively used as an effective method in analytical organic chemistry because of a lack of specificity in differentiation and because of the uncertainty in defining the baseline, which is the undeveloped part of the plain curve under the Cotton band.

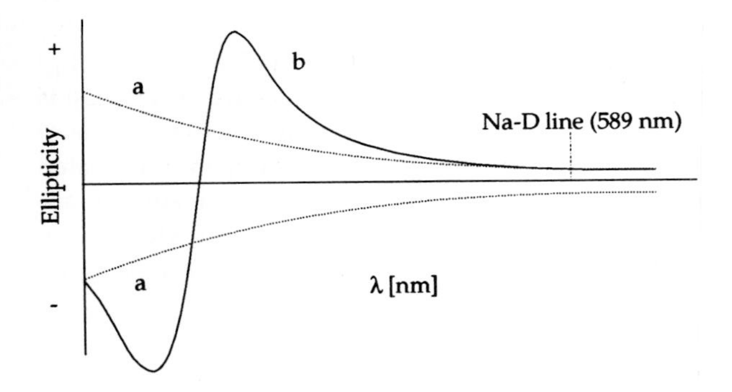

Figure 3.1.6 Typical ORD curves. (a) Plain ORD curve, (b) ORD curve with a single cotton effect (for ellipticity, cf. Section 3.1.4).

3.1.4 *Circular dichroism (CD)*

Circular dichroism (CD) is the most sophisticated of the three chiroptical methods in that the rotation and absorbance measurements are made simultaneously. Linearly polarized light consists of two beams of circularly polarized light propagating in phase but in opposite rotational senses. In chiral media the beams are phase differentiated because they 'see' two different refractive indexes and, consequently, travel at different speeds, a phenomenon that produces the rotation effect.

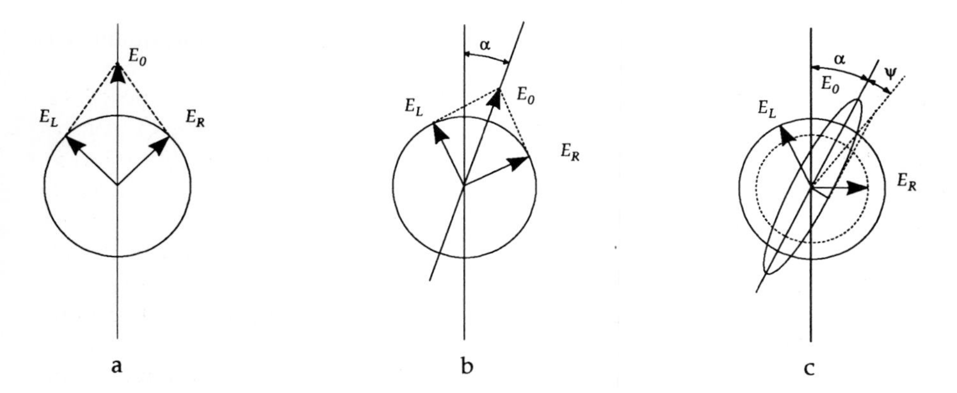

Figure 3.1.7 Phase relations associated with the passage of circularly polarized light through different media [30]; a-c, cf. explanations in the text.

CD represents the differential absorption of left circularly polarized (LCP) and right circularly polarized (RCP) light. The effect of the differential absorption is that when the electric vector projections associated with the LCP and RCP light are recombined after leaving the chiral medium, they describe an ellipse whose major axis lies along the new angle of rotation. The measure of the eccentricity of this ellipse is defined as the ellipticity ψ. These relations are illustrated in Figure 3.1.7.

(a) For optically inactive media, the recombination of left (E_L) and right (E_R) circularly polarized light yields linearly polarized light whose electric vector (E_0) is unchanged with respect to the incident axis. (b) For optically active media outside of an absorption band, recombination of E_L and E_R yields a resultant E_0 vector rotated from the incident axis by the angle α. (c) For optically active media inside an absorption band, recombination after differential absorption of E_L and E_R yields a resultant E_0 vector rotated from the incident axis by the angle α, and describing an ellipse with ellipticity ψ [30]. Consequently, it follows:

$$\psi = [\pi \, d \, (k_L - k_R)] \; / \; \lambda$$

where d is the path length in centimeters, k_L the absorption coefficient of the left and k_R the absorption coefficient of the right circularly polarized light, and λ is the particular wavelength at which ψ is measured [3]. If c is the concentration of absorbing chiral solute in moles per liter, then the mean molar absorptivity, ε, is derived from the absorption coefficient by:

$$\varepsilon = (4 \, \pi \, k) \; / \; (2.303 \, \lambda \, c)$$

In this case, the ellipticity (still in radians) becomes:

$$\psi = [2.303 \, c \, d \, (\varepsilon_L - \varepsilon_R)]$$

The expression of ellipticity in radians is cumbersome; consequently, this quantity is converted into degrees by the relation

$$\theta = \psi \, (360 \; / \; 2 \, \pi)$$

from which follows:

$$\theta = (\varepsilon_L - \varepsilon_R) \, c \, d \, (32.90)$$

The molar ellipticity is an intrinsic quantity which is calculated from

$$[\theta] = (\theta \, FW) \; / \; (l \, c' \, 100)$$

where

FW = the formula weight of the solute in question;

l = the medium path length in decimeters; and
c' = the solute concentration in units of grams per milliliter.

The molar ellipticity is related to the differential absorption by:

$$(\varepsilon_L - \varepsilon_R) = \Delta\varepsilon = [\theta] / 3298.$$

In the application of CD to analytical problems, the two physical properties of chirality and absorption provide the information necessary for qualitative and quantitative determinations, respectively. Because the change in molar absorptivity, $\Delta\varepsilon$, is defined as $(\varepsilon_L - \varepsilon_R)$, CD spectra can have positive and negative variations from the baseline as well as wavelengths where $\Delta\varepsilon = 0$ (crossover points), as shown in Figure 3.1.8. Because there is no CD signal at wavelengths where there is no absorption by the analyte, the baseline is easily defined. Thus CD is superior to ORD for analytical applications.

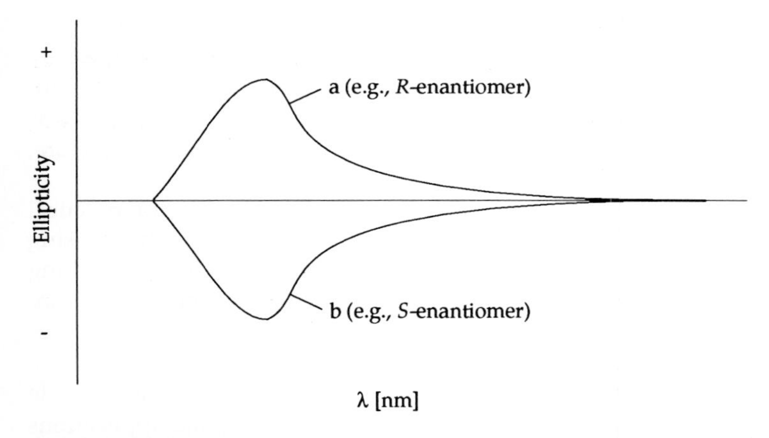

Figure 3.1.8 Typical CD curves with a single Cotton effect and (a) a positive and (b) a negative maxmum.

In the earliest CD instruments the experimental parameter measured was, in fact, the ellipticity of the transmitted beam. Modern instruments, however, are modified absorption spectrophotometers that measure the differences in the absorbances of the two beams as a function of wavelength. Because it is an absorption that is being measured, the data conform to the Lambert-Beer law.

The necessity for a chiral center in the analyte molecule makes CD a much more selective method than absorption spectrophotometry. Broad bands in absorption spectra are often separated into more than one Cotton band in a CD spectrum, and the bands are frequently of opposite polarity, enhancing the identification capabilities of CD over absorption. In addition, because the direct correlation is between the difference in the molar absorptivity, $\Delta\varepsilon$, and the molar ellipticity, θ_M, for a given analyte, the strongest CD signals are not necessarily associated with strong absorbers. This is especially useful at wavelengths > 240 nm, where molar absorbances are small compared with the strong bands observed in the far UV; CD bands are still sufficiently large. Analyte concentrations are typically 10^{-4} mol or less for wavelengths > 240 nm. CD signals in the far UV can be very large, but the signal-to-noise quality is poor whenever strong absorbers are present.

Because $\Delta\varepsilon$ is much smaller than the average ε value, the CD signal is a very small millivolt quantity riding on top of a relatively large value. Despite the large difference in signal size, detection limits are 0.1 µg/ml for analytes with θ_M values of approximately 200 m°/mol cm at the band maxima. This value can be improved by introducing fluorescence detection [31]. The use of fluorescence, however, introduces the need for a third structural requirement in the analyte molecule, which in effect decreases the applicability of CD. Instrument operating conditions and solution concentration variables are chosen to give the optimum ratio of CD to the total absorption, although one has to consider that a mixture may contain several absorbing species. When fluorescence is the detector of choice, more serious interference from the emissions of CD-passive fluorophores can be expected.

To predict whether an analyte will be optically active and can be determined by CD, the presence of chirality and absorbance must be confirmed. One must remember, however, that although the molecular structure may suggest that chirality is present, the substance may only be available as a racemic mixture and, therefore, undetectable.

Although these requirements may seem to make CD too selective for practical analytical use, the applicability of CD can be increased by adding the missing molecular property by in situ derivatization. One can make an achiral absorbing analyte CD-active by reacting it with a chiral partner, preferably one that is nonabsorbing, and a chromophore can be introduced in a way that either does or does not affect the overall chirality of the molecule.

These CD induction reactions should not be considered exclusively as possible precolumn or postcolumn derivatization reactions in chromatographic applications using CD detection. They are, instead, regarded to be so specific that they can be used for the analysis of unseparated mixtures.

Most instrumentation suitable for measuring CD is based on the design of Grosjean and Legrand [32]. A block diagram of their basic design is shown in Figure 3.1.9. Linearly polarized light is passed through a dynamic quarter wave plate, which modulates it alternatively into left and right circularly polarized (LCP and RCP) light. The quarter wave plate is a piece of isotropic material rendered anisotropic through the external application of stress. The device can be a Pockels cell (in which stress is created in a crystal of ammonium dideuterium phosphate through the application of alternating current at high voltage), or a photoelastic modulator (in which the stress is induced by the piezoelectric effect). The light leaving the cell is detected by a photomultiplier tube whose current output is converted to voltage and then split. One signal consists of an alternating signal proportional to the CD; it is due to the differential absorption of one circularly polarized component over the other. This signal is amplified by means of phase-sensitive detection. The other signal is averaged and is related to the mean light absorption. The ratio of these signals varies linearly as a function of the CD amplitude, and is the recorded signal of interest. The small signal intensity requires that the incident power be very large; a 500 W Xe lamp is usually employed. This source must be water-cooled and oxygen must be removed from the instrument to reduce the production of ozone, which is detrimental to the optics. The volume of a typical 1-cm path length cell is about 3.5 ml; smaller path length cells, which may require focusing of the incident beam, are available for analyses requiring smaller volumes.

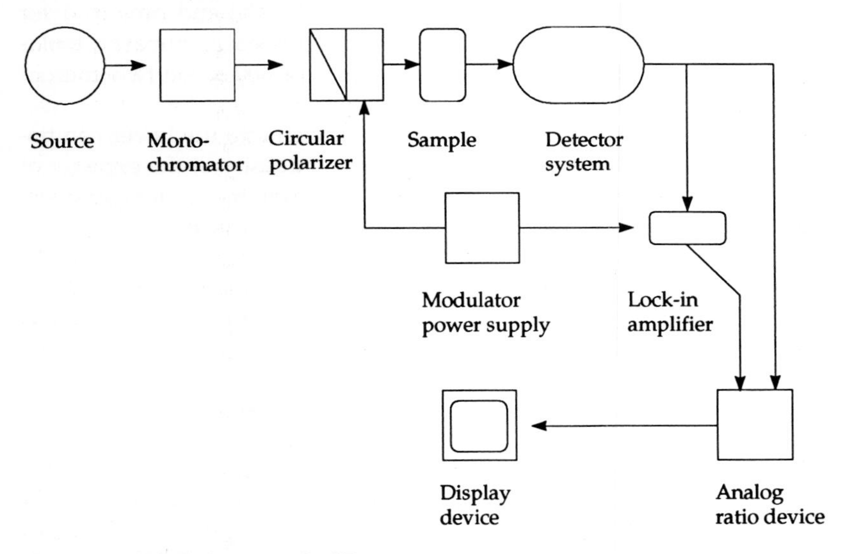

Figure 3.1.9 Block diagram of a CD spectrometer.

3.1.5 *Magnetic circular dichroism (MCD) and magnetic optical rotatory dispersion (MORD).*

Magnetic circular dichroism (MCD) is induced in all matter by a uniform magnetic field applied parallel to the direction of propagation of the measuring light beam. Although phenomenologically similar to natural CD, the molecular origin of the effect (called 'Faraday effect') is different. Both MCD and CD can be present in an optically active molecule in a magnetic field. The two effects are additive. Within the last years an increasing number of experimental MCD data has been collected and the theoretical analysis is well founded. MCD spectroscopy has found interest for applications in chemistry, physics and biochemistry [34].

The origin of the 'Faraday effect' depends on two facts: (i) Degenerate electronic states are split in the presence of a magnetic field (the first-order Zeeman effect) to yield a set of sub-levels called Zeeman components. All states are mixed together by an applied magnetic field (the second-order Zeeman effect). (ii) Electronic transitions from the Zeeman sub-levels of the ground state to those of an excited state are circulary polarized if the magnetic field is parallel (or anti-parallel) to the direction of the light beam.

The origin of MCD can be illustrated by the following example. Let us consider an electronic transition from a ^2S ground state, with spin S = 1/2 and zero orbital moment, to a $^2P_{1/2}$ excited state as shown in Figure 3.1.10. When a magnetic field is applied the degeneracies are lifted by the Zeeman effect. The selection rules for electric-dipole allowed transitions between the Zeeman sub-levels of the ground and excited states are $\Delta m_L = +1$ for the absorption of LCP light and $\Delta m_L = -1$ for RCP: $\Delta m_S = 0$. Thus the left- and right circularly polarized photons impart angular moments of opposite sign to the system. At temperatures of 300 K the two components of the ^2S ground state are almost equally populated.

When the temperature is lowered, the population is frozen into the lower component of the ground state and the LCP transition gains in intensity at the expense of the RCP intensity. Therefore, the MCD intensity of a paramagnet is temperature- (and field-) dependent, increasing in intensity as the temperature is lowered.

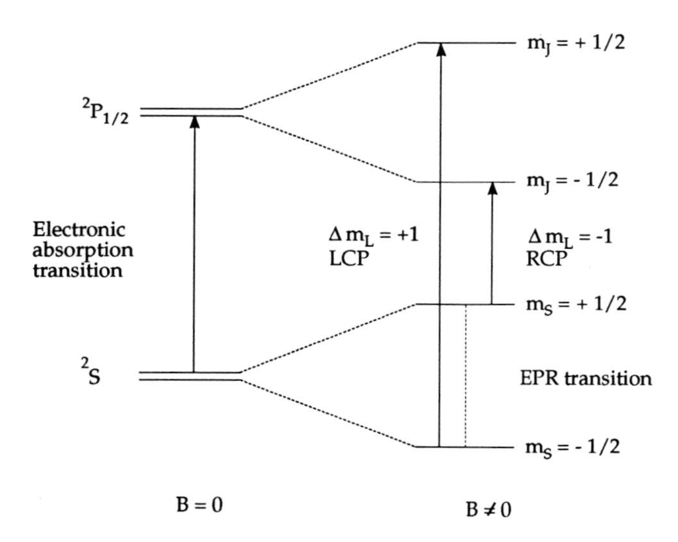

Figure 3.1.10 The origin of the MCD in the atomic transition ^2S to ^2P. The figure shows the allowed optical transitions between the Zeeman sub-levels of the ground and excited states [33].

3.1.6 *Vibrational optical activity (VOA) [33]*

Vibrational optical activity (VOA) comprises both vibrational circular dichroism (VCD) and Raman optical activity (ROA). VCD measures the difference in absorbance of LCP and RCP infrared light in the region of vibrational absorption bands of optically active molecules. ROA measures the difference in scattered intensity of LCP and RCP incident laser radiation. Vibrational optical activity is becoming a powerful tool for determining the stereochemistry of chiral molecules - both the conformation and the absolute configuration. Unlike electronic CD spectroscopy, which provides information only about chromophores and their immediate environments, in VCD every part of a molecule can contribute to the spectrum. Theoretically, it should be possible to determine both absolute configuration and conformation from the VCD or ROA alone. The primary experimental difficulty is that VOA is very weak, with signals being four or five orders of magnitude smaller than the parent effects, i.e. infrared absorption and Raman scattering. However, within the last years great progress has been made in both areas [35-39].

The highest and lowest frequencies at which VCD has been reported using dispersive instruments are approx. 6000 cm^{-1} and 900 cm^{-1}, respectively. Over this range the sensitivity limit in terms of absorbance $A = \varepsilon\ c\ d$ (where c is the concentration and d the path-length) can be $\Delta A = 10^{-5} - 10^{-6}$ at a bandwith of 5 - 10 cm^{-1}, sufficient to resolve most room temperature liquid-phase VCD spectra. The low-frequency limit of VCD measurements, using Fourier transformation instrumentation, is now approximately 600 cm^{-1}. Stephens and Lowe [38] have described a general theory of VCD, the so-called non-adiabatic theory, permitting prediction of vibrational rotational strengths and spectra. There are also a variety of heuristic models, including coupled oscillators and fixed partial charges.

For ROA no such frequency limits exist, although the largest effects commonly occur at frequencies below 1000 cm^{-1}. Furthermore, the difficulty of obtaining measurable signals in ROA has limited experiments to very concentrated samples, either pure liquids or saturated solutions, where intermolecular effects may be dominant. VCD, on the other hand, has been measured on sample concentrations of 0.1 - 0.01 mol. Since the demonstration that VCD spectra can be measured with high reliability, theoretical analysis has advanced rapidly. The field is entering a phase of collecting spectra and making comparisons with theoretical models.

3.1.7 *Detectors used in liquid chromatography*

The chiroptical detectors used in liquid chromatography (LC) are primarily single-wavelength detectors. Because of the constraints of both signal size and small elution volumes, lasers are the most suitable light sources for these detectors. Yeung and co-workers have described both polarimetric and CD detectors for LC [31,40]. Stopped-flow CD spectral detection for LC has been described both by development engineers from JASCO, Inc. [41] and by Westwood et al. [42].

Chiroptical detectors are particularly useful in studying substances of natural origin, and their use can complement the more common chromatographic detectors in investigating complex mixtures. The majority of the applications developed to date have involved laboratory preparations; the number of real samples investigated is very small.

If total separation of a mixture is possible, polarimetry is the chiroptical detector of choice. It has been used, for example, to identify and quantitate structurally related carbohydrates in a mixture [43]. Polarimetric instrumentation is inexpensive both to purchase and to operate, and the detector responds equally well to absorbing and nonabsorbing analytes. Furthermore, the effectiveness of a polarimetric detector has been demonstrated in both the direct mode, where the rotation caused by the analyte is measured [40], and in the indirect mode, where the change in the measured background rotation for an optically active mobile phase is used to quantitate the analyte [44].

A detector for high performance liquid chromatography (HPLC) based on optical activity would seem to possess several advantages in many attractive areas of organic analysis. Since most chromatographic eluents are not optically active, one is not limited in the choice of eluents or gradients. Such a detector is extremely selective, so that complex samples can be analyzed. The availability of a sensitive micropolarimeter will, therefore, benefit organic analysis when coupled to HPLC, and will broaden the applicability of spectropolarimetry in general.

For over a century, mechanical polarimeters have been constructed with sensitivities on the order of 0.01°. An example of this class of polarimeter is the model 241 LC from Perkin Elmer. Monochromatic light is passed through the polarizer, the flow cell (40 or 80 µl) with the sample and through the analyzer to a photomultiplier as detector (cf. Figure 3.1.5). The polarizer, which also means the polarization plane of the light, is modulated at 50 Hz at an inclination of 0.7° around the optical axis of the system. In the unbalanced state of the system a 50 Hz signal is produced in the photomultiplier which is intensified and transmitted with the correct sign to a servomotor. This motor turns the mechanically connected analyzer until the 50 Hz signal is reduced to zero. Thus, the system becomes balanced (optical zero balance) and the polarization planes of polarizer and analyzer form an angle of exactly 90°. An optically active sample placed into the light beam rotates the polarization plane; the analyzer is again balanced by the servomotor (at a speed of approximately 1.3°/s). The rotation of the analyzer is transformed into electric impulses by an optical encoder and the impulses are evaluated. When mechanical polarimeters are used the possibilities of chromatographic loss of resolution power caused by the finite balance velocity has to be taken into consideration. In addition, the adjustment of the flow cell is very time-consuming.

Currently available commercial polarimeters use the technique of Faraday compensation, resulting in sensitivities on the order of 0.001°. Examples of this class of instrument are the ChiraMonitor (ACS - Applied Chromatography Systems, UK) and the Chiralyser (IBZ, Hannover and Knauer, Berlin; both Germany). How they function is shown below, using the ChiraMonitor as example (Figure 3.1.11). The instrument consists of a solid state near-infrared laser (820 nm) chosen so that there is very little interference from the absorbing characteristics of compounds. The radiation from the laser passes, after being focused to less than 1 mm in diameter, through a polarizing prism to a calibrator/modulator with the Faraday rod (made from a special glass material) which gives rise to a 1 kHz (f_1) polarisation modulation of the laser beam varying about 1° around the polarizer/analyzer cross point (Figure 3.1.12; angle α).

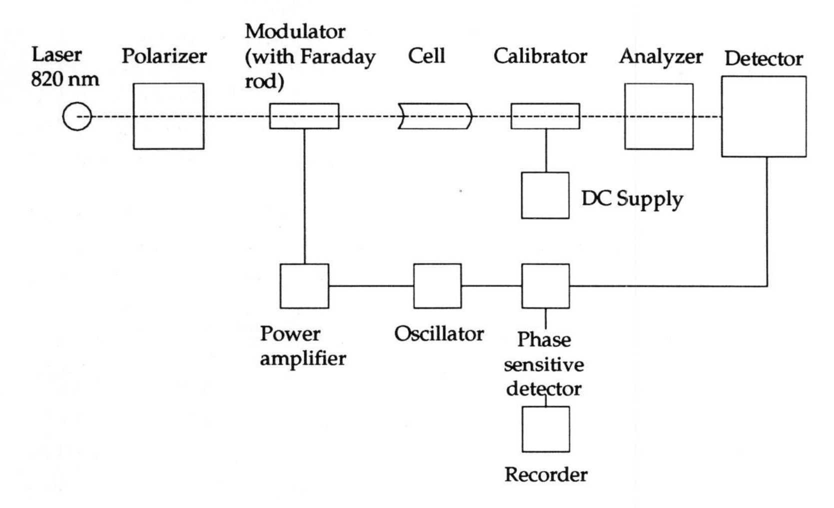

Figure 3.1.11 Schematic diagram of the polarimetric LC detector ChiraMonitor (ACS, UK).

This 1 kHz signal is also the reference signal for the phase sensitive detector (PSD). The calibrator is controlled by a DC power supply and it feeds a signal to the detector to check the calibration of the detector. After passing through the flow cell the light enters the analyzer. With polarizer and analyzer crossed at exactly 90° the exciting light is a pure 2 kHz (2 f_1) amplitude-modulated carrier signal of constant polarization.

Any optical rotation due to the sample (Figure 3.1.12; angle δ) deflects the system away from the cross point and there is a resultant 1 kHz amplitude modulation of the carrier. This signal is then recovered using a phase-sensitive detector. The phase anomaly generates a compensation current in the Faraday coil via the power amplifier which creates an electrical compensation field and compensates the influence of the optically active substance until the phase anomaly disappears.

The flow cell is the most critical component for optimization of the signal-to-noise ratio. The dimensions of the cell always represent a compromise between having a long light path and a small volume and must allow a laminar flow distribution. Whenever there is a sudden, large change in the refractive index, such as when a high concentration of material passes through, the laser beam is distorted and will not pass through the analyzer and the apertures properly. A similar problem exists when absorption processes cause thermal lensing. This can, in principle, be avoided by choosing an appropriate laser wavelength. Bubbles trapped inside the cell may also be a problem, particularly when the cell is used initially.

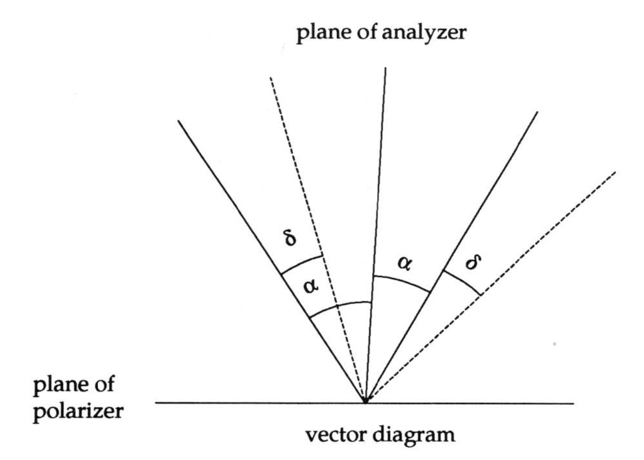

Figure 3.1.12 Function principle of the polarimetric LC detector ChiraMonitor (ACS, UK).

Scattering and reflections of the cell walls can also increase the noise level substantially. A flow direction against the light source can act as a hydrostatic lens to improve the signal/noise ratio of the instrument. Using the Drude relationship, it is possible to compare as follows the rotations measured at 820 nm (wavelength of the laser in the ChiraMonitor) with tabled data at 589 nm.

$$\alpha_D / \alpha_M = (\lambda_M^2 - \lambda_A^2) / (\lambda_D^2 - \lambda_A^2)$$

α_D = rotation at Na-D-line (589 nm)
α_M = rotation at 820 nm
λ_M = wavelength of the laser used (820 nm)
λ_D = wavelength of the Na-D-line (589 nm)
λ_A = absorption wavelength of the chromophore.

This relationship only holds for molecules with one optically active chromophore, or where the major contribution to a substance's optical activity is from one chromophore. More complex molecules can not be so easily described.

Example: A typical saturated lactone shows a λ_A around 200 nm. Using the Drude equation the quotient α_D/α_M = 2.06 results. This means the rotation measured at 820 nm is only half the rotation at 589 nm.

The differences between the ChiraMonitor and the Chiralyser are mainly in the construction of the flow cell and the light source. The cell of the ChiraMonitor consists of a stainless steel block through which a 1 mm diameter sample cell (opening out to 1.5 mm) is machined. The volume of the cell is approximately 20 µl. The flow cell of the Chiralyser consists of a glass-coated stainless steel tube with an optical length of 200 mm and a volume of approximately 40 µl. The reflecting surface of the glass coating of the flow cell allows optimal light transmission and a laminar flow distribution within the cell.

In contrast to the ChiraMonitor, which uses a laser diode, the light source of the Chiralyser consists of a halogen lamp allowing only polychromatic determination of the optical rotations. The advantage of this technique is the increased sensitivity compared with the older 'Chiraldetector' distributed by the same companies.

Limiting CD detection to a single-wavelength measurement reduces it to no more than a very expensive polarimeter with a much smaller range of application, since the nonabsorbing chiral compounds are now transparent to the detector. But if separation is not complete, then the differentiation capability of the full-range CD detector becomes necessary. As the need arises, convenient instrumentation for fast-scan measurements may become available. Early attempts to develop such devices have been described in the literature [45,46].

3.1.8 *Enantiomeric differentiaton*

Enantiomeric differentiation is a two-level problem. If the identity of the substance is known and only one isomer is present, then the sign of the rotation easily establishes its stereochemical identity, in which case polarimetric detection is sufficient. If both enantiomers are present, which is normally the case, the analysis takes on a different dimension; one has to determine the ee value or optical purity.

Polarimetry is the best choice to determine the enantiomeric enrichment at the exploratory level, where eluted volumes are small. When chromatographic procedures are developed to the point where large-scale separations are possible, CD is the better detector, because differences in the full spectrum of the analyte compared with that of the standard signify the presence of a coeluted chiral interference.

Because nonderivatized racemic mixtures coelute from conventional LC columns, neither a conventional detector nor a chiroptical detector alone is adequate to determine the ee. If the detectors are placed in series, however, a quantitative distinction can be made. Data from either an absorbance or RI detector provides the sum of the concentrations of the two isomers and the signal from the chiroptical detector (which is either the rotational difference, $[\alpha_{(+)} - \alpha_{(-)}]$, for the polarimetric detector, or the difference in ellipticity, $[\psi_{(+)} - \psi_{(-)}]$, for CD) provides information to calculate the concentration difference [47,48]. The concentration of each isomer is then readily obtained from the simultaneous solution of these equations.

In many situations where CD is the detector of choice, its selectivity is so great that it can be used as a stand-alone detector, providing the concentration difference information without separation. This is especially important whenever the eluted volumes are small, because of the small CD signal. An aliquot is injected onto a conventional column and the total concentration of both enantiomers is measured using absorption detection. The concentration difference is calculated simultaneously from the CD spectrum of another aliquot of the unseparated mixture. The ee is then calculated as described earlier.

Whichever method is used to determine the ee, the quality of the results depends decisively on the optical purity of the standard materials. These can never be considered to be optically pure, since the instrumental or separation methods are limited by their resolution capabilities. Theoretically, for spectroscopic analysis it is necessary to have only one of the isomers for instrument calibration, provided that diastereoisomerization is not a prerequisite for the determination, as it is in NMR. To have both isomers of equivalent optical purity as an internal check of the calibration in chiroptical methods is an unrealistic expectation. Reports of enantiomer ratio determinations should emphasize the fact that the ratio is relative to the purity of the best available standard reference material.

The detection limits obtained with chiroptical detectors are equivalent to those obtained with absorbance detectors for bulk measurements. In conventional chromatographic systems, nanogram detection limits are usual; in state-of-the-art detector development, picogram or even femtogram limits have been reported [48]. The ability to detect such small quantities is of critical importance only when the physical size of the sample is limited, as is the case in microbore LC, where the eluted volumes are very small. If the sample size is not limited, the simple alternative is to scale up the experiment.

Chiroptical methods are not yet competitive within the lowest of these achievable detection ranges, showing a typical cutoff in the nanogram-per milliliter range, unless fluorescence is used for signal enhancement or laser sources are used [40]. When sample sizes are not a problem and a typical working sample volume is a few milliliters, detection limits on the order of micrograms per milliliter are readily achieved using CD.

3.1.9 *Analytical applications*

Polarimetry and ORD have no potential as selective detectors; they are functional only when all interference has been removed. CD is in the same category as long as its use is limited to single-wavelength chromatographic detection. In many cases the most useful wavelength range is from 240 to 400 nm, which comprises the transitions from the aromatic ring and unsatured ketone chromophores. Few substances are CD-active in the visible range; at wavelengths < 240 nm signal-to-noise ratios are significantly decreased because of the extremely intense absorption bands. In addition, CD bands are observed to be broad and featureless and usually of only one sign, resulting in spectra that are the same as the corresponding absorption spectra.

The need for two detectors in the determination of ee or optical purity was discussed above. Some early examples combined UV or RI with polarimetric detection; for instance, cocaine and codeine [49], epinephrine [50], and D- and L-penicillamine [51] were investigated in this manner.

UV and CD were successfully used in series for prepared mixtures of *R*- and *S*-nicotine in which solutions of the natural isomer were spiked with aliquots of the other [1]. Subsequently, leaf extracts were spiked with the unnatural isomer and the ee was determined using conventional LC. The total nicotine concentration was measured by LC using an absorbance detector, and the CD spectrum of either an aliquot of the unseparated mixture or eluate from a conventional LC column yielded the data from which the concentration difference was calculated. The concentration of each was obtained by the simple solution of the simultaneous equations.

The assay of glycosides is a relatively unexplored area with several attractive possibilities for the application of chiroptical detectors. Structurally, these compounds fulfill the requirements for CD activity by having the chiral center in the sugar moiety and an aromatic chromophore in close juxtaposition; the connection between the two parts is through either carbon (cyanogenics), nitrogen (nucleosides and nucleotides), oxygen (saponins and flavonoids), or sulphur (glucosinolates). The magnitude of the CD signal will depend on how adjacent the nearest chiral center on the sugar is to the chromophore.

Nakanishi and coworkers have developed - based on CD exciton chirality - a microscale method for characterizing the structures of monosaccharides and their linkages in oligosaccharides. Sugar components were identified by UV and CD spectroscopy of chromophoric degradation products. CD spectral data of approximately 150 different reference glycopyranosides have been published [52].

The exciton chirality method is also useful for the stereochemical analysis of the aglycone part in glycosides. Recently, the determination of the absolute stereochemistry of natural 3,4-dihydroxy-ß-ionone glycosides has established the configuration as being 3ß,4ß (3*S*,4*R*) [53].

Furthermore, a lot of ORD and CD spectroscopy has been done in the field of carotenoid chemistry. In a fundamental paper by Klyne's and Weedon's groups [54] ORD spectra of carotenoids have been studied systematically. Extensive CD studies of carotenoids have been performed by Noak, Liaaen-Jensen and others [55-57]. Theoretical predictions and experimental data were shown by Sturzenberger et al. [58] to conform to the 'C_2-rule' [59].

References

[1] Purdie, N.; Swallows, K. A. *Anal. Chem.* (1989), *61*, 77A

[2] Crabbe, P. *ORD and CD in Chemistry and Biochemistry: An Introduction*; Academic Press: New York, 1972

[3] Charney, E. *The Molecular Basis of Optical Activity*; Wiley: New York, 1979

[4] Mason, S. F. *Q. Rev. Chem. Soc.* (1961), *15*, 287

[5] Snatzke, G. *Chemie in unserer Zeit* (1981), *3*, 78; (1982), *5*,160

[6] Purdie, N. *Prog. Anal. At. Spectrosc.* (1987), *10*, 345

[7] Drake, A. F. *Eur. Spectrosc. News* (1986), *69*, 10

[8] Harada, N.; Nakanishi, K. *Circular Dichroism Spectroscopy-Exciton Coupling in Organic Chemistry*; University Science Books: Mill Valley, 1983; Nakanishi, K.; Berova, N.; Woody, R.W. *Circular Dichroism-Principles and Applications*, VCH Verlagsgesellschaft: New York, 1994

[9] Schurig, V. *Kontakte (Darmstadt)* (1985), *1*, 54

[10] Eliel, E. L. *Stereochemistry of Carbon Compounds*; McGraw-Hill: New York; 1962

[11] Lyle, G. G.; Lyle, R. E. *Asymmetric Synthesis* (J. D. Morrison, ed.) Vol. 1, 13 Academic Press: New York, 1983
[12] *Beilsteins Handbuch der Organischen Chemie*, 4. Auflage, Band 3; Springer: Berlin, 1921
[13] Consiglio, G.; Pino, P.; Flowers, L. I.; Pittman jr.; C. U. *J. Chem. Soc., Chem. Commun.* (1983), 612
[14] Heller, W.; Curme, H. G. *Physical Methods of Chemistry* (Weissgerber, A.; Rossiter, B. W., eds.), Wiley: New York, part III C, 51
[15] Raban, M.; Mislow, K. *Top. Stereochem.* (1967), *1*, 1
[16] Horeau, A. *Tetrahedron Lett.* (1969), 3121
[17] Hoeve, W. T.; Wynberg, H. *J. Org. Chem.* (1980), *45*, 2754
[18] Horeau, A. *J. Am. Chem. Soc.* (1964), *86*, 3171
[19] Horeau, A. *Bull. Soc. Chim. Fr.* (1964), 2673
[20] Schoofs, A. R., Guetté, J.-P. *Asymmetric Synthesis* (J. D. Morrison, ed.); Academic Press: New York, 1983, Vol. I, 29
[21] Weber, R.; Schurig, V. *Naturwissenschaften*, (1984)
[22] Mori, K. *Technique of Pheromone Research* (Hummel, H. E.; Miller, T. A., eds), Springer: New York, 1984, 323
[23] Jaques, J.; Collet, A.; Wilen, S. H. *Enantiomers, Racemates and Resolutions*, Wiley: New York, 1981
[24] Leitich, J. *Tetrahedron Lett.* (1978), 3589
[25] Horn, D. H. S.; Pretorius, Y. Y. *J. Chem. Soc.* (1954), 1460
[26] Klyne, W.; Buckingham, J. *Atlas of Stereochemistry*, Vol. I, Chapman Hall: London, 1974
[27] Plattner, P. A.; Heusser, H. *Helv. Chim. Acta* (1944), *27*, 748
[28] Schurig, V. *Asymmetric Synthesis* (J. D. Morrison, ed.), Vol 1, 59, Academic Press: New York, 1983
[29] Cotton, A. *Compt. Rend.* (1895), *120*, 989
[30] Brittain, H. G. *Spectrosc. Int.* (1991), *3*, 12
[31] Synovec, R. E.; Yeung, E. S. *J. Chromatogr.* (1986), *368*, 85
[32] Velluz, L.; Legrand, M.; Grosjean, M. *Optical Circular Dichroism: Principles, Measurement, and Application*; Verlag Chemie: Weinheim, 1965
[33] Thomson, A. J. *Perspectives in Modern Chemical Spectroscopy* (Andrews, D. L., ed.), 255, Springer: Berlin, New York, 1990
[34] Piepho, S.; Schatz, P. N. *Group Theory in Spectroscopy*, Wiley: New York, 1983
[35] Osborne, G. O.; Cheng, J. C.; Stephens, P. J. *Rev. Sci. Inst.* (1973), *44*, 10
[36] Nafie, L. A.; Keiderling, T. A.; Stephens, P. J. *J. Am. Chem. Soc.* (1976), *98*, 2715
[37] Annamalai, A.; Keiderling, T. A. *J. Am. Chem. Soc.* (1984), *106*, 6254
[38] Stephens, P. J.; Lowe, M. A. *Ann. Rev. Phys. Chem.* (1985), *36*, 213
[39] Nafie, L. A.; Diem, M. *Acc. Chem. Res.* (1979), *12*, 296
[40] Synovec, R. E.; Yeung, E. S. *Anal. Chem.* (1986), *58*, 1237A
[41] Takakuwa, T.; Kurosu, Y.; Sakayanagi, N.; Kaneuchi, F.; Takeuchi, N.; Wada, A.; Senda, M. *J. Liquid Chromatogr.* (1987), *10*, 2759
[42] Westwood, S. A.; Games, D. E.; Sheen, L. *J. Chromatogr.* (1981), *204*, 103
[43] DiCesare, J. L.; Ettre, L. S. *Chromatogr. Rev.* (1982), *220*, 1
[44] Yeung, E. S. *J. Pharm. Biomed. Anal.* (1984), *2*, 255

[45] Anson, M.; Bayley, P. M. *J. Phys. E* (1974), *7*, 481
[46] Hatano, M.; Nozawa, T.; Murakami, T.; Yamamoto, T.; Shigehisa, M.; Kimura, S.; Kakakuwa, T.; Sakayanagi, N.; Yano, T.; Watanabe, A. *Rev. Sci. Instrum.* (1981), *52*, 1311
[47] Boehme, W. *Chromatogr. Newsl.* (1980), *8*, 38
[48] Meinard, C.; Bruneau, P.; Perronnett, J. *J. Chromatogr.* (1985), *349*, 109
[49] Palma, R. J.; Young, J. M.; Espenscheid, M. W. *Anal. Letters* (1985), *18*, 641
[50] Scott, B. S.; Dunn, D. L. *J. Chromatogr.* (1985), *319*, 419
[51] DiCesare, J. L.; Ettre, L. S. *J. Chromatogr.* (1962), *251*, 1
[52] Wiesler, W.T.; Berova, N.; Ojika, M.; Meyers, H.V.; Chang, M.; Zhou, P.; Lo, L.C.; Niwa, M.; Takeda, R.; Nakanishi, K. *Helv. Chim. Acta* (1990), *27*, 748
[53] Humpf, H.U.; Zhao, N.; Berova, N.; Nakanishi, K.; Schreier, P. *J. Nat. Prod.* (1994), *57*, 1761
[54] Bartlett, L.; Klyne, W.; Mose, W.P.; Scopes, P.M.; Galasko, G.; Mallams, A.K.; Weedon, B.C.L.; Szabolcs, J.; Toth, G. *J. Chem. Soc. C* (1969), 2527
[55] Noak, K. In *Carotenoid Chemistry and Biochemistry*, Britton, G.; Goodwin, T.W., Eds., Pergamon: Oxford, 1982, p. 135
[56] Liaaen-Jensen, S. In *Proc. Intern. Conference on Circular Dichroism*, Bonn, 1991, p. 47
[57] Buchecker, R.; Marti, U.; Eugster, C.H. *Helv. Chim. Acta* (1980), *65*, 896
[58] Sturzenberger, V.; Buchecker, R.; Wagnière, G. *Helv. Chim. Acta* (1980), *63*, 1074
[59] Wagnière, G.; Hug, W. *Tetrahedron Letters* (1970), 4765; Hug, W.; Wagnière, G. *Helv. Chim. Acta* (1971), 54, 633; Hug, W.; Wagnière, G. *Tetrahedron* (1972), 28, 1241

3.2 Nuclear magnetic resonance

Nuclear magnetic resonance (NMR) does not allow to differentiate between enantiomers, since the resonances of enantiotopic nuclei are isochronous. The determination of enantiomeric compositions by NMR spectroscopy, therefore, requires the conversion of the enantiomers into diastereomers by means of a chiral auxiliary. The chemical shift nonequivalence of diastereotopic nuclei in diastereoisomers in which the stereogenic centers are covalently linked in a single molecule was first noted by Cram [2]. Under appropriate experimental conditions the chemical shift nonequivalence provides a direct measure of diastereomeric composition which can be related directly to the enantiomeric composition of the original mixture.

Three types of chiral auxiliary are used. (i) Chiral lanthanide shift reagents (CLSR) [3,4] and (ii) chiral solvating agents (CSA) [5,6] form diastereomeric complexes in situ with substrate enantiomers and may be employed directly. (iii) Chiral derivatizing agents (CDA) [7] require the separate formation of discrete diastereoisomers prior to NMR analysis. With CDA it has to be ensured that neither kinetic resolution nor racemization of the derivatizing agent occurs during derivatization.

3.2.1 *Chiral derivatizing agents (CDA)*

Derivatization of enantiomers with an enantiomerically pure compound (CDA) is the most widely used NMR technique for the assay of enantiomeric purity. In contrast to chiral lanthanide shift reagents (CLSR) and chiral solvating agents (CSA), which form diastereomeric complexes that are in fast exchange on the NMR time scale, derivatization yields discrete diastereomers for which the observed chemical shift nonequivalence $\Delta\delta$ is typically five times greater than for related complexes with a CSA.

Several prerequisites exist for the CDA method: The derivatizing agent must be enantiopure; the presence of a small amount of the enantiomeric compound reduces the enantiomeric purity. During the formation of diastereomers racemization must be excluded. For instance, racemization during ester formation had been observed by Raban and Mislow [8] as they first reported the chemical shift nonequivalence in the ^{1}H NMR spectra of diastereomeric 2-phenylpropionic acid esters of 1-(2-fluorophenyl)ethanol. In addition, the possibility of kinetic resolution due to differential reaction rates of the substrate enantiomers must be excluded. This danger can be minimized by using an excess of the derivatizing agent.

3.2.1.1 ^{1}H and ^{19}F NMR analysis

Alcohols and amines

Selected examples of useful CDAs for ^{1}H and/or ^{19}F analysis are represented in Table 3.2.1. The most widely used is α-methoxy-α-(trifluoromethyl)phenylacetic acid (MTPA) (**1**), introduced by Mosher in 1969 [9,10]. Since there is no hydrogen atom at the chiral center, racemization during derivatization is excluded. MTPA is available commercially in enantiomerically pure form, either as the acid or the acid chloride. Reaction with primary and secondary alcohols or amines forms diastereomeric amides or esters that may be analyzed by ^{1}H or ^{19}F NMR [9-11]. In ^{1}H NMR analysis chemical shift nonequivalence is typically 0.1 to 0.2 ppm ($CDCl_3$; 298 K). Problems with kinetic resolution have been reported [12,13], however, NMR analysis after MTPA derivatization remains the method of choice for simple chiral amines and alcohols [14-16]. Often the diastereomers can be separated by GC or HPLC as well (see Sections 3.4 and 3.5), permitting independent verification of enantiomeric purity.

Table 3.2.1 Selected chiral derivatizing reagents for ^{1}H and ^{19}F NMR analysis

R-O-Acetylmandelic acid
R,*R*-2,3-Butanediol
Camphanic acid
R-2-Fluoro-2-phenylethylamine
S-O-Methylmandelic acid
S-α-Methoxy-α-(trifluoromethyl)phenylacetic acid (MTPA)
S-α-Naphthylethylamine
S-α-Phenylethylamine

The accuracy of the measured values depends upon the instrumental conditions, the methods of data handling, and the size of the shift nonequivalence. The error can be estimated to be +/-1%. Although several analogues of MTPA have been studied, e.g., **2a-e** [17] and **3a-d** [18], they suffer from racemization under the forcing conditions required to form ester derivatives of sterically hindered alcohols.

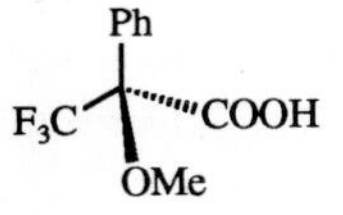

Ph, F_3C, COOH, OMe — **1**

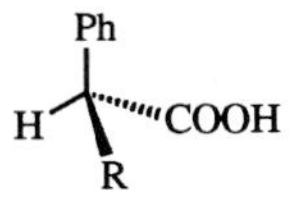

Ph, H, COOH, R — **2a-e**

R: a) OMe
b) t-Bu
c) CF_3
d) OH
e) Cl

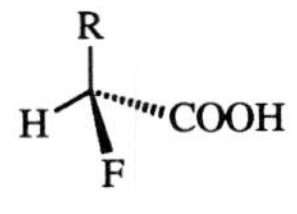

R, H, COOH, F — **3a-d**

R: a) SPh
b) Ph
c) OPh
d) CH_2Ph

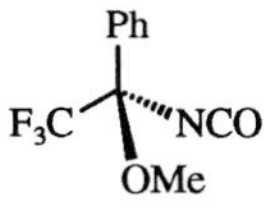

Ph, F_3C, NCO, OMe — **4**

Some success has been achieved with the isocyanate **4** [19]. The isocyanate does not react with hindered alcohols, but with primary and secondary chiral amines it yields diastereomeric ureas that show higher chemical shift nonequivalence than the corresponding MTPA derivatives. With improvements in the synthesis of esters (also with hindered alcohols [20]) or amides under nonracemizing conditions, derivatizing agents other than MTPA may be used. Thus, in the analysis of chiral alcohols, O-methylmandelic acid (**5**) and O-acetylmandelic acid (**6**) [21,22] can also be used. Often higher values of $\Delta\delta$ for the diastereomeric derivatives will be obtained.

Ph
MeO–C–COOH
H
5

Ph
AcO–C–COOH
H
6

In the past, NMR configurational correlation schemes have been developed that permit the assignment of the absolute configuration of alcohols and amines in MTPA and related mandelate derivatives [10,23]. In the MTPA model, the α-trifluoromethyl and carbonyl oxygen are eclipsed, as supported by chiroptical measurements [24], so that the preferred conformation has the carbonyl hydrogen eclipsed with the carbonyl group. Assuming an extended trans-ester or amide conformation, an extended Newman projection of the preferred conformation places one of the groups (R_1 in Figure 3.2.1) close to the phenyl ring.

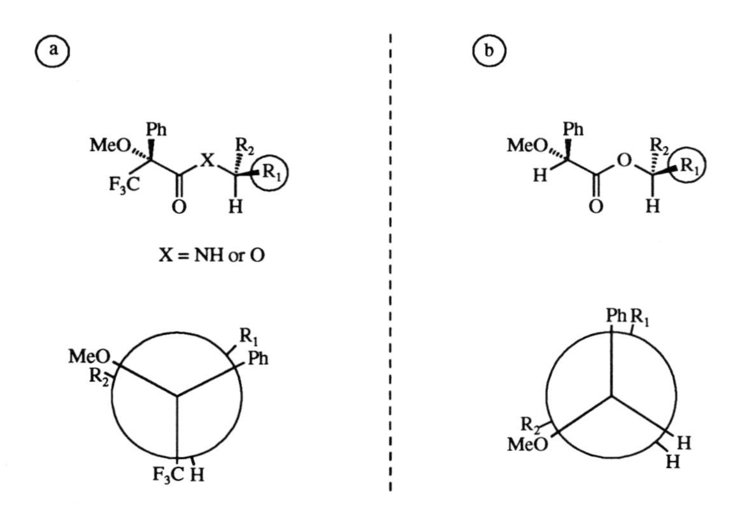

Figure 3.2.1 *R*-MTPA (a) and *R*-O-methylmandelic acid derivatives (b).

Being close to the shielding influence of the magnetically anisotropic aromatic group, this group, R_1, resonates to lower frequency (i.e., to higher field) of R_1 in the alternative diastereomer (where R_1 and R_2 exchange sites). The model extends well to MTPA derivatives of α-hydroxy esters [25] or α-amino esters [26]. In these derivatives it was observed that the chemical shift nonequivalence of the diastereotopic α-methoxy group was much larger ($\Delta\delta_H$ = 0.2 ppm ($CDCl_3$)) than it is usually the case in simple MTPA amides or esters ($\Delta\delta_H$ = 0.04 ppm). Another method for the determination of the absolute configuration of secondary alcohols, amines and thiols uses *R*-2-phenylpropionic acid (cf. 'Practical examples') [27].

The magnitude of chemical shift nonequivalence with MTPA and related derivatizing agents may often be enhanced by the addition of an achiral lanthanide shift reagent, such as $Eu(fod)_3$ ('fod' = 6,6,7,7,8,8,8-heptafluoro-2,2-dimethyl-3,5-octanedione). In MTPA derivatives, induced shifts have been found whose magnitude was generally greater for the methoxy singlet in the *R,R* diastereomer than in the *R,S* isomer [28]. Similarly, the lanthanide-induced shift was greater for the o-phenyl protons in the *R,R* isomer as compared with the *R,S* diastereomer. This effect has been observed in a high number of cyclic and acyclic secondary alcohols, permitting the determination of absolute configuration and enantiomeric purity [29,31]. The method is also operative with MTPA derivatives of α- and β-amino acid esters [26] and α- and β-hydroxy acid esters [32]. It has also been extended to atropisomers, such as the bisnaphthole **7** [33].

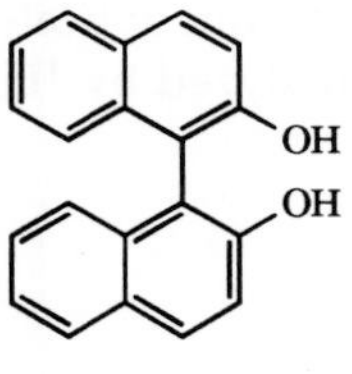

OH
OH

7

Another useful chiral derivatizing agent is camphanic acid [34]. It was originally used by Gerlach [35] to determine the enantiomeric purity of α-deuterated primary alcohols. Later, it was employed in the chiral analysis of amines [22] and β-amino alcohols [36]. The methyl singlets in the camphanoyl moiety are useful ^1H NMR reporter groups, and anisochronous resonances have been observed (typically, $\Delta\delta_H$ = 0.06 ppm in $CDCl_3$ or deuterated benzene) for a high number of substrates.

Carboxylic acids

There are only few useful chiral derivatizing agents for carboxylic acids. *R*-α-Phenylethylamine has been used in the analysis of compounds **8** [37] and **9** [38], the latter supported by the addition of $Eu(fod)_3$. A very useful CDA for acids is *S*-methyl mandelate [22,39]. Derivatization works with chiral carboxylic acids using dicyclohexyl carbodiimide (DCC) as a coupling agent in the presence of the acyl transfer catalyst 4-(dimethylamino)pyridine (DMAP). The mandelate methine proton resonates at 6.05 ppm in typical esters; a shift nonequivalence of up to 0.2 ppm is commonly observed. This agent has also been used in the determination of the enantiomeric purity and absolute configuration of chiral α-deuterated primary carboxylic acids [22,40,41].

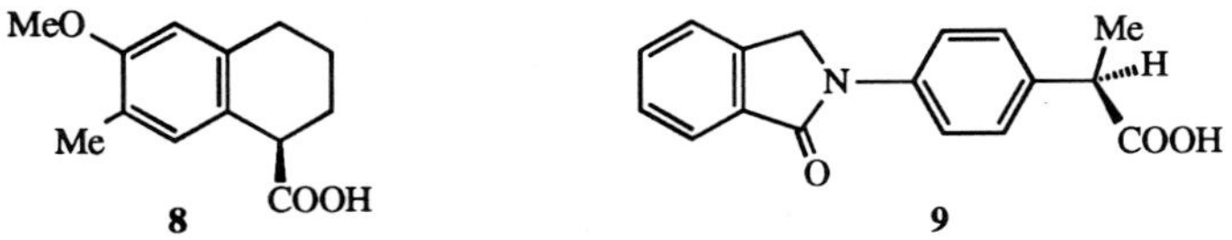

MeO Me COOH **8**

N O Me H COOH **9**

Carbonyl compounds

For the analysis of carbonyl compounds CDAs are not commonly used. Usually, the carbonyl is reduced under nonracemizing conditions, e.g., using $LiAlH_4$. The corresponding alcohol is then derivatized with MTPA or another CDA. Derivatization of carbonyls can be performed by forming chiral 1,3-dioxolanes under acid catalysis with enantiomerically pure diols [42] such as *R,R*-butane-2,3-diol or *R,R*-pentane-2,4-diol [43-45]. Similarly, chiral oxazolidines [46] have been derived from 1*R*,2*S*-ephedrine and imidazolidines have been prepared from optically pure 1,2-diaryldiamines [47,48] (cf. Figure 3.2.2); they have been analyzed by ^{1}H and ^{19}F NMR.

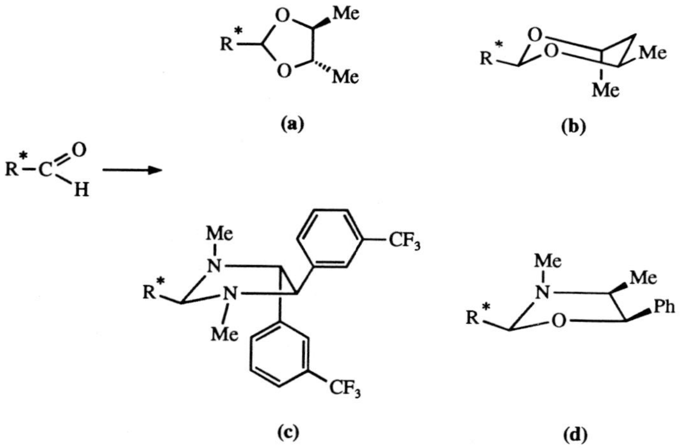

R*–C(=O)H →

(a) R* O Me O Me

(b) R* O O Me Me

(c) Me N R* N Me CF_3 CF_3

(d) Me N R* O Me Ph

Figure 3.2.2 Derivatizing scheme of chiral aldehydes using (a) *R,R*-butane-2,3-diol, (b) *R,R*-pentane-2,4-diol, (c) N,N'-dimethyl-di-(3-trifluoromethylphenyl)-aminoethane, (d) 1*R*,2*S*-ephedrine.

The imidazolidines exhibited reasonable $\Delta\delta_F$ values for a wide range of chiral aldehydes. Excess diamine should be used in order to ensure quantitative reaction and avoid kinetic resolution. This step is chemoselective, since ketones do not react under the mild derivatization conditions.

3.2.1.2 Nuclear magnetic resonance with other nuclei

For the application of NMR in the analysis of chiral molecules ^{31}P is an attractive nucleus. The chemical shift dispersion is large and simple spectra are recorded under broad-band proton decoupling conditions. Several chiral phosphoryl or thiophosphoryl chlorides have been studied as chiral derivatizing agents for alcohols and amines.

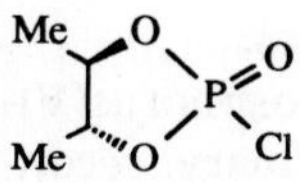

10 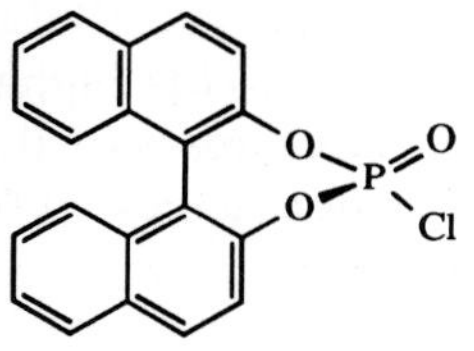 11

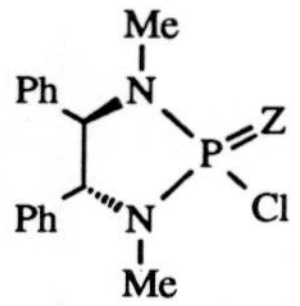

12 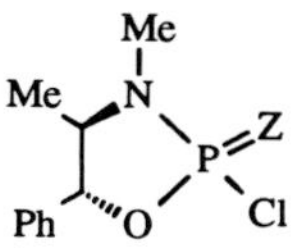 13

Z = O or S

The chlorodioxaphospholane **10** reacts with chiral primary and secondary alcohols, in the presence of a base, to give diastereomeric phosphates for which $\Delta\delta_P$ is small, (0 to 0.13 ppm ($CDCl_3$) [49]. The binaphthyl CDA **11** yields larger $\Delta\delta_P$ values; it reacts easily with a variety of chiral alcohols in the presence of 1-methylimidazole to give the diastereomers [50]. In these reagents, as in the chiral diamine-derived CDA **12** [51], the phosphorus atom is not chiral. Thus, inversion or retention of configuration at phosphorus during derivatization of an enantiopure alcohol results in a single diastereomer. The reduced electrophilicity of the phosphorus atom in **12**, together with the presence of two P-N bounds, requires forcing conditions (NaH, THF, reflux). However, larger $\Delta\delta_P$ values were obtained: the phosphates derived from 2-butanol gave $\Delta\delta_P$ values of 0.006 ppm for **10** but 0.454 ppm for **12** (Z = oxygen) and 0.20 ppm for the diastereomers derived from **13** (cf. Table 3.2.2).

Table 3.2.2 ^{31}P shift nonequivalence of derivatized chiral amines/alcohols R-X using CDA **13** [1].

R-X	$\Delta\delta_P$ [ppm, $CDCl_3$]
α-phenylethylamine	0.175
sec-butylamine	0.628
1,3-dimethylbutylamine	0.843
1-methyl-2-phenoxy-ethylamine	0.347
2-octanol	0.307
α-phenylethanol	0.111
2-methyl-2-butanol	0.167
4-methyl-2-pentanol	0.301

The latter CDA, derived from 1*R*,2*S*-ephedrine [52], reacts with chiral amines (THF, Et_3N, 24h, 65°C) but requires formation of the alkoxide (BuLi, Et_2O) in order to form derivatives of chiral alcohols. The thio analogue yields higher $\Delta\delta_P$ values.

The chiral phosphorus(III) CDA **14** is more reactive than the phosphorus(V)-based CDAs **10-13**, forming diastereomeric derivatives with chiral primary, secondary and tertiary alcohols (C_6D_6, 20°C).

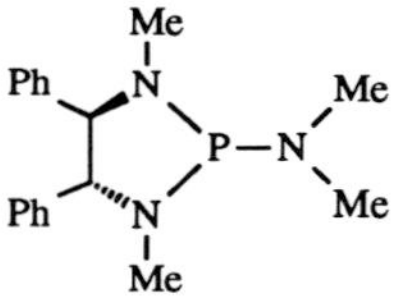

Me
Ph N Me
P—N
Ph N Me
Me

14

Further reaction with sulphur yields the more stable thiophosphates, which can also be analyzed by ^{31}P NMR. Representative $\Delta\delta_P$ values are given for a broad range of chiral alcohols (Table 3.2.3).

Attempts to extend the principle of ^{31}P NMR to the analysis of chiral ketones were only partially successful. Condensation of the chiral hydrazine **15** with sterically demanding ketones or α,β-unsaturated ketones was not satisfactory; the use of this method was restricted to some chiral monosubstituted cyclohexanones [53].

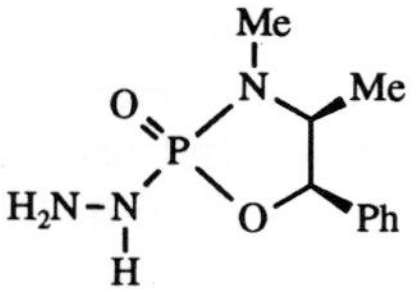

Me
O N Me
P
H_2N-N O Ph
H

15

An organometallic CDA has been recommended for ^{31}P NMR to determine the enantiomeric purity of chiral alkenes, alkynes and allenes. By using the biphosphine

2,3-O-isopropylidene-2,3-dihydroxy-1,4-bis(diphenylphosphino)butane (DIOP), the zerovalent platinum and palladium ethene complexes **16** were studied [54,55]. Displacement of ethene with alkynes, allenes, and electron-poor or strained alkenes (e.g., enones or norbornenes) proceeds easily in situ (THF or C_6D_6) and the resultant diastereomeric complexes show good ^{31}P NMR chemical shift differences.

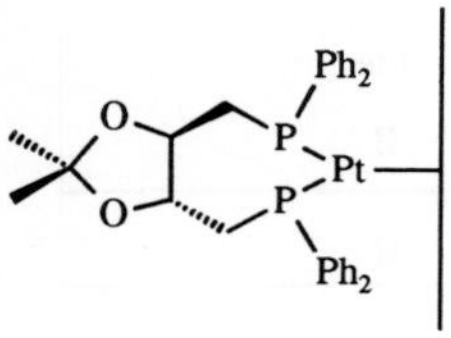

16

^{29}Si NMR can also be applied in the analysis of the enantiomeric purity of chiral alcohols. Successive reaction of diphenyldichlorosilane with an enantiomerically pure alcohol e.g., menthol, quinine or methyl mandelate followed by the alcohol to be determined (pyridine, 4.5 h, 60°C) yielded diastereomeric silyl acetals with small chemical shift nonequivalence ($\Delta\delta_{Si}$ = 0.053 ppm ($CDCl_3$, 298 K) for **17** [56].

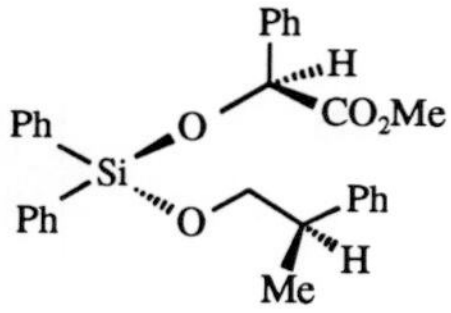

17

Selenium-77 is a relatively sensitive NMR nucleus, possesses a large chemical shift range (about 3400 ppm) and is particularly sensitive to an electronic environment. The selenocarbonyl group itself displays a chemical shift range of 2600 ppm; accordingly, the use of ^{77}Se NMR has been described for the assay of the enantiomeric purity of, e.g., racemic 5-methylheptanoic acid with the optically pure selone **18** (DCC-DMAP, CH_2Cl_2, 0°C, 1h). The derived diastereomeric N-acyl selones gave $\Delta\delta_{Se}$ values of 0.1 ppm ($CDCl_3$, 293 K) [57]. For ^{77}Se NMR spectroscopy see also literature [57a,b].

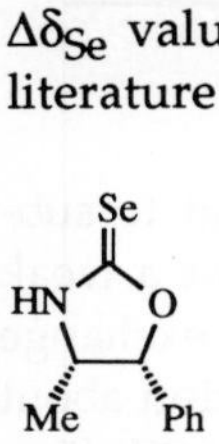

18

Table 3.2.3 ^{31}P shift nonequivalence of several derivatives using CDA **14** [1].

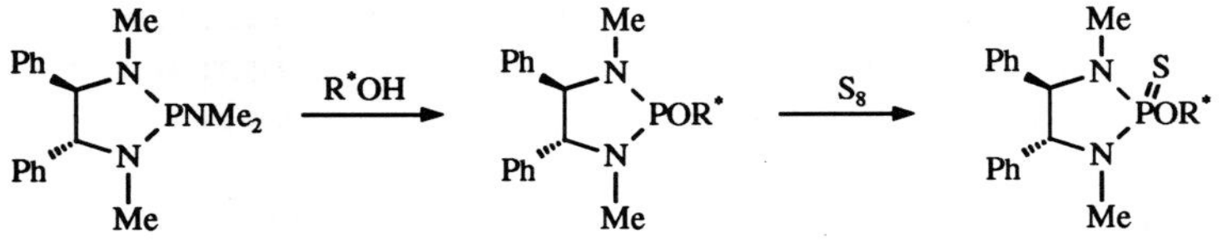

R*OH	I	$\Delta\delta_P$ [ppm, C_6D_6]	II
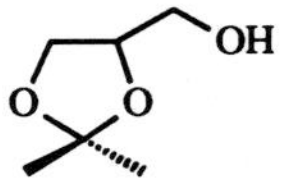	0.34		0.004
	0.67		0.09
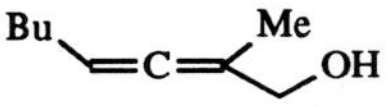	0.337		0.125
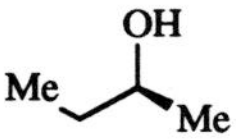	3.70		0.275
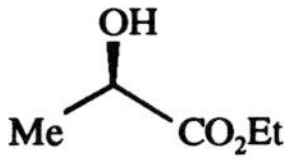	3.73		0.34
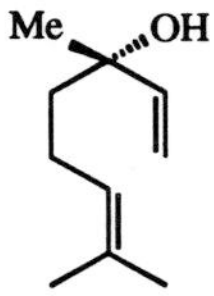	0.61		0.00

3.2.2 *Chiral lanthanide shift reagents (CLSR)*

Addition of a lanthanide shift reagent to an organic compound may lead to substantial shifts of resonances. The six-coordinated lanthanide complex forms a weak addition complex with a high number of organic compounds that is in fast exchange with the unbound substrate on the NMR scale. Based on the first information about the simplification of NMR spectra by the use of lanthanide shift reagents (LSR) [58], it has been realized that the rapid and reversible diastereomeric association between a chiral organic substrate and an optically active (nonracemic) LSR can result into efficient chiral recognition [59-61]. While chemical shift differences caused by hydrogen-bonding association are usually small and sometimes require the addition of achiral LSR to induce an increase of shift differences, chiral and paramagnetic LSR exhibit the abilities to discriminate between enantiomers and to induce large shifts. The chemical shift non-equivalence decreases in magnitude with increasing distance of the enantiotopic nuclei from the site of the chiral center.

Whitesides and Lewis [59] have reported a chemical shift non-equivalence for the externally enantiotopic protons of racemic α-phenylethylamine in the presence of the chiral LSR tris(3-tert-butyl-hydroxymethylene-1*R*-camphorato)-europium(III) [$Eu(pvc)_3$] (Table 3.2.4). Following this early work, several other chiral shift reagents were introduced (Table 3.2.4). The dicampholyl reagent $Eu(dcm)_3$ exhibited the best differential shift dispersion, and $Eu(hfc)_3$ gave particularly large $\Delta\delta$ values in ^{13}C NMR analysis for its diastereomeric complexes with chiral substrates. The praseodymium complex $Pr(hfc)_3$ worked better than $Eu(hfc)_3$ in ^1H NMR, giving the largest $\Delta\delta$ values at lowest concentrations of added shift reagents [62,63], while $Yb(hfc)_3$ has been shown to be superior to $Pr(hfc)_3$ in the analysis of a series of chiral sulphoxides [64]. The praseodymium shift reagents are in so far advantageous, e.g., in the analysis of diastereotopic methyl groups, that induced shifts are to lower frequency rather than to higher frequency, as been observed for the europium and ytterbium complexes.

Table 3.2.4 The most-used chiral lanthanide shift reagents [1].

L in LnL_3	Lanthanon	Abbreviation
Dicamphoyl-d-methanato-	Eu	$Eu(dcm)_3$
Heptafluorohydroxymethylene-d-camphorato-	Eu	$Eu(hfc)_3$
	Pr	$Pr(hfc)_3$
	Yb	Yb(hfc)3
Pivaloyl-d-camphorato-	Eu	$Eu(pvc)_3$
Trifluorohydroxymethylene-d-camphorato-	Eu	$Eu(tfc)_3$
	Pr	$Pr(tfc)_3$
	Yb	$Yb(tfc)_3$

Whitesides et al. [65] and Goering et al. [66] have also prepared chiral LSR derived from terpene ketones other than camphor which are available in an optically active form such as menthone, pulegone, carvone and nopinone. They do not seem to be superior to the classical camphor-related LSR.

In several reviews details of the application of chiral lanthanide shift reagents have been provided [67-69]. Most applications involve ^1H NMR analysis, but ^{13}C,

^{19}F, and ^{31}P NMR are also used. Despite the distinct advantages of $Pr(hfc)_3$ and $Yb(hfc)_3$, $Eu(tfc)_3$ and $Eu(hfc)_3$ were almost exclusively used [70-73].

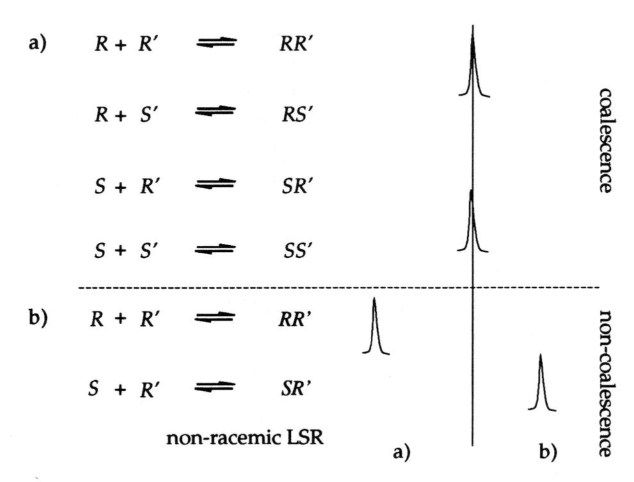

Figure 3.2.3 Associations between a racemic substrate and racemic and non-racemic CLSR.

It has to explained why the chiral LSR must be non-racemic in order to effect chemical shift differences for enantiotopic nuclei. Let us study the equilibria formed between the enantiomers ***R*** and ***S*** of the substrate and the enantiomers *R'* and *S'* of the LSR, assuming that the LSR is present as a monomer and that only a mono-adduct is formed, i.e. ***R*** + *R'* = ***R****R'*, ***R*** + *S'* = ***R****S'*, ***S*** + *R'* = ***S****R'* and ***S*** + *S'* = ***S****S'*.

In the presence of a racemic LSR four configurational isomers, i.e. two diastereomers, which are enantiomeric (***R****R'*/***S****S'* and ***R****S'*/***S****R'*), are formed. In the absence of additional elements of chirality only the diastereomers but not the enantiomers produce different resonance signals. At very low temperatures two resonance absorptions for the diastereomeric adducts may be expected. At ambient temperatures, however, only an average shift will be observed because the association and dissociation equilibria become rapid on the NMR time scale, leading to peak coalescence (see Figure 3.2.3). However, peak coalescence cannot occur when the racemic substrate is allowed to interact with only one enantiomer of the chiral LSR.

A consequence of the foregoing for enantiomer ratio determination is that the chiral LSR need not be enantiomerically pure in order to induce chemical shift non-equivalence. This is an important advantage over the above-mentioned method of converting enantiomers into diastereomers. The chemical shift difference with a chiral LSR increases as the enantiomeric purity of the lanthanide chelate increases and it will be zero when the LSR is racemic. The relative intensities of the signals, which reflect the enantiomeric composition of the substrate, are not affected by the enantiomeric purity of the chiral LSR [73].

There are two important requirements to select the lanthanide ion, i.e. (i) large shifts $\Delta\delta$; (ii) low line-broadening. No reliable guidelines are yet available; thus, the selection of the chiral LSR for producing spectral non-equivalence in chiral substrates is still a matter of trial and error. Nevertheless, there are certain trends and the interested reader is referred to references on this subject [69,74].

Most determinations of the enantiomeric excess are performed in $CDCl_3$ or CCl_4, whereby the latter solvent generally causes larger shift differences $\Delta\delta$. n-Pentane may also be used, although solubility problems and the solvent's interfering resonances may restrict its use. The spectrometer should be run at its highest sensitivity and the amount of substrate should be minimized. The more concentrated the substrate, the broader will be the signals. At a given molar ratio, δ as well as $\Delta\delta$ are quite insensitive to changes in concentration and tend to increase with a decrease in concentration. Convenient substrate concentrations are 0.1-0.25 mol.

Usually it is sufficient to add the solid LSR in incremental portions to the solution of the substrate. Solution is then effected by agitating or gentle warming. Low substrate concentrations are recommended in order to keep the amount of LSR to be added to a minimum. Some researchers prefer to add the LSR as a filtered solution. If this procedure is adopted the substrate concentration will be changed after subsequent additions of LSR.

It is essential to dry the shift reagent prior to use, as hydrolysis leads to formation of Eu_2O_3 and severe line broadening, although sublimation (200 °C, 0.05 mm Hg) is preferred. Provided that instrument handling and data aquisition and manipulation are performed carefully, accurate values of enantiomeric purity will be obtained [75]. In the range 40 to 60% ee, the best reported deviation is +/- 2% [76]; for ee values > 90 %, the error in measurement was reported to be on the order of 10% [77]. Polar substrates such as chiral 1,2- and 1,3-diols can be handled in acetonitrile-D_3 as NMR solvent. Chiral carboxylic acids are not usually amenable to direct analysis in this manner. They may either be converted into their corresponding tertiary amides (amides are good σ-donors for europium or ytterbium) [78] or they may be examined directly in aqueous solution [79].

3.2.3 *Chiral solvating agents (CSA)*

Chiral solvating agents form diastereomeric solvation complexes with solute enantiomers via rapidly reversible equilibria in competition with the bulk solvent. There are two reasons for the chemical shift anisochrony in this method. The first is the relative position of magnetically anisotropic groups (e.g., phenyl, carbonyl) in the low-energy solution conformers with respect to other substituents in the diastereomeric complexes. Exchange between chiral and achiral solvates is rapid on the NMR time scale and the observed resonance signals derived from each enantiomer δ_R(obs) and δ_S(obs) represent population weighted averages of the chemical shifts for the discrete chiral and achiral solvates δ_R, δ_S and $\delta_{achiral}$, respectively. This method is rapid and simple to perform, with no problems of kinetic resolution or sample racemization, provided that the complexes remain in solution. The enantiomeric purity of the CSA is not critical. If it is < 100%, then the size of the chemical shift nonequivalence will be reduced; only for a racemic CSA is $\Delta\delta$ zero. The main drawback of the method is that $\Delta\delta$ values tend to be small, but with high-field NMR instrumentation this point is not critical. In addition, only a limited number of cosolvents may be used. Nonpolar solvents ($CDCl_3$, CCl_4, C_6D_6) tend to maximize the observed anisochrony between the diastereomeric complexes, while more polar solvents preferentially solvate the solute and $\Delta\delta$ falls to zero.

Distinct ^{19}F NMR resonances for the enantiomers of 2,2,2-trifluoro-1-phenylethanol in the presence of *R*-α-phenylethylamine have been observed first by Pirkle [80]. By using *R*-2-naphthylethylamine, the size of the shift nonequivalence increased. These amines have also been used in the analysis of chiral carboxylic acids that form diastereomeric salts [81-84]. The analysis of chiral amines and amino alcohols using an enantiopure carboxylic acid CSA, i.e. the reciprocal experiment, has been more thoroughly investigated [85-88]. The chiral derivatizing agent MTPA (see Table 3.2.1) has also been studied but its use is restricted by the tendency of its salts to precipitate in $CDCl_3$ and C_6D_6. Pyridine-D_5 offers an alternative in these solutions [85]. In the case of the diastereomeric salts formed by *R*-O-acetylmandelic acid and α-phenylethylamine, $\Delta\delta_H$ for the C-Me doublet varies with both temperature and enantiomeric composition. Concentration and the CSA/solute ratio also effect $\Delta\delta_H$. For example, lowering the temperature increases $\Delta\delta_H$, while high concentrations (greater than 0.3 mol) lead to ion pair aggregation and a diminution in $\Delta\delta_H$. As expected, $\Delta\delta_H$ reaches a maximum value at 1:1 stoichiometry, when salt formation is complete. The dissociation equilibrium constants for salt formation of the two diastereomers must be nonequivalent, so that as the enantiomeric composition of the solute changes $\Delta\delta_H$ also varies [87].

The most commonly used CSA is 1-(9-anthryl)-2,2,2-trifluoroethanol (TFAE) (**19**) which has been applied for the determination of the enantiomeric purity of a high number of compounds, including lactones [89,90], ethers [91], oxaziridines [92] and sulphinate esters [93]. A solvation model has been proposed for this hydrogen bond donor, involving a secondary interaction between the methine hydrogen of the CSA and the less basic of the two basic sites in a given solute. By using this model the absolute configuration of a series of dibasic solutes, e.g., lactones [89] (see 'Practical examples') may be deduced from the sense of the observed shift. Although hydrogen-bonding interactions are the most common diastereoselective inter-

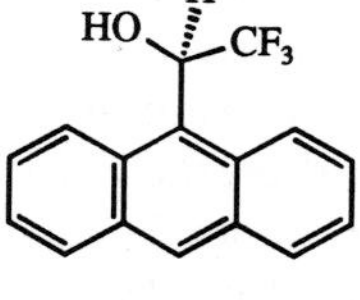

H
HO CF3

19

actions found in CSA-solute complexes, other interactions such as π-stacking (π-acid-π-base) are regarded to also contribute.

3.2.4 *Practical examples*

3.2.4.1 Problem 1: Determination of the absolute configuration of Z-4-hepten-2-ol [94].

The determination of the enantiomeric composition and absolute configuration of biologically active compounds is of great importance in flavour analysis [95]. The major aliphatic secondary alcohols among banana volatiles are heptan-2-ol and Z-4-hepten-2-ol. The enantiodifferentiation of these two substances was achieved via their isopropylcarbamates by means of multidimensional gas chromatography (MDGC) using a combination of an achiral (J&W DB-5) and chiral (Chrompak XE-60-*S*-VAL-*S*-PEA) capillary column (see Section 3.5). No information was available about the absolute configuration of Z-4-hepten-2-ol; thus, it was impossible to determine the exact order of elution during MDGC separation on the chiral phase .

Solution: Chiral derivatization of Z-4-hepten-2-ol (CDA)

Racemic Z-4-hepten-2-ol was kindly provided by Oril SA, Paris. According to the methodology of Gerlach et al. [96] optically enriched Z-4-hepten-2-ol was synthesized.

Description of the experiment: Five mmol racemic alcohol was esterified with 5 mmol dodecanoic acid in 20 ml heptane (bidistilled) under lipase (porcine pancreatic lipase, PPL; Sigma) catalysis at 40°C. After 48 h a yield of 53% was reached and the reaction stopped by filtration. After concentration in vacuo to 0.5 ml the remaining alcohol was separated by liquid chromatography on silica gel (10 x 1 cm) using a pentane-diethyl ether gradient (fraction I, 50 ml 9+1, discarded; fraction II, 50 ml 100% diethyl ether, dried over Na_2SO_4 and concentrated in vacuo to 0.5 ml). Ten microliter of the concentrated eluate was subjected to derivatization for GC analysis on the chiral phase XE-60, revealing an enantiomeric excess of 81.2%. After concentration to dryness under a stream of nitrogen and dilution to 2 ml ethanol (abs.), evaluation of the optical rotation value on a Perkin Elmer 241 MC polarimeter (589 nm) revealed the enrichment of the (-)-enantiomer. The determination of the configuration of the levorotatory alcohol was performed via its *R*-2-phenylpropionic acid ester by ^1H NMR according to the method of Helmchen et al. [97].

Description of the experiment: A 2 mmol equivalent portion of *R*-(-)-2-phenylpropionic acid was transferred to the corresponding acid chloride with oxalyl chloride (5 min, 60°C). Three times 1 ml benzene was added and distilled off, leading to optically pure *R*-2-phenylpropionic acid chloride used for esterification of 1 mmol *R,S*- and (-)-alcohol (3 d, 60°C). After concentration, the solutions were stirred with 2 ml dioxan/water (1+1) and extracted with diethyl ether/benzene (7+3). The organic layers were treated with 1 N NaOH, washed with water until neutral reaction and dried over Na_2SO_4. Purification after concentration was achieved by preparative thin layer chromatography on silica gel (dichloromethane).

According to Helmchen [97], the ^1H NMR spectrum of the diastereomeric esters of *R*-2-phenylpropionic acid shows distinct chemical shifts at constitutionally equivalent (=externally diastereotope) protons that can be decisively correlated with the absolute configuration of the alcohol moiety. The reason for this is the diamagnetic shield (=upfield shift) effected by the phenyl group on the alkoxy residue of the ester. Provided that the ester group exists in a planar trans-configuration, the ^1H NMR signals of the substituents of the alcohol moiety that are arranged on the same side as the phenyl group of *R*-2-phenylpropionic acid (R_1 in Figure 3.2.4) show a pronounced upfield shift relative to the proton signals of a group arranged at the opposite side (R_2 in Figure 3.2.4).

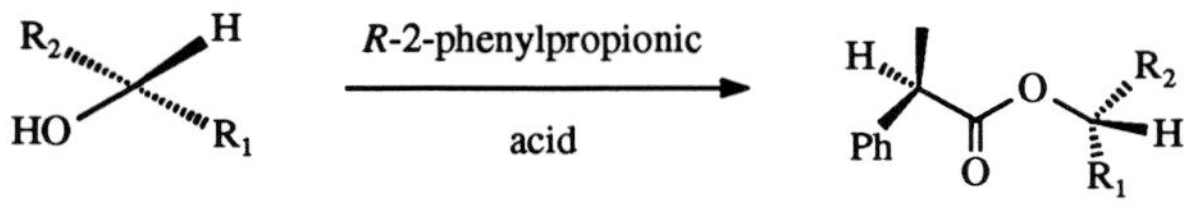

Figure 3.2.4 The principle of the 'Helmchen model' [97] for the evaluation of the absolute configuration of secondary alcohols.

The ^1H NMR spectra of the *R*-2-phenylpropionic acid esters of *R,S*- (**20**) and (-)-Z-4-hepten-2-ol are outlined in Table 3.2.5.

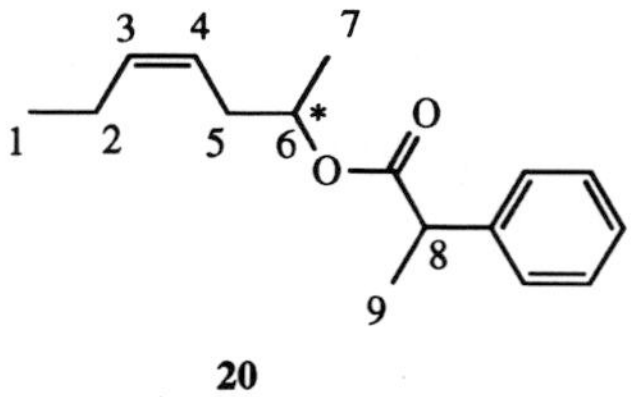

20

Table 3.2.5 The ^1H NMR data of the *R*-2-phenylpropionic acid derivatives of *R,S*- and (-)-Z-4-hepten-2-ol [94]

Alcohol	7[1] (d)	6 (q)	5 (m)	4 (d-t)	3 (d-t)	2 (d-q)	1 (t)
R,S-	1.17	4.85	2.23	5.42	5.26	1.92	0.95
	1.08			5.23	5.14		0.89
(-)-	1.10	4.88	2.30	5.43	5.28	2.00	0.96

[1] Number of the C-atom, the protons are bound to.

As seen from the data outlined in Table 3.2.5 (cf. upfield shifted signal of C_7-methyl protons), ^1H NMR spectroscopy revealed *S*-configuration for the levorotatory enantiomer, i.e. *S*-(-)-Z-4-hepten-2-ol was assigned.

3.2.4.2 Problem 2: Determination of the absolute configuration of γ-pentadecalactone [98].

In many countries legislation restricts the use of coumarin and/or coumarin-containing plants such as sweet clover (*Melilotus officinalis* L. Lam.). In order to check its illegal use it is essential to define an appropriate 'key substance' that is characteristic for the coumarin-containing plant. For sweet clover, γ-pentadecalactone (**21**) has been found to be a suitable indicator, since this lactone does not occur in other plant tissues.

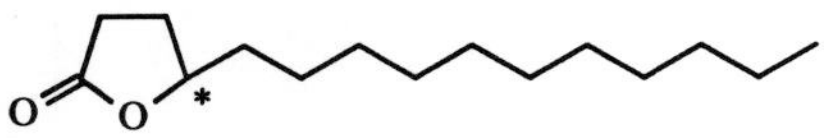

O O *

21

For the enantiodifferentiation of the naturally occurring **21**, it was necessary to synthesize optically pure γ-pentadecalactone and to determine the absolute configuration of its enantiomers.

Solution: Addition of a chiral solvating agent (CSA)

γ-Pentadecalactone (**21**) was synthesized according to the literature [99]. HPLC enantioseparation was achieved by using an optically active polyacrylamide-*S*-phenylalanine-ethylester phase (ChiraSpher, cf. Section 3.4) On-line polarimetric detection revealed a negative rotation for the first eluted enantiomer.

Description of the experiment: Synthesized **21** was dissolved in hexane and separated by preparative HPLC as follows: pump, UV-detector (220 nm) and Chira-Monitor (Zinsser) were connected in series; column: ChiraSpher 250 x 4 mm (Merck); eluent: n-hexane (bidistilled), 1.1 ml/min. The retention times were 11.2 and 12.0 min for the (-)- and (+)-enantiomer, respectively.

Although the order of elution was coincident with that of homologous γ-lactones studied earlier [100], the determination of the absolute configuration of the enantiomers was necessary to confirm the results. Assignment was achieved by ^1H NMR spectroscopy [90] using diastereomeric solvate complexes with the optically pure shift reagent TFAE (**19**).

The anthryl group of TFAE shows a different screening effect on the substituents of the lactone ring. This effect is caused by the aromatic system; depending on the position of the substituent, i.e. cis or trans to the ring system, an upfield shift or downfield shift, respectively, of the resonance signal will be observed.

Since in the above-mentioned example *S*-TFAE was used, the proton at the asymmetric C-atom is in cis-position for the *S*-lactone, leading to an upfield shift (relative to the *R*-enantiomer). As shown from the proton signal at the asymmetric C-atom in Figure 3.2.5, a stronger upfield shift was observed for the (-)-lactone than for the (+)-enantiomer, enabling the assignment of the (-)-enantiomer to be *S* and that of the (+)-enantiomer to be *R*.

Description of the experiment: Each of the γ-pentadecalactone (**21**) enantiomers (0.08 mmol; obtained in pure form by preparative HPLC), together with 3 mol equivalents *S*-TFAE (Sigma) in $CCl_4/CDCl_3$ (3+1), was investigated by ^1H NMR (TMS as standard; 200 MHz).

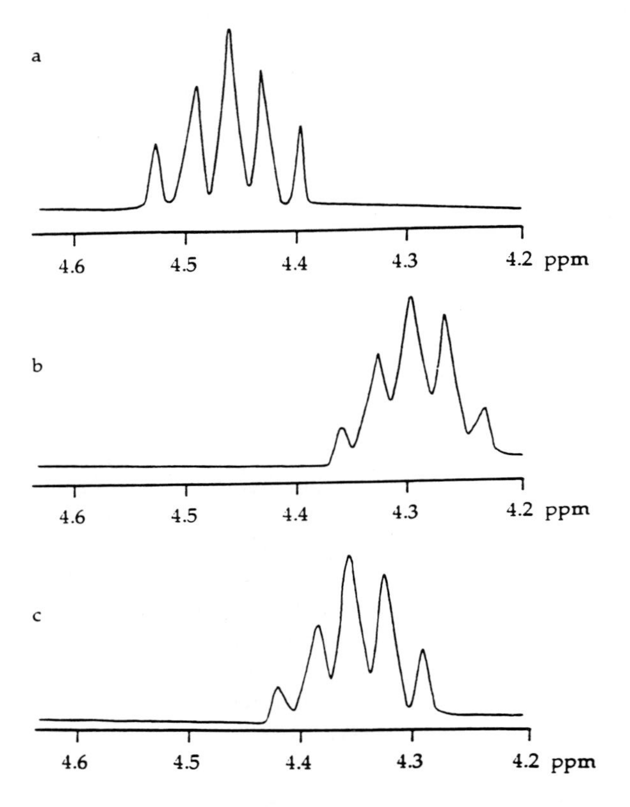

Figure 3.2.5 Determination of the absolute configuration of γ-pentadecalactone with ^1H NMR spectroscopy using the chiral solvating agent *S*-TFAE. (a) lactone without TFAE, (b) (-)-lactone with TFAE, (c) (+)-lactone with TFAE.

Enantiodifferentiation of **21** from sweet clover herb performed by liquid chromatography and gas chromatography (cf. Sections 3.4 and 3.5) on two chiral phases revealed an ee of 42 % *S* [98].

References

[1] Parker, D. *Chemical Reviews* (1991), *91*, 1441

[2] Cram, D. J.; Mateos, J. L. *J. Am. Chem. Soc.* (1959), *81*, 5150

[3] Fraser, R. R. In *Asymmetric Synthesis*; Morrison, J. D., Ed., Vol. 1, Chapter Academic Press: New York, 1983, p 173

[4] *Methods in Stereochemical Analysis*; Morrill, T. C., Ed.; VCH Publishers: New York, 1986; Vol. 5

[5] Weisman, G. R. In *Asymmetric Synthesis*; Morrison, J. D., Ed.; Vol.1, Chapter 8, Academic Press: New York, 1983, p 153

[6] Pirkle, W. H.; Hoover, D. J. *Top. Stereochem.* (1982), *13*, 263

[7] Yamaguchi, S. In *Asymmetric Synthesis*; Morrison, J. D., Ed.; Vol. 1, Chapter 7, Academic Press: New York, 1983, p 125

[8] Raban, M.; Mislow, K. *Tetrahedron Lett.* (1965), 4249

[9] Dale, J. A.; Dull, D. L.; Mosher, H. S. *J. Org. Chem.* (1969), *34*, 2543

[10] Dale, J. A.; Mosher, H. S. *J. Am. Chem. Soc.* (1973), *95*, 512

[11] Sullivan, G. R.; Dale, J. A.; Mosher, H. S. *J. Org. Chem.* (1973), *38*, 2143

[12] Hietaniemi, L.; Pohjala, E.; Malkonen, P.; Riekkola, M. L. *Finn. Chem. Lett.* (1989), *16*, 67

[13] Dutcher, J. S.; MacMillan, J. G.; Heathcock, C. H. *J. Org. Chem.* (1970), *41*, 2663

[14] Williams, R. M.; Glinka, T.; Ewa, K.; Hazeol, C.; Stille, J. K. *J. Am. Chem. Soc.* (1990), *112*, 808

[15] Kitamura, M.; Ohkuma, T.; Takunaga, M.; Noyori, R. *Tetrahedron Asymmetry* (1990), *1*, 1

[16] Nieduzak, T. R.; Carr, A. A. *Tetrahedron Asymmetry* (1990), *1*, 535

[17] Dale, J. A.; Mosher, H. S. *J. Am. Chem. Soc.* (1968), *90*, 3732

[18] Takeuchi, Y.; Ogura, H.; Ishii, Y.; Kaizumi, T. *J. Chem. Soc., Perkin Trans.* (1989), *1*, 1721

[19] Nabeya, A.; Endo, T. *J. Org. Chem.* (1988), *53*, 3358

[20] Neises, B.; Steglich, W. *Angew. Chem., Int. Ed. Engl.* (1978), *17*, 522

[21] Trost, B. M.; Mignani, S.; Acemoglu, M. *J. Am. Chem. Soc.* (1989), *111*, 7487

[22] Parker, D. J. *J. Chem. Soc., Perkin Trans.* (1983), *2*, 83

[23] Trost, B. M.; Belletire, J. L.; Godleski, S.; McDougal, P.; Balkovec, J. M.; Baldwin, J. J.; Christy, M. E.; Ponticello, G. S.; Varga, S. L.; Springer, J. P. *J. Org. Chem.* (1986), *51*, 2370

[24] Barth, G.; Voeter, W.; Mosher, H. S.; Bunnenberg, E.; Djerassi, C. J. *J. Am. Chem. Soc.* (1970), *92*, 875

[25] Yasuhara, F.; Kabuto, K.; Yamaguchi, S. *Tetrahedron Lett.* (1978), 4289

[26] Yasuhara, F.; Yamaguchi, S. *Tetrahedron Lett.* (1980), 2827

[27] Helmchen, G. *Tetrahedron Lett.* (1974), *16*, 1527

[28] Dale, J. A.; Mosher, H. S.; Yamaguchi, S. *J. Org. Chem.* (1972), *37*, 3174

[29] Mosher, H. S.; Yamaguchi, S. *J. Org. Chem.* (1973), *38*, 1870

[30] Kabuto, K.; Yamaguchi, S. *Bull. Chem. Soc. Jpn.* (1977), *50*, 3074

[31] Bouman, T. D.; Gawronski, J. K.; Lightner, D. A. *J. Am. Chem. Soc.* (1980), *102*, 1983

[32] Kabuto, K.; Yasuhara, F.; Yamaguchi, S. *Tetrahedron Lett.* (1978), 4289

[33] Miyano, S.; Tobita, M.; Hashimoto, H. *Bull. Chem. Soc. Jpn.* (1981), *54*, 3522

[34] Gerlach, H. *Helv. Chim. Acta* (1966), *49*, 2481

[35] Gerlach, H.; Zagalak, B. *J. Chem. Soc., Chem. Commun.* (1973), 274

[36] Williams, R. M.; Sinclair, P. J.; Ahari, D.; Chen, D. *J. Am. Chem. Soc.* (1988), *110*, 1547

[37] Feringa, b.; Wynberg, H. *J. Org. Chem.* (1981), *46*, 2547
[38] Munari, S. D.; Marazzi, G.; Forgione, A.; Lango, A.; Lombard, P. *Tetrahedron Lett.* (1980), 2273
[39] Baker, K. V.; Brown, J. M.; Cooley, N. A.; Hughes, G. D.; Taylor, R. J.; *J. Organometal. Chem.* (1989), *370*, 397
[40] Brown, J. M.; Parker, D. *Tetrahedron Lett.* (1981), 2815; 4994
[41] Brown, J. M.; Parker, D. *J. Org. Chem.* (1982), *97*, 2722
[42] Lemiere, G. L.; Dommisse, R. A.; Lepoivre, J. A.; Alderweireldt, F. C.; Hiemstra, H.; Wynberg, H.; Jones, J. B.; Toone, E. J. *J. Am. Chem. Soc.* (1987), *109*, 1363
[43] Fujiwara, J.; Fukutani, Y.; Hasagawa, M.; Maruoka, K.; Yamamoto, H. *Tetrahedron Lett.* (1984), 5004
[44] Maruoka, K.; Yamamoto, H. *Angew. Chem., Int. Ed. Engl.* (1985), *24*, 668
[45] Mangeney, P.; Alexakis, A.; Normant, J. F. *Tetrahedron Lett.* (1987), 2363
[46] Agami, A.; Meynier, F.; Berlan, J.; Besace, Y.; Brochard, L. *J. Org. Chem.* (1986), *51*,73
[47] Cuvinot, D.; Mangeney, P.; Alexakis, A.; Normant, J. F.; Lellouche, J. P. *J. Org. Chem.* (1989), *54*, 2420
[48] Mangeney, P.; Alexakis, A.; Normant, J. F. *Tetrahedron Lett.* (1988), 2677
[49] Anderson, R. C.; Shapiro, M. J. *J. Org. Chem.* (1984), *49*, 1304
[50] Kato, N. J. *J. Am. Chem. Soc.* (1990), *112*, 254
[51] Alexakis, A.; Mutti, S.; Normant, J. F.; Mangeney, P. *Tetrahedron Asymmetry* (1990), *1*, 437
[52] Johnson, C. R.; Elliott, R. C.; Penning, T. D. *J. Am. Chem. Soc.* (1984), *106*, 5019
[53] Dehmlow, E. V.; Sauerbier, C.; Z. *Naturforsch.* (1989), 240
[54] Taylor, R. J.; Parker, D. *J. Chem. Soc., Chem. Commun.* (1987), 1781
[55] Parker, D.; Taylor, R. J. *Tetrahedron* (1988), *44*, 2241
[56] Chan, T. H.; J-Peng, Q.; Wang, D.; Guo, J. A. *J. Chem. Soc., Chem. Commun.* (1987), 325
[57] Silks, L. A.; Dunlop, R. B.; Odan, J. D. *J. Am. Chem. Soc.* (1990), *112*, 4979; (a) Duddeck, H.; Wagner, P.; Biallaß, A. *Magn. Reson. Chem.* (1991), *29*, 248; (b) Duddeck, H.; Wagner, P.; Rys, B. *Magn. Reson. Chem.* (1993), *31*, 736
[58] Hinckley, C. C. *J. Am. Chem. Soc.* (1969), *91*, 5160
[59] Whitesides, G. M.; Lewis, D. W. *J. Am. Chem. Soc.* (1970), *92*, 6979
[60] Fraser, R. R.; Petit, M. A.; Saunders, J. K. *J. Chem. Soc., Chem. Commun.* (1971), 1450
[61] Schurig, V. *Tetrahedron Lett.* (1972), 3297
[62] Fraser, R. R.; Petit, M. A.; Miskow, M. *J. Am. Chem. Soc.* (1972), *94*, 3253
[63] Kainisho, M.; Ajisaka, K.; Pirkle, W. H.; Beare, S. D. *J. Am. Chem. Soc.* (1972), *94*, 5924
[64] Tangermann, A.; Zwanenburg, B. *Recl. Trav. Chim. Pays-Bas* (1977), *96*, 196
[65] McCreary, M. D.; Lewis, D. W.; Wernick, D. L.; Whitesides, G. M. *J. Am. Chem. Soc.* (1974), *96*, 1038
[66] Goering, H. L.; Eikenberry, J. N.; Koermer, G. S.; Lattimer, C. J. *J. Am. Chem. Soc.* (1974), *96*, 1493
[67] Parker, D. *Chem. Rev.* (1991), *91*, 1441
[68] Fraser, R. R. In *Asymmetric Synthesis*; Morrison, J. D., Ed.; Academic Press: New York, 1983; Vol. 1, Chapter 9, 173
[69] Sullivan, G. R. *Top. Stereochem.* (1976), *10*, 287
[70] Guanti, G.; Banfi, L.; Narisano, E. *Tetrahedron Asymmetry* (1990), *1*, 721

[71] Baldenius, K. U.; Kagan, H. B. *Tetrahedron Asymmetry* (1990), *1*, 597
[72] Rabiller, C.; Maze, F. *Magn. Reson. Chem.* **1989**, *27*, 582
[73] Raban, M.; Mislow, K. *Top. Stereochem.* **1967**, *2*, 199
[74] McCreary, M. D.; Lewis, D. W.; Wernick, D. L.; Whitesides, G. M. *J. Am. Chem. Soc.* (1974), *96*, 1038
[75] Peterson, P. E.; Stephanian, M. *J. Org. Chem.* (1988), *53*, 1907
[76] Bucciarelli, M.; Forni, A.; Moretti, I; Torre, G. *J. Org. Chem.* (1983), *48*, 2640
[77] Crans, D. C.; Whitesides, G. M. *J. Am. Chem. Soc.* (1985), *107*, 7019
[78] Brown, J. M.; Parker, D. J. *J. Chem. Soc., Chem. Commun.* (1980), 342
[79] Kabuto, K.; Saskai, Y. *J. Chem. Soc., Chem. Commun.* (1984), 316
[80] Pirkle, W. H. *J. Am. Chem. Soc.* (1966), *88*, 1837
[81] Mamiok, L.; Marquet, A.; Lacombe, L. *Tetrahedron Lett.* (1971), 1093
[82] Baxter, C. A. R.; Richards, H. C. *Tetrahedron Lett.* (1972), 3357
[83] Mikolajczyk, M.; Ormelonczuk, J.; Leitloff, M.; Drabrowicz, J.; Ejchart, A.; Jurzak, J. *J. Am. Chem. Soc.* (1978), *100*, 7003
[84] Aitken, R. A.; Gopal, J. A. *Tetrahedron Asymmetry* (1990), *1*, 517
[85] Villani, F. J.; Costanzo, M. J.; Inners, R. R.; Mutter, M. S.; McLure, D. E. *J. Org. Chem.* (1986), *51*, 3715
[86] Benson, S. C.; Cai, P.; Colon, M.; Haiza, M. A.; Tokles, M.; Snyder, J. K. *J. Org. Chem.* (1988), *53*, 5335
[87] Shapiro, M. J.; Archinal, A. E.; Jarema, M. A. *J. Org. Chem.* (1989), *54*, 5826
[88] Parker, D.; Taylor, R. J. *Tetrahedron* (1987), 5451
[89] Pirkle, W. H.; Adams, P. E. *J. Org. Chem.* (1980), *45*, 4111; (1980), *45*, 4117
[90] Pirkle, W. H.; Sikkenga, D. L.; Pavlin, M. S. *J. Org. Chem.* (1977), *42*, 384
[91] Pirkle, W. H.; Boeder, C. W. *J. Org. Chem.* (1977), *42*, 3697
[92] Pirkle, W. H.; Rinaldi, P. L. *J. Org. Chem.* (1977), *42*, 3217; (1978), *43*, 4475
[93] Pirkle, W. H.; Hoekstra, M. S. *J. Am. Chem. Soc.* (1976), *98*, 1832
[94] Fröhlich, O.; Huffer, M.; Schreier, P. *Z. Naturforsch.* (1989), *44c*, 555
[95] Mosandl, A. *Food Rev. Int.* **1988**, *4*, 1
[96] Gerlach, D.; Schneider, S.; Göllner, T.; Kim, K. S.; Schneider, P. *Bioflavour '87*; Schreier, P., Ed.; W. de Gruyter: Berlin, New York, 1988, 543
[97] Helmchen, G.; Schmierer, R. *Angew. Chem.* (1976), *13*, 770
[98] Wörner, M.; Schreier, P. *Z. Lebensm. Unters. Forsch.* (1990), *190*, 425
[99] Heiba E. I.; Dessau, R. M.; Rodewald, P. G. *J. Am. Chem. Soc.* (1974), *96*, 7977
[100] Huffer M.; Schreier, P. *J. Chromatogr.* (1989), *469*, 137

3.3 General aspects of chromatography

To facilitate a better understanding of technical descriptions and chromatographic evaluation formulas the most important terms used in this text are explained below.

3.3.1 *Definitions and formulas used in chromatography*

If a stationary phase A (selector) exhibits chemical affinity for a component B (selectand) to be separated, the chromatographic retention behavior of B will be influenced both by the physical distribution balance between the mobile and stationary phase and by the chemical balance A + B = AB. If A and B are chiral and if A is optically active and B a racemic mixture, then a chromatographic separation of enantiomers is possible without isolation of diastereomers. Energetically different diastereomeric associates, A_RB_R and A_RB_S, will be rapidly and reversibly formed [1]. Essential peak parameters for chromatographic enantiomer separations are given in Table 3.3.1.

Table 3.3.1 Chromatographic parameters for enantiomer separation [2].

Peak Parameter	Definition
Peak retention	Chemoselectivity [$-\Delta G^0$]: thermodynamic measure for selective association between selectand and selector
Peak separation	Enantioselectivity [$-\Delta_{R,S}$ (ΔG^0)]: thermodynamic measure for chiral recognition between a racemic selectand and an optically active selector
Peak area ratio	Enantiomeric ratio (enantiomeric excess, ee): precise quantitative measure for the enantiomeric composition of the selectand
Peak relation	Order of elution (*R*; *S*): correlation of the retention of the selectand and molecular configuration (only indications)
Peak coalescence	Enantiomerization barrier [$\Delta G^{\#}$]: kinetic measure of the enantiomerization (racemization) of the selectand during the separation

Besides the physical partition balance *peak retention* also will be determined by the chemical association balance.

Peak separation of racemic selectands on optically active selectors is caused exclusively by the difference of the free association enthalpy $\Delta_{R,S}(\Delta G0)$. But the conception that an effective enantiodifferentiation can occur only within a strong chemical interaction has been proven to be wrong in many cases. In some instances, the hindrance of molecular association, e.g., owing to steric effects, seems to be even advantageous for the discrimination of configurational isomers.

Peak area ratio is a direct measure of the quantitative ratio of an enantiomeric pair. The measured enantiomeric excess corresponds to the real enantiomeric excess of the sample if all chemical, physical and analytical manipulations are performed in an achiral environment.

Peak relation (order of elution) can be determined only empirically, e.g., by coinjection of a racemic selectand with one of the two enantiomers of known configuration. In liquid chromatography it is possible to determine the order of elution by using an on-line polarimetric detector (see Section 3.1). Earlier conclusions based on the retention behavior related to the absolute configuration cannot be valid according to the current state of scientific knowledge (For information on the inversion of elution orders see Section 3.5).

For *peak coalescence* (peak fusion) there are three different explanations:

(i) First kind (GC and LC): this occurs if instead of an optically active selector, a racemic one is used. In this way enantiomers can be differentiated from diastereomers or other isomers, since only enantiomers are able to coalesce.

(ii) Second kind (mainly GC): this occurs if, during the process of chromatographic separation, enantiomers with unstable configurations are inverted. Enantiomerization of the selectand will lead to peak interconversion profiles from which kinetic inversion parameters can be calculated as a reversible first order reaction. While during a slow inversion a plateau between the terminal peaks will be formed which is a result of inverting molecules, a fast inversion of the configuration will lead to disappearance of the separation.

(iii) Third kind (only GC): this occurs, if a racemic mixture that can be resolved into its enantiomers at low temperature cannot be separated after an increase in temperature. If the temperature is increased further, separation with an inversion of the order of elution might become possible again [3].

The following retention parameters are important for gas chromatographic enantiomer separation (according to ASTM [4]):

a) Net retention time t'_R [min] (= adjusted retention time). Equivalent to the residence time in stationary phase; this is determined by substracting the gas holdup volume from the solute retention time:

$$t'_R = t_R - t_M$$

t_R = retention time [min] (= total retention time); time (or distance) from the point of injection to the point of peak maximum.

t_M = gas holdup time [min]; time (or distance) required for elution of a nonretained substance (e.g., mobile phase).

b) Capacity ratio k (= partition ratio). Ratio of the amounts of a solute in stationary phase and mobile phase, which is equivalent to the ratio of the times the solute spends in the two phases: $k = t'_R / t_M$

c) Separation factor α (according to IUPAC) (= relative retention; selectivity). This is the ratio of the adjusted (net) retention times of both enantiomers, measured under identical conditions:

$$\alpha = t'_{R2} / t'_{R1}$$

t'_{R2} = net retention time of the more retained peak [min]
t'_{R1} = net retention time of the less retained peak [min].

d) Peak resolution, R_s (= separation efficiency). Measure of separation as evidenced by both the distance between the peak maximum and the peak widths (Figure 3.3.1):

$$R_s = 2\,(t_{R2} - t_{R1}) / (w_{b1} + w_{b2})$$

w_{b1} = peak width at base of the less retained peak [min]
w_{b2} = peak width at base of the more retained peak [min].

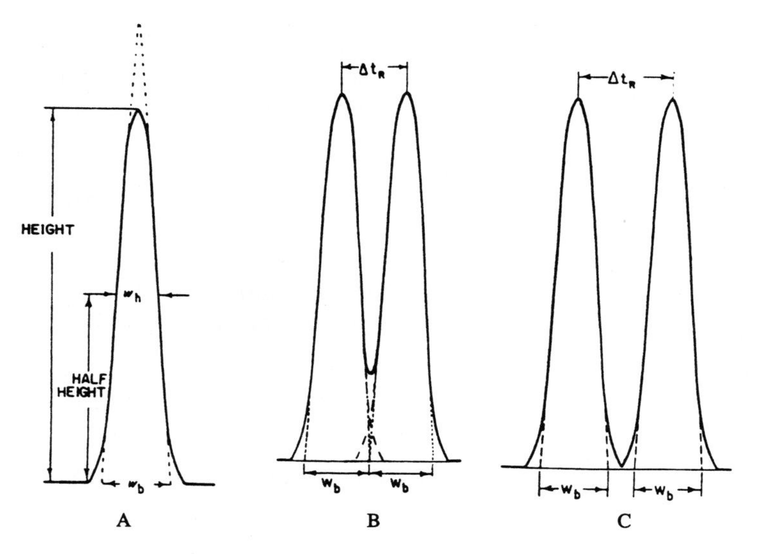

Figure 3.3.1 Graphic presentation of the resolution [5]. A = Half height and baseline peak width. B = sufficient separation (R_s = 1). C = baseline separation (R_s = 1.5).

If the peaks are assumed to be Gaussian:

$$R_s = 1.18\,(t_{R2} - t_{R1}) / (w_{h1} + w_{h2})$$

w_{h1} = peak width at half height of the less retained peak [min]. This is determined by measuring the length of baseline defined by intercepts extrapolated from the points of inflection of the peak. Equivalent to four standard deviations (σ) in a Gaussian peak.
w_{h2} = peak width at half height of the more retained peak [min]. Measured across the peak halfway between the baseline and the peak maximum. In a Gaussian peak, equal to 2.35 standard deviations (σ).

e) Valley size ϑ [%] (= percentage of baseline separation). This is the measure of incomplete separation, especially if peak tailing occurs:

$$\vartheta = [(h_1 + h_2 - 2\, h_{1,2}) \,/\, (h_1 + h_2)]\ 100\ \%$$

h_1 = peak height (maximum) of the less retainded peak [mm]
h_2 = peak height (maximum) of the more retainded peak [mm]
$h_{1,2}$ = peak minimum between both peak maxima [mm].

3.3.2 *Sources of error in the determination of enantiomeric compositions (ee values) by chromatographic methods*

Despite the great success of chromatographic methods in enantiomer analysis, there exist potential errors in the determination of ee. The following sources of error can be discerned [6,7]:

a) Decomposition of the selectand on achiral parts of the column: The enantiomer that remains longer in the column will be lost preferentially, causing an error in ee; i.e. the ee value will be shifted in favour of the first-eluted enantiomer.

b) Decomposition of the selectand on the chiral stationary phase: This chemical change may proceed enantioselectively, causing the preferential loss of one enantiomer. The ee value will be shifted in favour of the first-eluted enantiomer. But the opposite phenomenon has been observed as well [6], when the decomposition of the selectand is mediated by the chiral stationary phase in an enantioselective fashion in favour of the first-eluted enantiomer.

c) Contamination of peaks with impurities, increasing peak areas: This error may be recognized and eliminated by changing the chromatographic parameters, such as temperature, carrier gas flow (GC), mobile phase type, use of detectors to provide additional information (eg., MS, FTIR for GC; photo-diode array detector, polarimetric detector for LC) or use of multidimensional chromatographic methods.

Errors due to a), b) and c) may be recognized by determining the ee value with stationary phases of opposite chirality [8-11]. To be on the safe side, verification of the expected 1:1 ratio on enantiomer separation is strongly recommended in enantiomer analysis.

d) Fractionation of enantiomers by the so-called 'ee-effect' [12,13] during sample manipulation. The scalar physical properties of enantiomers may vary with the composition deviations from the original ee of a mixture of enantiomers. This effect may lead to accidental fractionation of enantiomers during sample preparation, isolation or chromatographic separation. It may be explained - in a quite over-simplified manner - by assuming that the enantiomers in excess function as a chiral environment for the enantiomer separation of the residual racemic mixture. Whether or not an 'ee-effect' is important in sampling procedures with splitting devices in GC is open to discussion. Let us consider the determination of 0.1 % of an *R*-enantiomer in a mixture with 99.9 % *S* and let us assume that *R* preferentially combines with the excess of *S* to give *R,S*, while *S* combines with *S* to give *S,S*. If *R,S* is more volatile than *S,S*, the former will be preferentially lost on evaporation in the injector through the splitting device and the ee of *S* will, therefore, be overestimated. Clearly, on-column injection will eliminate this source of error.

e) Enantiomerization of configurationally labile solutes, producing peak distortions due to inversion of configuration during enantiomer separation.

f) Non-linear detector response; it should be clear that in enantiomer analysis a linear detector response is indispensable.

g) Peak distortions; these can be caused by inadequate instrumentation (e.g., peak splitting = 'christmas-tree'-effect caused by cold-spots in the transfer-line of a multi-dimensional gas chromatographic system [14].

References

[1] Schurig V. *Angew.Chem.* (1984), *96*, 733
[2] Schurig V., Bürkle W. *J. Am. Chem. Soc.* (1982), *104*, 7573
[3] Schurig V., Ossig J., Link R. *Angew. Chem.* (1989), *101*, 197
[4] Leibnitz E., Struppe H.G. *Handbuch der Gaschromatographie* 3. Ed., Geest & Portig: Leipzig, 1984
[5] Jennings W. *Analytical Gas Chromatography,* Academic Press: Orlando, London, 1987
[6] Schurig V. *J. Chromatogr.* (1988), *441*, 135
[7] Schurig V. *Kontakte (Darmstadt)* (1986), *1*, 3
[8] Gil-Av E., Feibush B., Charles-Sigler R. In *6th Int. Symp. on Gas Chromatography and Associated Techniques, Rome 1966* Littlewood A.B., Ed., Inst. of Petroleum: London, 1967, p. 227
[9] Lochmüller C.H., Souter R.W. *J. Chromatogr.* (1973), *87*, 243
[10] Bayer E., Allmendinger H., Enderle G., Koppenhoefer B. Z. *Anal. Chem.* (1985), *321*, 321
[11] Schurig V. *Angew. Chem.* (1977), *89*, 113

[12] Tsai W.-L., Hermann K., Hug E., Rohde B., Dreiding A.S. *Helv. Chim. Acta* (1985), *68*, 2238
[13] De Min M., Levy G., Micheau J.C. *J. Chim. Phys.* (1988), *85*, 603
[14] Bernreuther A. Doctoral Thesis, University of Würzburg, Germany 1992

3.4 Liquid chromatography

High performance liquid chromatography (HPLC) is a powerful separation technique, especially for non-volatile substances such as most pharmaceuticals. The resolution of enantiomers by HPLC usually proceeds rapidly in the presence of an efficient chiral stationary phase. Due to the highly selective chiral recognition of liquid chromatography, a great variety of chiral phases is needed. The best suited chiral phase for a defined problem must still be choosen empirically by 'trial and error'. In recent years, efforts have been directed towards finding new types of chiral stationary and mobile phases on the basis of stereochemical principles and employing the technical advances in modern liquid chromatography.

Basically, there are two approaches to the separation of an enantiomeric pair by liquid chromatography. In the indirect way, the enantiomers are converted into covalent diastereomeric compounds by reaction with a chiral reagent. These diastereomers are routinely separated on a conventional achiral stationary phase. In the direct approach, several variations are possible; (i) the enantiomers (or their derivatives) are applied to a column containing a chiral stationary phase; (ii) the solutes are passed through an achiral column using a chiral solvent or; (iii) more commonly, the solutes are passed through a mobile phase containing a chiral additive.

3.4.1 *Covalent derivatization with chiral reagents to form diastereomers*

Diastereomer formation is used for such purposes as preparative isolation of enantiomers, evaluation of optical purity, determination of absolute configuration and enantiodifferentiation in biological samples. These applications emphasize various methodological aspects. While sensitivity and selectivity are essential for enantiodifferentiation in biological systems, these parameters are not critical in preparative chromatography. Careful cleavage for the recovery of enantiomers from separated diastereomers is important in preparative work, but less essential in analytical applications, where ease of diastereomer formation is more critical.

The diastereomers differ in their physico-chemical properties in an achiral environment and are, therefore, separable on conventional achiral columns. For the reagent *R*-A and an enantiomeric mixture of B it will be obtained:

R-A + *R*,*S*-B -----> [*R*-A,*R*-B] + [*R*-A,*S*-B]

The degree of separation of [*R*-A,*R*-B] and [*R*-A,*S*-B] depends on the chiral structures of A and B and the linkage between the chiral centres, as well as on the chromatographic system employed. The order of elution of the derivatized enantiomers will be reversed if the other reagent enantiomer, *S*-A, is used. Thus, the order of elution of the B enantiomers can be directed by selection of the reagent enantiomer. Access to both enantiomeric forms of the reagent is desirable. For the investigation of the degree of chromatographic resolution of a pair of enantiomers, e.g., when different reagents are evaluated, it is sufficient to use a racemic mixture of the reagent. To determine the order of elution and the enantiomeric composition of a chiral substance an optically pure reagent is required. Nonetheless, there are certain drawbacks which require careful interpretation of the results It is important, for example, to know the optical purity of the derivatization reagent. Only if this is 100% optically pure the results will be representative of the enantiomer composition. The reason for this is shown in Figure 3.4.1.

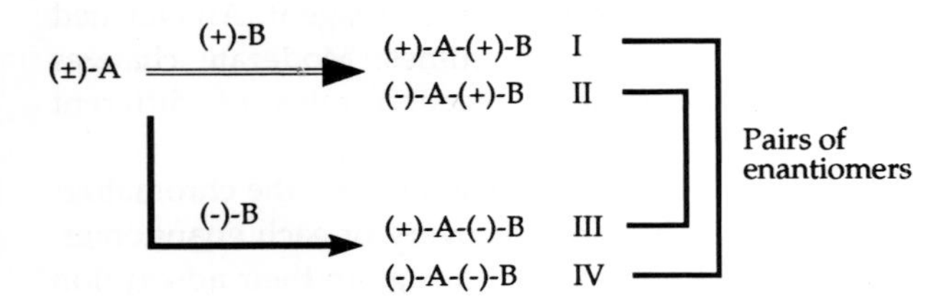

Figure 3.4.1 The effect of the use of a chiral derivatization reagent with < 100% optical purity.

If the enantiomers of A are to be separated and determined as their diastereomeric derivatives I and II, obtained by reaction with *R*-B, any contamination with *S*-B will be detrimental. The products III and IV produced with *S*-B form enantiomeric pairs with the two main products (IV with I and III with II) and are therefore added to the corresponding peaks. The effect is shown by the following example. Assuming that A consists of the *R*-form in 99% and the reagent *R*-B in 97% optical purities, 99.5% of *R*-A, 0.5% of *S*-A, 98.5% of *R*-B and 1.5% of *S*-B will occur in the reaction mixture. If complete reaction is supposed, the proportions of I-IV will then be 98.0075%, 0.4925%, 1.4925% and 0.0075%, respectively. Since I and IV are superimposed in the chromatogram, as well as II and III, the two peaks will show the proportions 98.015% and 1.985%, respectively. Therefore, if the optical purity of the reagent is not considered, A will be found to contain 98% of the *R*-form, which corresponds to an optical purity of 96%.

3.4.1.1 Practical considerations of diastereomer formation

For the determination of unknown concentrations of enantiomers, the derivatization should be quantitative, i.e. 90-100%, in order to ensure high accuracy. A high yield also ensures that the enantiomeric ratio has not been changed by kinetic resolution. The reagent should be in excess of the substrate. For small quantities of the substrate, the ratio between amounts of reagent and substrate is not important. Instead, the absolute concentration of the reagent in the reaction mixture is essential, as it determines the rate of diastereomer formation. When the reagent is not continuously consumed by, e.g., solvolytic reactions and its concentration (and other parameters such as pH) remains approximately constant, the formation of diastereomers often follows pseudo first-order kinetics, i.e. the substrate concentration declines exponentially with time. This means that the half-life is independent of the initial substrate concentration. Lower reagent concentrations require longer reaction times. For a reaction in which the reagent is continuously consumed by solvolysis, it is essential that the initial rate of diastereomer formation be high enough so that quantitative conversion is obtained while the reagent concentration is still high. If not, prolonging the reaction time will contribute little to the reaction yield. The recovery of slow-reacting substrates can be increased by using a higher reagent concentration. Therefore, when using reagents such as isocyanates or acylating reagents in aqueous media, it is helpful to know the actual rate of hydrolysis of the reagent. An elevated temperature may be used as well to shorten reaction times. Moderate changes normally have marginal influence on the relative reaction rates of different competing reactions.

If extensive clean-up is carried out after derivatization and before the chromatographic determination, it is recommended to check the recovery of each enantiomer. The diastereomers formed may exhibit different properties, e.g., in their adsorption characteristics.

It is obvious to exclude racemization or epimerization. In particular, compounds with asymmetric carbon centres which could change configuration through carbanion or carbocation intermediates have to be checked carefully as they may not exhibit the necessary configurational stability under basic or acidic conditions.

For the separation of diastereomers the distance between the two chiral centres in the derivatives is also important. In general, the centres should be as close as possible to each other in order to maximize the difference in chromatographic properties. Normally, a maximum of three bonds should separate the two centres.

As mentioned above, requirements for chromatographic resolution depend on whether the work is analytical or preparative. Factors that affect the chromatographic resolution R_S are the relative retention of the two diastereomers (the separation factor α), the partition ratio k and the actual column efficiency N ($N= [t'_R/\sigma]^2$; cf. Section 3.3). In preparative chromatography high α-values are desirable as they allow greater sample loading on the column, without peak overlapping due to reduction in N. For analytical purposes, columns with high separation efficiency are used to separate the diastereomers. In these cases, a separation factor of 1.1-1.2 is sufficient to obtain adequate resolution. When one enantiomer is to be determined in the presence of a large excess of the other enantiomer, higher α-values may be needed. It is desirable that the smaller peak elute before the larger one because of the often observed peak tailing. The desired order of elution can be obtained by choosing the proper enantiomer of the reagent.

The chiral substances most commonly resolved as diastereomers are amines, alcohols and carboxylic acids. The chiral derivatization reagents for these compounds and other substances are the same as used in gas chromatography (see Section 3.5). The main types of derivatives formed are outlined in Figure 3.4.2.

Since amines are the easiest to derivatize a large variety of reagents is available. In bifunctional compounds such as amino acids and amino alcohols the amino group is usually the site for derivatization. Chiral amines, alcohols and carboxylic acids are also used as substrates for chiral derivatizing reagents.

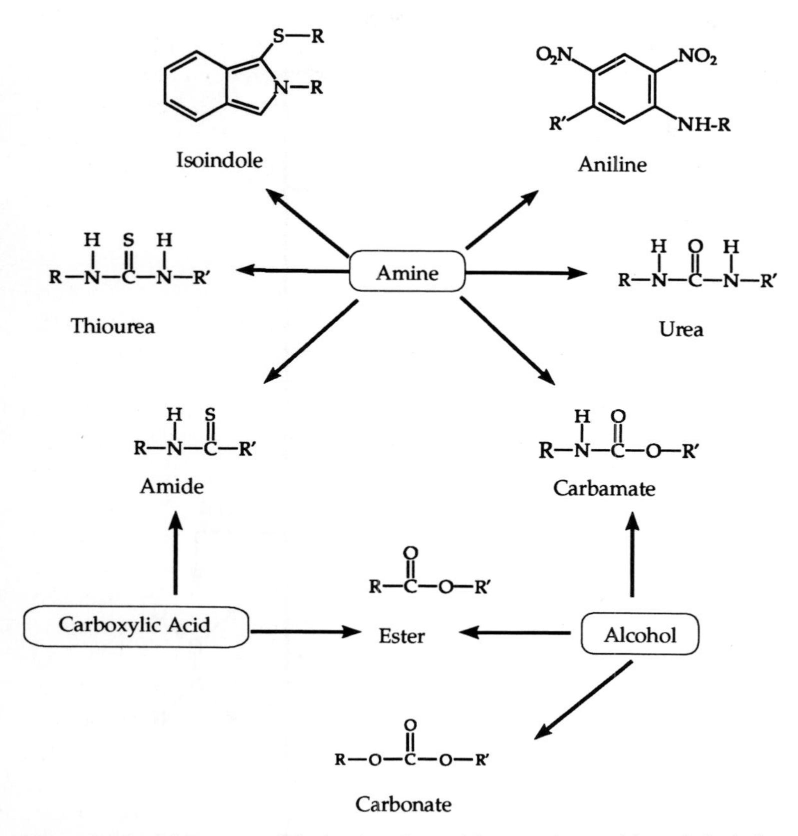

Figure 3.4.2 Main types of derivatives formed from amines, acids and alcohols.

In gas chromatography (cf. Section 3.5) the formation of esters is preferred to that of amides to separate carboxylic acids, because of the less favourable chromatographic properties of amides. The opposite situation arises in liquid chromatography where diastereomeric amides are easily resolved due to hindered rotation around the amide bond: In HPLC, the majority of carboxylic acids are, therefore, separated as amides. Various ways of achieving the formation of amides are outlined in Figure 3.4.3.

3.4.1.2 The background of the chromatographic separation of diastereomers

Differential elution of diastereomers in a chromatographic system is based on interactions of the sample molecules with both the stationary phase and the solvent.

Systematic studies of the chromatographic behaviour of diastereomers may be favourable for predicting optimal chiral reagents and allowing the assignment of the absolute configuration of chiral compounds.

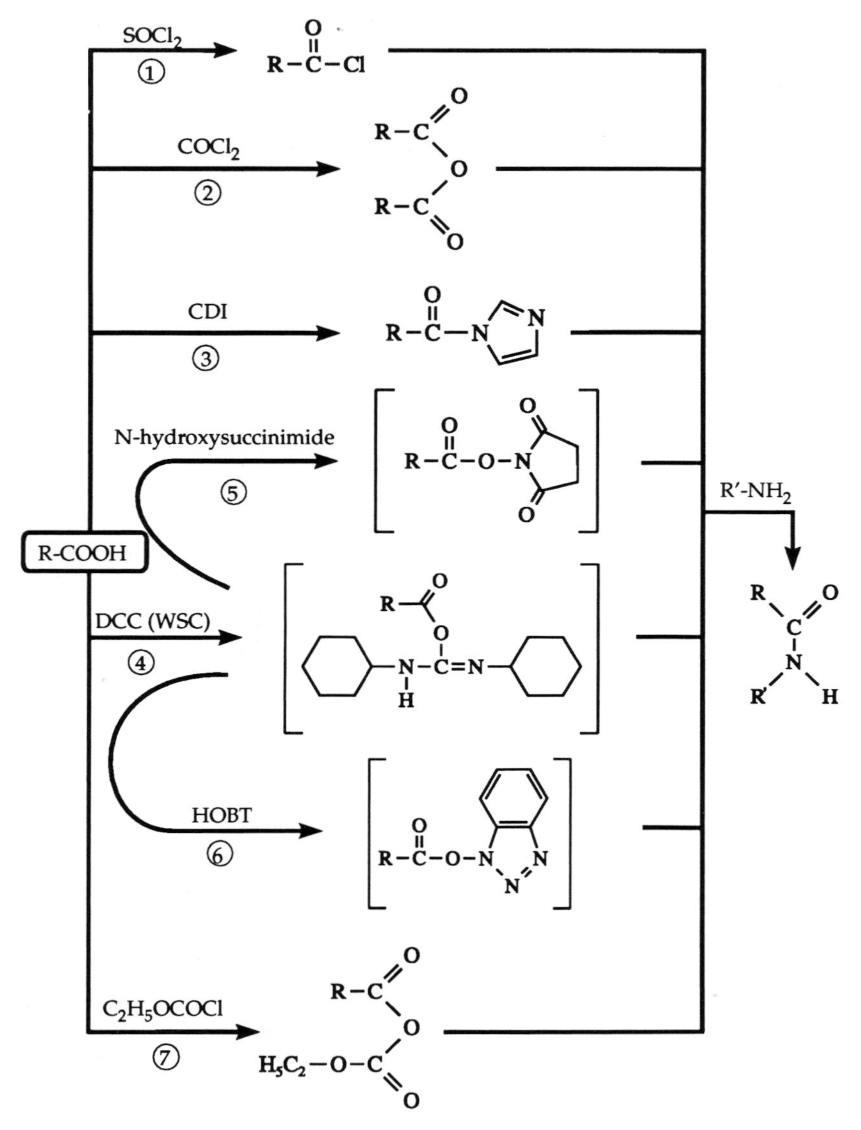

Figure 3.4.3 Formation of amides to separate carboxylic acids.

The nature of the interactions involved in the retention differs in the various modes of chromatography. Nonetheless, several common features such as distance of the chiral centres and conformational immobility around the centres are considered to enhance the resolution of diastereomers. The effect on the resolution of increasing the distance between the chiral centres has been demonstrated by studying the separation of acyclic isoprenoid acid enantiomers derivatized as p-nitrophenylethylamines [2]. The separation factor dropped from 2.2, with the secondary methyl chiral centre in the α-position, to 1.0 when it was at carbon 5.

Many separations of diastereomers have been performed by reversed-phase chromatography. However, a systematic study of the parameters is still lacking. Detailed studies have been made of the relative retention of diastereomers on silica and alumina. Separation of more than 100 pairs of diastereomers on silica have been discussed previously [3,4]. The order of elution was characterized by preferential retention of molecules with the least steric hindrance of the adsorbing polar groups. The data showed significant influence of the mobile phase composition on separation.

The retention was explained in terms of solvent-solute localization, which refers to the tendency of a solute (solvent) to attach to specific sites on the adsorbent. The results also indicated that a localizing solvent (e.g., acetonitrile) preferentially reduces the retention of the strongly retained solute. Maximum separation factors were, therefore, obtained with non-localizing mobile phases, e. g., dichloromethane.

The bonds in amides [5-7] and carbamates [8,9] are planar. Provided that the chiral substituents are non-polar, adsorption to the stationary phase is assumed to take place between the chiral moieties primarily by hydrogen bonding of the acyl oxygen and - to a lesser extent - of the hydrogen on the amide nitrogen. Helmchen et al. [5-7] found that the degree of shielding by the non-polar groups in secondary amides governed the order of elution. Better shielding and less retention resulted when the substituents most efficient in steric hindrance were situated on opposite faces of the common plane formed by the amine bond. Polar groups such as hydroxyls in close location to the carbonyl group were found to exhibit a favourable effect on the resolution [7].

Poor separation was observed for carbamate-derivatives of secondary amines (methyl group on the nitrogen) [8,9]. The same effect was recognized in separating amphetamine from methamphetamine derivatized with 4-nitrophenylsulphonyl-*S*-propyl chloride, with no resolution of methamphetamine [10]. Reversed-phase separation of the derivatives showed the opposite behaviour, with successful resolution of only methamphetamine.

Various normal- and reversed-phase systems have been compared for the separation of epimers of corticosteroids [11]. As normal-phase supports were checked nitrile, silica and an amino-phase. The separation factors increased in the order just given, which parallels the increased polarity of the supports.

Comparison with the reversed-phase mode showed that an increased length of the alkyl group located at the centre of chirality (C-22) yielded improved separation in the reversed mode, while the resolution in the normal-phase mode remained unaffected.

3.4.1.3 The detection properties of diastereomers

As a result of derivatization the detection properties of enantiomers may change. Chiral reagents often exhibit conjugated structures or are created when new bonds are formed, and may contribute not only to the separation, but also to the detection of the diastereomers formed. Chiral reagents such as naphthyl propionic acid and naphthylethylamine yield derivatives which may be detected by UV absorption at, e.g., 254 or 280 nm, or by fluorescence emission at 320-350 nm. The thiourea structure formed by reaction between isothiocyanates and amines exhibits a molar absorptivity of about 12,000 at 250 nm [12]. This means that such diastereomers can be detected at very low concentration levels. Enantiomers with native UV absorption can normally be expected to have equally good or even better detectability after derivatization.

Certain chiral reagents have been designed to allow sensitive detection. Anthracene has an extremely high molar absorptivity, i.e. 2×10^5 mol^{-1} cm^{-1} at 254 nm, and shows high fluorescence quantum efficiency. Anthrylethylamines were introduced by Goto et al. [13] for derivatization of N-protected amino acids; detection limits were reported to be 100 fmol. Fluorene has a molar absorptivity of 2×10^4 at 260 nm and a high quantum yield with an emission maximum at 315 nm. Fluorenylethyl chloroformate was introduced by Einarsson et al. [14]. This reagent was used for derivatization of primary and secondary amino acids and amines.

Detection principles other than UV absorption and fluorescence do not play an important role, however, electrochemical detection should be mentioned. Ferrocene compounds show interesting electrochemical properties. They can be oxidized at a low potential and thereafter reduced at a second downstream electrode, in some cases with very high sensitivity and selectivity [15]. The chiral ferrocene reagents ferrocenylethylamine and ferrocenylpropylamine were introduced for electrochemical detection by Shimada et al. [16].

When enantiomers with favourable native fluorescence or UV properties are to be determined at low concentrations in, e.g., biological samples, it may be advantageous to use so-called transparent reagents. Such reagents and their degradation products do not themselves fluoresce or absorb light, and no fluorophores/chromophores are introduced into other reactive components in the sample.

An example of the use of transparent reagents is the method of Björkman [17] for determination of indoprofen in plasma after derivatization with ethyl chloroformate and leucinamide. For fluorescence detection, excitation and emission wavelengths are 275 and 433 nm, respectively, which results in very little interference from other co-extracted sample components.

The spectral properties of the two diastereomers formed by covalent derivatization with a chiral reagent are not necessarily identical. Therefore, a slight difference in measured peak areas for a derivatized racemate may be due to different detector response for the two diastereomers.

3.4.2 *Addition of chiral reagents to the mobile phase*

Modern reversed-phase HPLC sorbents (C_2-C_{18} silicas) are suitable substances for achieving a high column efficiency. Today, mobile phase additives in reversed-phase liquid chromatography are widely used in order to regulate the retention behaviour of an analyte. Basically, the technique uses ion-pair formation in organic media, where the partition of a positively charged analyte such as protonated amine is strongly influenced by the nature of the counter-ion.

By using an optically active counter-ion diastereomeric ion-pairs are formed, which may be successfully separated with an achiral reversed-phase column. The technique of using amphiphilic additives can be regarded as a dynamic coating of the achiral sorbent, i.e. a physical immobilization of the amphiphile. Thus, if this additive is optically active, the achiral sorbent will be converted into a chiral one.

The ionically-bound Pirkle phases (cf. Subsection 3.4.3.1) are excellent chiral stationary phases as long as they are used in non-polar solvents where the mobile phase has very little tendency to displace the selector from its adsorption sites. Under these conditions, the mobile phase can be used without any added selector. In cases where the selector is immobilized by strong hydrophobic interaction to an alkyl-silica or other hydrophobic matrix, the selector will require to be present in the mobile phase.

Many of the principles based on covalently-bound chiral phases can also be applied to the technique of adding the chiral selector to the mobile phase. They can be divided into three categories; (i) the metal complexation used in chiral liquid exchange chromatography; (ii) the use of various uncharged additives; and (iii) the ion-pairing techniques used for charged analytes.

3.4.2.1 Chiral additives at metal complexation

Metal complexation for chiral liquid exchange chromatography (CLEC) applications was first used by Karger [19,20]. Employing chiral triamines (L-2-ethyl- and L-2-isopropyl-4-octyldiethylenetriamine) as mobile phase additives in the presence of Zn(II) and other transition metal ions, a series of dansyl amino acids was resolved on C_8-reversed-phase. A similar system, i.e. L-prolyl-N-octylamide and Ni(II), was successfully applied to the same type of resolutions with a C_{18}-column [21].

A detailed study was performed by Davankov et al. [22] to achieve a successful analytical and preparative resolution of free amino acids. N-alkyl-L-hydroxy-prolines (C_7, C_{10} and C_{16} chain length) were used to coat a C_{18}-silica column and Cu(II) acetate (0.1 mmol) in methanol/water (15+85, v/v) was employed as the mobile phase. All additives were adsorbed by means of strong hydrophobic interaction with the C_{18} sorbent. In general, decreasing alkyl chain length has been found to result in larger separation factors. The reason for this is unclear as the amounts of chiral additive adsorbed could not be determined.

The following effects caused by a variation of the mobile phase composition can be mentioned; (i) an increase in pH (above 5.5) causes increased retention and larger separation factors; (ii) a decrease in Cu(II) concentration results in a small increase in retention, but no significant effects on separation factors; (iii) increasing the ammonium acetate concentration strongly decreases the retention.

The relative stabilities of the mixed ligand sorption complexes formed in CLEC are strongly dependent on the method used for immobilization. When the selector is physically immobilized by hydrophobic interaction, the order of elution of all amino acids is $k(\mathrm{L}) < k(\mathrm{D})$. This experimental result, combined with the mobile phase effects, suggests the enantioselective mechanism outlined in Figure 3.4.4.

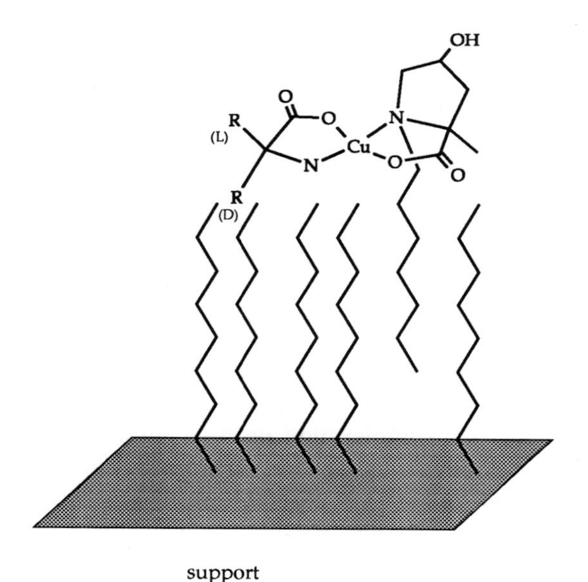

Figure 3.4.4 Proposed interactions between the C_{18}-silica, N-alkyl-L-hydroxyproline, Cu^{2+} and an amino acid [22].

The N-alkyl chains of the selector are thought to be oriented parallel to the C_{18}-chains. By co-ordination with a Cu(II) ion this fixed ligand adopts a conformation such that the hydropyrrolidine ring and its N-alkyl group extend in a direction opposite to the main co-ordination plane of the Cu(II) chelate. Therefore, a mixed ligand sorption complex formed by the D-enantiomer of the analyte, the α-substituent of this enantiomer will be directed towards the hydrophobic (C_{18}) sorbent surface. Thus, an increased stabilization by means of hydrophobic interaction will be achieved. The L-enantiomer does not have this possibility and is, therefore, eluted faster than the D-enantiomer.

The same principle has been used in a technique based on coating alkyl-silicas with N-decyl-L-histidine [23]. The separation factors were generally lower than those found by employing N-alkyl-L-hydroxyproline selectors, but were still useful for the determination of enantiomer composition. As in the previous case, the highest enantioselectivity was found for the amino acids with the largest α-substituents (alkyl and aryl groups).

If the alkyl chain of the chiral selector is omitted, conditions are obtained under which there should no longer be any strong hydrophobic interactions with the alkyl silica and no actual physical immobilization of the selector. Consequently, the chromatographic process might then best be understood as *in situ* generation and separation of diastereomeric complexes in a reversed-phase mode.

A variety of methods based on this principle of chiral metal complexation in the mobile phase has been developed. Chiral metal complexes used as mobile phase additives for the optical resolution by chiral ligand exchange chromatography are summarized in Table 3.4.1. Applications for amino acids, hydroxy acids and hormones have been described.

Table 3.4.1 Chiral metal complexes used as mobile phase additives for the optical resolution by chiral ligand exchange chromatography [18].

Selector	Metal ion	Stationary phase	Analyte
L-proline	Cu^{2+}	octyl-silica	amino acids
L-proline	Cu^{2+}	octyl-silica	dansyl amino acids
L-proline	Cu^{2+}	silica	thyroid hormones
L-proline	Cu^{2+}	cation exchanger	amino acids
L-histidine	Cu^{2+}	octyl-silica	amino acids
L-histidine methyl ester	Cu^{2+}	octadecyl-silica	amino acids
L-arginine	Cu^{2+}	octyl-silica	amino acids
L-phenylalanine	Cu^{2+}	octadecyl-silica	aromatic amino acids
L-phenylalanine	Cu^{2+}	octadecyl-silica	mandelic acids
L-aspartic acid monoalkylamides	Cu^{2+}	octadecyl-silica	amino acids
L-aspartyl-L-phenylalanine methyl ester (Aspartame)	Cu^{2+}, Zn^{2+}	octadecyl-silica	amino acids
N,N-dipropyl-L-alanine	Cu^{2+}	octadecyl-silica	dansyl amino acids methylamino acids
N,N-dialkyl-L-amino acids	Cu^{2+}	octadecyl-silica	amino acids
N-(p-tosyl)-L-(and D-) phenylalanine	Cu^{2+}	octadecyl-silica	amino acids
L-amino acids	Cu^{2+}	octadecyl-silica	hydroxy acids
R,R-tartaric acid monooctylamide	Cu^{2+},Ni^{2+}	octadecyl-silica	amino acids

3.4.2.2 Uncharged chiral mobile phase additives

Many chiral phases of the diamide type give rise to enantioselective hydrogen bond interactions. These bonded selectors should therefore also be useful as additives in a non-polar mobile phase in normal-phase chromatography. Under these conditions they are adsorbed rather strongly on a silica surface, which can then be regarded as coated with a chiral stationary phase (CSP). In particular, N-acetyl-L-valine tert-butylamide (**1**) and *R,R*-di-isopropyl tartaric diamide (DIPTA, **2**), have been found suitable for the optical resolution of a number of polar solutes [24,25].

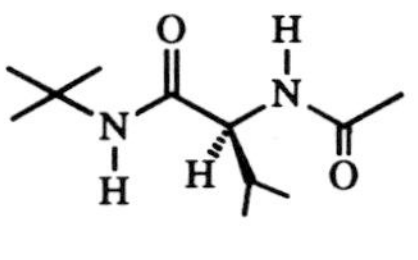

1 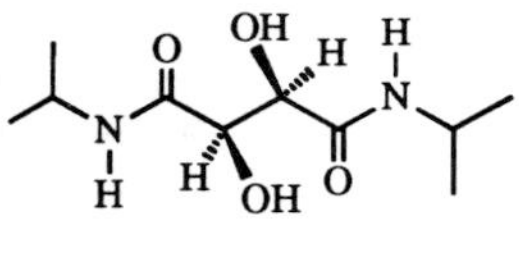 **2**

As to chiral recognition, the behaviour shown by the *threo-* and *erythro*-forms of solutes containing a 1,2-diol structure (cf. 1-phenyl-1,2-diol in Figure 3.4.5) is relevant.

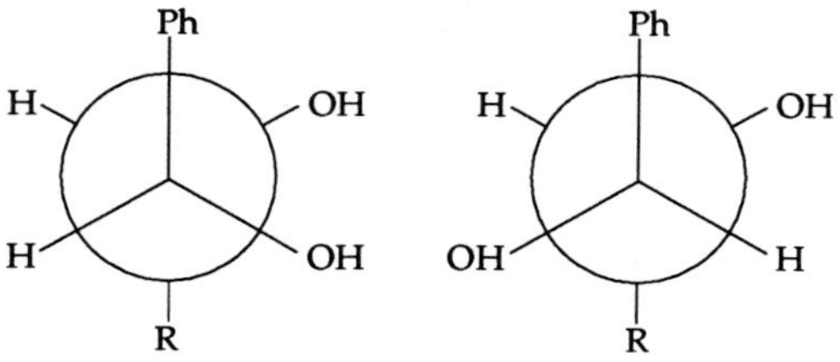

a b

Figure 3.4.5 The preferred conformations of the *threo* (a)- and *erythreo* (b)-isomers of a 1-phenyl-1,2-diol.

Whereas the *threo*-compounds (a) yield larger separation factors with increasing bulkiness of the substituent R, the reverse is found for the *erythro*-forms (b). This can be explained in view of the preferred conformations of the two forms, which become more populated as the steric requirements of R increase (Figure 3.4.5). Because a two-point hydrogen bond formation between the CSP and the solute will require a *gauche* conformation of the two hydroxyl groups in this case, the experimental results strongly support such a mechanism of formation.

Another type of uncharged chiral selector which has been used as a mobile phase additive is cyclodextrin (CD). CD, mainly in the β-form (β-CD), is used in a reversed-phase system employing C_{18}-silica and an aqueous buffer system [26]. The first studies were performed on substituted mandelic acids. Large substituent effects were observed and retention decreased with increasing pH or increasing ß-CD concentration. Complete optical resolution (α = 1.8) was achieved for o-chloromandelic acid at pH 2.1 using 14.4 mmol β-CD in buffer; otherwise, separation factors were low and decreased with increasing pH.

N-methyl barbituric acid derivatives and other pharmaceuticals have been separated [27] using an achiral RP-18 phase and several cyclodextrins as mobile phase additives. By equilibrating the stationary phase with the mobile phase (0.025 mol sodium phosphate buffer [pH 2.5] and ethanol were mixed [8+2; v/v]; to 1+l of the resulting solution 25 mmol of α-, β-, or γ-CD was added), the CDs were adsorbed onto the RP-18 modified silica surface. In a dynamic process, the adsorption and desorption of cyclodextrin molecules from the surface determine the chromatographic properties of the system. In equilibrium, the concentration of adsorbed and desorbed CD is constant.

Table 3.4.2 Separation values (α) of different pharmaceuticals obtained by the addition of β-CD or γ-CD to the mobile phase [27].

Racemate	β-Cyclodextrin	γ-Cyclodextrin
Prominal	1.086	1.000
5-Cyclopenten-1-yl-5-ethyl-1-methyl barbituric acid	1.092	1.000
N-Methylcyclohexylethyl barbituric acid	1.225	1.063
1,5-Dimethylphenyl barbituric acid	1.073	1.000
Hexobarbital	1.112	1.287
*MTH-tyrosine	1.197	1.000
**PTH-phenylalanine	1.000	1.029
PTH-norvaline	1.039	1.019
Dansyl-tryptophan	1.000	1.192
***DNP-α-aminobutyric acid	1.028	1.019
Carvedilol	1.058	1.000
Chlorthalidone	1.165	1.000
Chlormezanone	1.030	1.000
Methaqualone	1.000	1.033
Bendroflumethiazid	1.000	1.071
Oxazepam	1.000	1.078

*MTH=Methylthiohydantoin; **PTH=Phenylthiohydantoin; ***DNP=Dinitrophenyl

Under these conditions, the stationary RP-18 phase is dynamically coated with a layer of chiral selectors exhibiting a high enantioselectivity. Interestingly, it was not possible to separate any racemate using α-CD. Apparently, the cavity of α-CD is too small to include the solutes. In Table 3.4.2 the α-values obtained for different substances using β- or γ-CD are compared. If one modifies the mobile phase by changing the organic solvent from acetonitrile to ethanol and methanol, the chiral separation will be strongly affected. Although methanol seems to be the best modifier for the mobile phase, it is, in comparison with ethanol and acetonitrile, less advantageous because of the poor solubility of β-cyclodextrin.

3.4.2.3 Charged additives used in ion-pairing techniques

As a logical consequence of the ion-pair chromatographic technique [28], a chiral counter-ion, (+)-10-camphorsulphonic acid, was introduced as a mobile phase additive for the optical resolution of several amino alcohols [29]. The amino alcohol in its protonated form combines with the camphorsulphonate by electrostatic interaction to form diastereomeric complexes. It is assumed that a second interaction occurs, a hydrogen bond between the keto group and the hydroxy group of the two partners, leading to the different chromatographic retentions observed. The separations were performed in normal-phase mode on silica or diol-silica by using dichloromethane with a small amount of a polar retention modifier (dichloromethane is an excellent ion-pair solvent). The enantiomer separation should be regarded as a separation of labile diastereomeric ion-pair complexes, i.e. the chiral discrimination occurs in the mobile phase. As expected, the capacity factors decrease considerably with increasing concentration of the polar component of the mobile phase.

The requirement of a two-point interaction for enantioselection is supported by results illustrating the structural effects of the analyte. No optical resolution occurs when the hydroxyl and amino groups are separated by more than two carbon atoms or if the hydroxyl group is absent. Furthermore, oxprenolol (**3**) is not resolved, probably due to an internal hydrogen bond between the hydroxyl group and the oxygen atom of the allyloxy group.

Improved results have been obtained using an N-protected dipeptide, N-carbobenzoxycarbonyl-glycine-L-proline, as the chiral counter-ion. In these cases α–values as high as 1.4 were obtained for some of the amino alcohols [30]. These results encouraged the use of the ion-pair system for the enantioseparation of racemic sulphonic and carboxylic acids with optically active amino alcohols as counter-ions. Alprenolol (**4**), however, with its binding groups located in a flexible alkyl chain, showed a low degree of stereoselectivity. Considerable improvement has been achieved using compounds with rigid ring systems such as quinine, quinidine and cinchonidine.

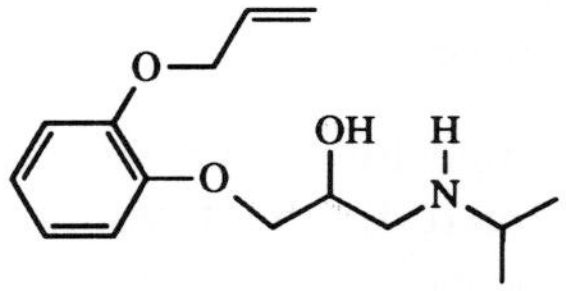

3 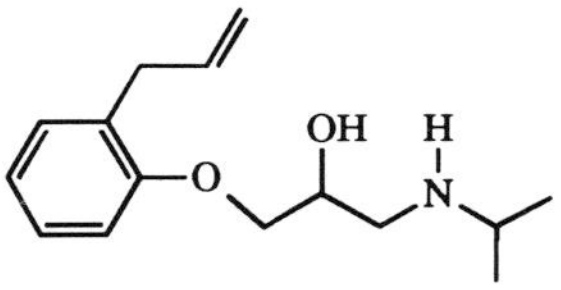 **4**

Although the use of chiral counter-ions for optical resolution by the technique described above has proved to be successful for several applications, the chromatographic system is complex and the effects on retention and resolution of many of the factors involved are not always easy to elucidate. A few of these factors are discussed below:

(i) *The chromatographic sorbent:* The surface properties of the strongly polar silica sorbent are critical. Diol-silica is generally preferred due to its better performance.

(ii) *The water content of the mobile phase:* This is a critical factor due to its distinct influence on retention (and resolution). A water content of 80-90 ppm has been recommended. Higher concentrations of water are detrimental. The small amount of water is probably essential in order to deactivate the silica surface, which otherwise might adsorb polar components too strongly.

(iii) *The capacity ratios of the enantiomers:* These usually decrease as the concentration of the counter-ion increases. This fact has been explained as due to competition of the counter-ion and the ion-pairs for the same adsorption sites on the sorbent.

(iv) *The polar components of the mobile phase:* These cause a drastic decrease in retention of the solute, which is usually accompanied by a decrease in optical resolution. The latter effect is regarded to result from competition for hydrogen-bonding groups in the ion-pair components, which lead to reduced stereoselectivity.

(v) *The optical purity of the mobile phase additive:* This influences the separation as follows:

$$\alpha_{obs} = \frac{\alpha P + (100 - P)}{\alpha(100 - P) + P}$$

where *P* is the fraction (%) of one enantiomer in the mobile phase additive, and α_{obs} is the observed separation factor (*P* is not equal to the optical purity, which is 0 for *P* = 50).

A different ion-pairing chromatographic technique, employing (+)-dibutyl tartrate (DBT) as chiral additive has also been described [31]. The principle of this technique originated in studies by Prelog et al. [32] of the liquid-liquid distribution of enantiomeric amino alcohols as ion-pairs in the presence of tartaric acid esters; an unequal distribution of the enantiomers was observed. In chromatography, DBT is used as a physically immobilized CSP, applied by adsorption from an aqueous phase onto a hydrophobic matrix (alkyl-silica). This system has permitted partial optical resolution of a series of amino alcohols into their ion-pairs with hexafluorophosphate in a reversed-phase system using a completely aqueous buffer. The mechanism of this enantioseparation is unclear.

3.4.3 *Chiral stationary phases (CSP)*

The separation of enantiomers by HPLC using chiral stationary phases (CSP) is based on the formation of transient diastereomeric complexes between the enantiomorphs of the solute and a chiral selector in the stationary phase. A diagram of a chiral resolution on a CSP is presented in Figure 3.4.6. In order to separate the two enantiomeric solutes, a difference in internal energy must be produced on the column. This is achieved by binding a chiral molecule to the chromatographic stationary phase which is able to form diastereomeric complexes with the solutes. The difference in stability between these complexes leads to a difference in retention time; the enantiomer forming the less stable complex is eluted first. In the example in Figure 3.4.6, the solute-CSP complex formed between the *S*-solute and the CSP has a greater energy content, i.e. is less stable than the solute-CSP complex formed by the *R*-solute.

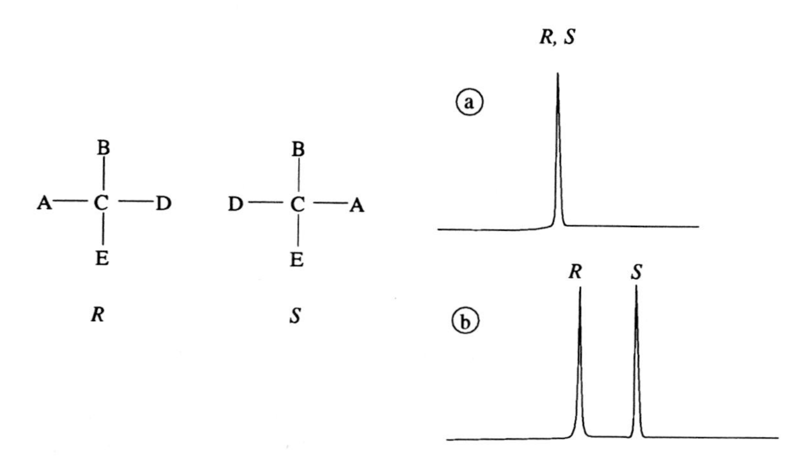

Figure 3.4.6 Chromatographic separation of two enantiomers; (a) achiral phase; (b) chiral phase [33].

The basic knowledge of chiral recognition. CSP must be able to form diastereomeric complexes that differ due to the stereochemistry of the enantiomeric solutes. Thus, the CSP must be able to recognize the stereochemical differences between the enantiomers. The first LC model for chiral recognition by a CSP was developed by Dalgliesh [34] explaining the resolution of aromatic amino acids by cellulose paper chromatography. A 'three-point' attachment between the solute and CSP based on attractive interactions has been proposed. These interactions comprised hydrogen bonding and adsorption of the aromatic moiety. In the 'three-point' model the determination of the configuration about a chiral carbon involves at least three of the four bonds attached to that center.

An example of chiral resolutions based on interactions between the solute and the CSP is presented in Figure 3.4.7. Enantiomer I interacts with the CSP at three sites: A-A', B-B' and C-C', whereas enantiomer II lacks the C-C' interaction. If the C-C' interaction is an attractive interaction, enantiomer I will be retained on the column longer than II. If the C-C' interaction is repulsive, however, then the diastereomeric complex with enantiomer II is more stable and enantiomer I will be eluted first.

In the following, several theoretical aspects of important binding types playing a role in enantioselective sorption processes are discussed.

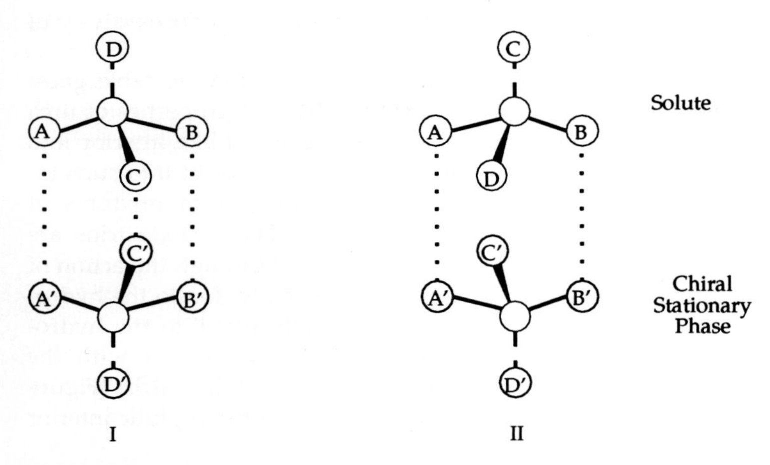

Figure 3.4.7 The 'three-point' interaction model for chiral recognition [33].

Co-ordination to transition metal. The transition metals are characterized by unfilled inner-shell d-orbitals. A transition metal complex is formed by ligands which may donate electrons to these unfilled orbitals. Such co-ordination complexes possess a well-defined geometry; the ligands can only occupy certain given positions in space. Thus, the donor ligand atoms in the complex are held at fixed distances from the metal atom and in definite orientations. This so-called co-ordination sphere is therefore packed with the ligands and with the solvent molecules. The latter also form a second (outer) highly organized sphere. Consequently, the stability of the complex is highly dependent on the stereochemical relationships, since in the presence of two or more chiral ligands in the coordination sphere their interaction - either directly or through the solvent molecules of the first or second sphere - will lead to differences in complex stability, thereby causing enantioselectivity. In both cases the principle is to use a chiral stationary phase composed of an immobilized chiral ligand forming a coordination complex with the transition metal ion. During passage of a suitable racemic mixture through the column, diastereomeric mixed-ligand sorption complexes are formed.

Charge-transfer (CT) interaction. This type of interaction requires π-electron systems. Stable CT-complexes are often formed between aromatic rings acting as donor and acceptor components. Such an aromatic π-π interaction, together with additional polar interactions (hydrogen bonding, dipole interactions), is the foundation for efficient chiral selectors used in LC.

Nitroaromatics are usually good π-acceptors because a negative charge is effectively delocalized by participation of the substituents in resonance stabilization. Good π-donors are also aromatics with electron-releasing substituents such as amino or alkoxy groups. Although π-π interaction as the sole source of retention has proven sufficient for the optical resolution of condensed aromatic hydrocarbons of planar chirality (helicenes), further bonding interactions increase the enantioselectivity of the selector.

Inclusion phenomena. The ability of certain compounds to include suitable guest molecules has long been known. Classical examples are the host properties of urea and starch. Urea forms spontaneously complexes with a channel-like interior into which unbranched alkanes fit well. Since branched alkanes do not fit into such interiors, this phenomenon can be used to separate n-alkanes from mixtures of isomers. Starch is well known for its inclusion of iodine. The cyclodextrins are crystalline degradation products of starch, which are obtained through the action of microorganisms. The formation of inclusion complexes is mainly due to the hydrophobic moiety of the solute molecule which must be 'tightly fitted' to the hydrophobic cavity; the polar substituents of the solute may then interact with the secondary hydroxyl groups on the hydrophilic 'mouth' of the cyclodextrin (Figure 3.4.8). Crown-ether is another type of host molecule, one with a hydrophilic interior and a hydrophobic 'mouth' (Figure 3.4.8).

The cavity as hydrophilic interior contains heteroatoms such as oxygen, where lone-pair electrons are able to participate in bonding to electron acceptors such as metal or organic cations, e.g., ammonium cations. In the latter case, the ammonium ion is held within the cavity by hydrogen bonds to the ether oxygen atoms.

Particular inclusion effects are observed in chiral matrices composed of swollen microcrystalline cellulose derivatives. The triacetate has been shown to be effective, in part, by steric exclusion. For instance, of a series of aromatic hydrocarbons, benzene is strongly retained, 2,3,5-trimethylbenzene (mesitylene) is less strongly retained and 1,3,5-tri-tert-butylbenzene is completely excluded. This phenomenon can be explained by the lamellar arrangement of the polysaccharide chains. These yield a kind of two-dimensional molecular sieve, permitting, in particular, inclusion of flat aromatic compounds, but excluding sterically more demanding structures.

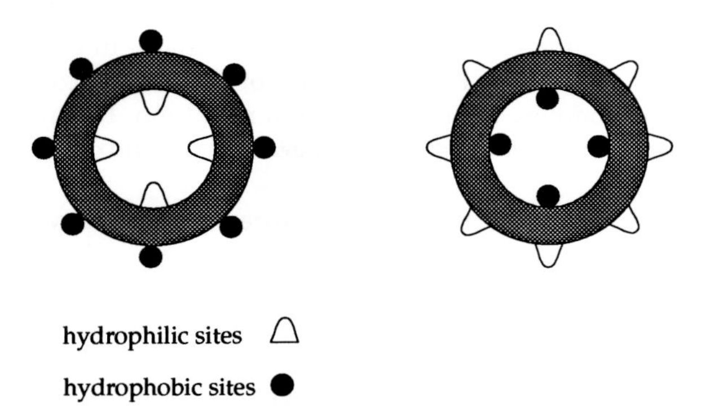

Figure 3.4.8 Different types of host molecules [38].

The interactions should be different in nature, e.g., a hydrogen bond, a π-π-interaction, and a dipole interaction. Otherwise, unfavourable associations might be formed. For example, if more than one π-π interaction is possible between an enantiomer and the receptor, the receptor loses its ability to perform molecular recognition. If A-A' and B-B' are both similar π-π interactions, A-B' and B-A' associations may also result, as shown in Figure 3.4.9; both enantiomers might be retained to the same extent, precluding resolution.

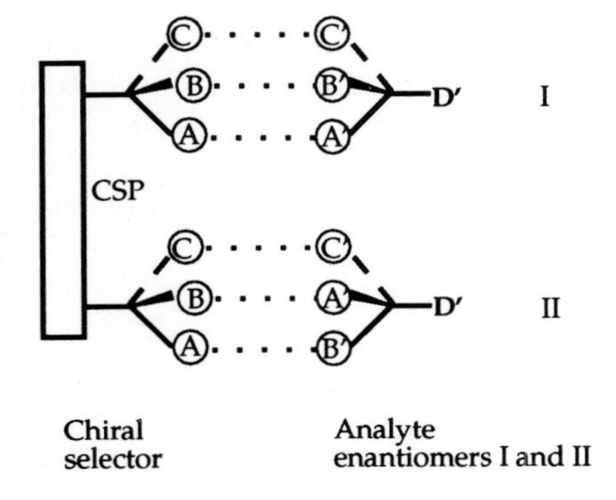

Figure 3.4.9 Similar π-π interactions (A-A', B-B') precluding resolution [35].

As already mentioned, chiral recognition is possible if one repulsive and two bonding interactions are involved. The repulsive interaction is often called steric. Typically, a bulky group is bonded to the chiral center. However, a bonding interaction is possible if one of the enantiomers forms a perfect steric fit and affords favourable van der Waals' interactions with the bulky group. For instance, Pirkle's [36] explanation of the stereoselectivity obtained with two attractive interactions at π-π (A-A') and H-bonding (B-B') sites, together with a steric interaction, should be considered (Figure 3.4.10).

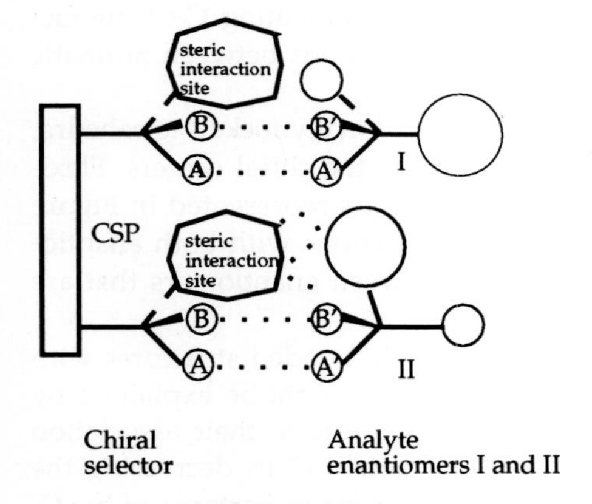

Figure 3.4.10 Stereoselectivity arising from one repulsive and two attractive interactions [35].

It is obvious that enantiomer II will be more strongly repulsed from the stationary phase due to the unfavourable position of its bulky group. A secondary effect is the weakening of the H-bonding (B-B') and the π-π (A-A') attractions. Enantiomer I develops stronger pseudo-bonds with the stationary phase than does enantiomer II; it is preferentially retained on the column.

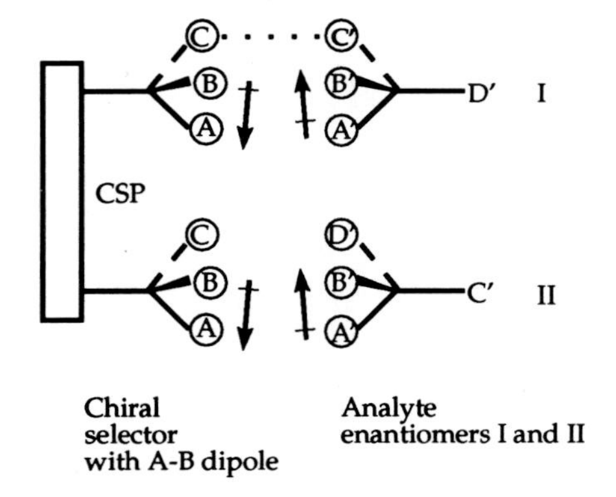

Figure 3.4.11 Discrimination through dipole stacking interaction [35].

Data obtained from intermolecular nuclear Overhauser effects in NMR and from molecular mechanics calculations supported this multiple interaction.

Interactions between selector and analyte are regarded to occur at the corners of two tetrahedra centered on the stereogenic centers of the two species. This illustration is a simplification of the molecular recognition process and it is potentially misleading. The 'three-point' rule does not imply that the interacting functions must be similarly arranged with respect to the stereogenic centers of the selector and analyte. Considering the case in which dipoles lie along the AB and A'B' vectors (Figure 3.4.11), stacking of these dipoles is functionally equivalent to two point-to-point interactions; chiral recognition results from the discriminating C-C' interaction. A similar argument can be applied to the π-π interactions between aromatic rings.

Thus, the species involved do not have to be conformationally locked tetrahedra, although there is typically some rigidity in the vicinity of the chiral centers. Flexibility can reduce a receptor's ability to be enantioselective, as represented in Figure 3.4.12. In this case, the receptor is capable of three interactions with both enantiomers. Similarly, rigid receptors may not distinguish between enantiomers that are highly flexible.

If the receptors are in close proximity (as they will be in bonded structures with high surface coverages), poor selectivities may result. This might be explained by the fact that the retention of the enantiomers is enhanced due to their association with more than one adjacent receptor molecule. With certain CSPs decreasing the concentration of receptors reduces this effect and increases the importance of the C-

C' interaction in chiral discrimination. Increased enantioselectivity is observed on less densely bonded silicas.

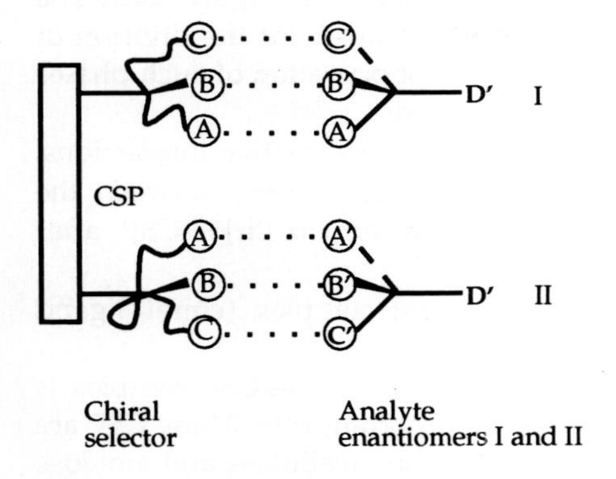

Figure 3.4.12 Nondiscrimination arising from flexibility of the chiral selector [35].

A single type of bonding interaction may be sufficient for optical resolution in certain LC modes of separation. The fact that enantiomers exhibiting only one hydrogen bonding substituent can be separated suggests that only one attractive force is necessary for chiral discrimination in chromatography. This might be explained by a difference in the equilibrium constants of the diastereomeric complexes between the chiral selector and the analyte. This difference is due to effects from the chiral binding site that force one of the enantiomers to take on an unfavourable conformation.

The question arises whether any postulated chiral recognition mechanism bears much resemblance to reality. Forst of all, the complex process of enantioseparation cannot be explained by simple models. Chiral recognition results from a weighted time-average of the contributions of many possible complexes, i.e. different directions of approach, different conformers, and different combinations of three (or more) simultaneous interactions are involved. Different models have to be used to explain the results of various experiments.

Studies involving the theory and mechanism of chiral recognition are increasing in number and sophistication. For example, as general criterion for chiral recognition the inequality of the distance matrices of complexes of different compounds with a resolving agent has been formulated [37]. Molecular modeling with energy calculations, NMR interaction studies, statistical mechanics, X-ray crystallography, and surface analysis are all being used in this area of research.

Classification of commonly used chiral stationary phases

Several authors have tried to categorize HPLC-CSP into classes. In one of these systems developed by Wainer [33], the chiral recognition process is divided up into two stages; (i) the formation of the solute-CSP complex, represented by interactions A-A' and B-B' in Figure 3.4.7; and (ii) the expression of stereochemical differences between the enantiomorphs represented by the interaction C-C' in Figure 3.4.7. The mechanisms involved in the first stage are used as a parameter for the division of HPLC-CSP into groups. However, while the mechanism of operation of such phases remains unclear, a simpler classification has been proposed:

Type 1. The solute-CSP complexes are formed by multiple attractive interactions, including hydrogen-bonding, π-π interactions, dipole stacking etc., between the solute and a low molecular-weight CSP. These are often termed Pirkle-CSP, after their principal inventor, or 'brush type' CSP.

Type 2. The solute is part of a diastereomeric metal complex (chiral ligand exchange chromatography [CLEC]).

Type 3. The primary mechanism for the formation of the solute-CSP complex is via attractive interactions, but inclusion also plays an important role. These CSP are high molecular mass polymers, often with a helical structure (cellulose and amylose derivatives; synthetic polymers).

Type 4. Interactions between the analyte and CSP follow from inclusion into chiral cavities within a CSP of relatively low molecular mass (cyclodextrins; crown-ether).

Type 5. The CSP is a protein and the solute-CSP complexes are based upon combinations of hydrophobic and polar interactions.

Type 6. The CSP is an optically active ruthenium(II)-complex coated onto spherical ceramics. Interactions follow from an ion exchange process.

The current commercially available CSPs can be differentiated using this classification and are presented in Table 3.4.3. The following Figure 3.4.13 gives some help in choosing the appropriate chiral stationary phase for a given problem. In the matter of chiral drug analysis, there is a database available (CHIRBASE; C. Roussel, Marseille) providing descriptions of 24,000 analytical separations; 8,000 structures of compound's enantiomers have been archived.

Table 3.4.3 Commercially available chiral stationary phases for HPLC (DNB = dinitrobenzoyl; cov. = covalent).

Type	Name	Chiral unit	Typical mobile phase	Supplier
Brush	Apex Prepsil L-Valyl-phenylurea	*S*-Valyl-phenylurea	Normal phase (Hexane/isopropanol)	Jones (Ercatech, Bern 31, Suisse)
	Bakerbond Chiral Covalent DNBLeu	*S*-DNB-Leucine (cov.)	-"-	Baker (Baker Chemicals, 64521 Groß-Gerau)
	Bakerbond Chiral Covalent DNBPG	*R*-DNB-Phenylglycine (cov.)	-"-	Baker (-"-)
	Bakerbond Chiral Ionic DNBPG	*R*-DNB-Phenylglycine (ionic)	-"-	Baker (-"-)
	Chiral D-DL	*R*-DNB-Leucine (cov.)	-"-	Serva (Serva,

Type	Name	Chiral unit	Typical mobile phase	Supplier
				69115 Heidelberg)
	Chiral L-DL	*S*-DNB-Leucine (cov.)	-"-	Serva (-"-)
	Chiral D-DPG	*R*-DNB-Phenylglycine (cov.)	-"-	Serva (-"-)
	Chiral L-DPB	*S*-DNB-Phenylglycine (cov.)	-"-	Serva (-"-)
	ChiRSil I	*R*-DNB-Phenylglycine (ionic)	-"-	RSL (Alltech Germany, 82008 Unterhaching)
	Chi-RoSil	*R*-DNB-Phenylglycine (ionic)	-"-	RSL (-"-)
	Covalent D-Leucine	*R*-DNB-Leucine (cov.)	-"-	Regis/Alltech (Promochem, 46485 Wesel)
	Covalent L-Leucine	*S*-DNB-Leucine (cov.)	-"-	Regis/Alltech
	Covalent D-Naphthyl-alanine	*R*-Naphthylalanine	-"-	Regis/Alltech
	Covalent L-Naphthyl-alanine	*S*-Naphthylalanine	-"-	Regis/Alltech
	Covalent D,L-Naph-thylalanine	*R,S*-Naphthylalanine	-"-	Regis/Alltech
	Covalent D-Phenyl-glycin	*R*-DNB-Phenylglycine (cov.)	-"-	Regis/Alltech
	Covalent L-Phenyl-glycin	*S*-DNB-Phenylglycine (cov.)	-"-	Regis/Alltech
	Covalent D,L-Phenyl-glycin	*R,S*-DNB-Phenylglycine (cov.)	-"-	Regis/Alltech
	ES D-DNB-Leu	*R*-DNB-Leucine (cov.)	-"-	ES (Macherey-Nagel, 52348 Düren)
	ES L-DNB-Leu	*S*-DNB-Leucine (cov.)	-"-	ES (-"-)
	ES D-DNB-PHGLY	*R*-DNB-Phenylglycine (cov.)	-"-	ES (-"-)
	ES L-DNB-PHGLY	*S*-DNB-Phenylglycine (cov.)	-"-	ES (-"-)
	ES R-PU	*R*-Phenylethylurea	-"-	ES (-"-)
	ES S-PU	*S*-Phenylethylurea	-"-	ES (-"-)
	Grom-Chiral-*R*-DNBPG-C	*R*-DNB-Phenylglycine (cov.)	-"-	Grom (M. Grom, 71083 Herrenberg)
	Grom-Chiral-*R*-DNBPG-I	*R*-DNB-Phenylglycine (ionic)	-"-	Grom (-"-)
	Grom-Chiral-*S*-DNBPG-C	*S*-DNB-Phenylglycine (cov.)	-"-	Grom (-"-)
	Grom-Chiral-*S*-DNBPG-I	*S*-DNB-Phenylglycine (ionic)	-"-	Grom (-"-)
	Grom-Chiral-*R*-DNBL-C	*R*-DNB-Leucine (cov.)	-"-	Grom (-"-)
	Grom-Chiral-*R*-DNBL-1	R-DNB-Leucine (ionic)	-"-	Grom (-"-)

Type	Name	Chiral unit	Typical mobile phase	Supplier
	Grom-Chiral-*S*-DNBL-C	*S*-DNB-Leucine (cov.)	-"-	Grom (-"-)
	Grom-Chiral-*S*-DNBL-I	*S*-DNB-Leucine (ionic)	-"-	Grom (-"-)
	Grom-Chiral-U	*R*-N-α-Phenylethylurea	-"-	Grom (-"-)
	Ionic L-Leucin	*S*-DNB-Leucine (ionic)	-"-	Regis/Alltech (Promochem, 46485 Wesel
	Ionic D-Phenyl Glycine	*R*-DNB-Phenylglycine (ionic)	-"-	Regis/Alltech
	Nucleosil Chiral-2	Mixture of tartrate + DNB-Phenylethylamine	-"-	Macherey-Nagel (52348 Düren)
	Nucleosil Chiral-3	-"-	-"-	Macherey-Nagel
	Spherisorb Chiral 1	*R*-N-α-Phenylethylurea	-"-	PhaseSep (Promochem, 46485 Wesel)
	Spherisorb Chiral 2	*R*-Naphtylethylurea	-"-	PhaseSep
	Sumichiral OA-1000	α-Naphthylethylamide	-"-	Sumika (Sumitomo, 40474 Düsseldorf)
	Sumichiral OA-2000	*R*-DNB-Phenylglycine (ionic)	-"-	Sumika (-"-)
	Sumichiral OA-2000A	*R*-DNB-Phenylglycine (cov.)	-"-	Sumika (-"-)
	Sumichiral OA-2100	Chlorophenyl-isovaleroyl-phenylglycine	-"-	Sumika (-"-)
	Sumichiral OA-2200	Chrysanthemoyl-phenyl-glycine	-"-	Sumika (-"-)
	Sumichiral OA-3000	tert-Butylaminocarbonyl-valine	-"-	Sumika (-"-)
	Sumichiral OA-4000	*S,S*-α-Naphthylethyl-aminocarbonylvaline	-"-	Sumika (-"-)
	Sumichiral OA-4100	*R,R*-α-Naphthylethyl-aminocarbonylvaline	-"-	Sumika (-"-)
	Supelcosil LC-*R*-Naphthylurea	*R*-Naphthylethylurea	-"-	Supelco (ict, 65929 Frankfurt)
	TSKgel Enantio P1	*S*-Aromat. amide	-"-	TosoHaas, 70567 Stuttgart
Ligand	Chiralpak MA(+)	Amino acid derivative copper	Aqueous buffers/Cu^{2+}	Daicel (Baker, ex-64521 Groß-change Gerau)
	Chiralpak WE	N-(2-Hydroxy-1,2-diphenylethyl)glycine-copper	-"-	Daicel (-"-)
	Chiralpak WH	Proline-copper	-"-	Daicel (-"-)
	Chiralpak WM	Amino acid-copper	-"-	Daicel (-"-)

Type	Name	Chiral unit	Typical mobile phase	Supplier
	Chiral Hypro-Cu	Hydroxyproline-copper	-"-	Serva (Serva, 69115 Heidelberg)
	Chiral ProCu	Proline-copper	-"-	Serva (-"-)
	Chiral ValCu	Valine-copper	-"-	Serva (-"-)
	Grom-Chiral-HP	Hydroxyproline-copper	-"-	Grom (M. Grom, 71083 Herrenberg)
	Grom-Chiral-P	Prolineamide	-"-	Grom (-"-)
	Grom-Chiral-PC	Proline-copper	-"-	Grom (-"-)
	Grom-Chiral-VC	Valine-copper	-"-	Grom (-"-)
	MCI Gel CRS10W	C_{18}-Silica gel with N,N-dioctyl-*S*-alanine	-"-	Mitsubishi (Mitsubishi, 40547 Düsseldorf)
	Nucleosil Chiral-1	Hydroxyproline-copper	-"-	Macherey-Nagel (52348 Düren)
	TSKgel Enantio L1	Aliphat. amino acid-copper	-"-	TosoHaas (70567 Stuttgart)
	TSKgel Enantio L2	Aromat. amino acid-copper	-"-	TosoHaas (-"-)
Helicity	Chiralcel CA-1	Cellulose triacetate	Aqueous buffers	Daicel (Baker, 64521 Groß-Gerau)
	Chiralcel OA	Cellulose triacetate	-"-	Daicel (-"-)
	Chiralcel OB	Cellulosetribenzoate	-"-	Daicel (-"-)
	Chiralcel OB-H	High resolution version of Chiralcel OB	Hexane/isopropanol	Daicel (-"-)
	Chiralcel OD	Cellulose-trisphenylcarbamate	Aqueous buffers	Daicel (-"-)
	Chiralcel OD-H	High resolution version of Chiralcel OD	-"-	Daicel (-"-)
	Chiralcel OD-R	Reversed phase version of Chiralcel OD	40% CH_3CN aq./ $NaClO_4$ aq.	Daicel (-"-)
	Chiralcel OF	Cellulose-tris-4-chlorphenylcarbamate	Aqueous buffers	Daicel (-"-)
	Chiralcel OG	Cellulose-tris-4-toluylcarbamate	-"-	Daicel (-"-)
	Chiralcel OJ	Cellulose-tris-4-toluylate	-"-	Daicel (-"-)
	Chiralcel OK	Cellulose-tricinnamate	-"-	Daicel (-"-)
	Chiralpak AD	Amylose-tris-3,5-dimethylphenylcarbonate	-"-	Daicel (-"-)
	Chiralpak AS	Amylose-tris-ethylbenzylcarbonate	-"-	Daicel (-"-)
	Chiralpak OT(+)	Poly(triphenylmethylmethacrylate)	Hexane/isopropanol	Daicel (-"-)
	Chiralpak OP(+)	Poly(2-pyridyl-diphenylmethyl-methacrylate)	-"-	Daicel (-"-)
	ChiraSpher	Poly-N-acryloyl-(S)-phenylalanineethylester	Normal phase (hexane/ether)	Merck (64293 Darmstadt)
	Triacetylcellulose	Cellulose triacetate (no carrier!)	Reversed phase (water/alcohol)	Merck (-"-)

Type	Name	Chiral unit	Typical mobile phase	Supplier
Cavity	ChiraDex	β-Cyclodextrin	Aqueous buffers/ methanol	Merck (-"-)
	Crownpak CR	Chiral 18-crown-6 ether	$HClO_4$ aq.	Daicel (Baker, 64521 Groß-Gerau)
	Cyclobond I	β-Cyclodextrin	Aqueous buffers/ methanol	Astec (ICT, 65929 Frankfurt)
	Cyclobond I DMP	β-Cyclodextrin-3,5-dimethylphenyliso-cyanate	Normal phase	Astec (-"-)
	Cyclobond I N	β-Cyclodextrin-naph-thylethylcarbonate	Reversed or normal phase	Astec (-"-)
	Cyclobond I PT	β-Cyclodextrin-4-toluylate	Normale phase	Astec (-"-)
	Cyclobond I RN	β-Cyclodextrin-*R*-naphthylethyl-iso-cyanate	Reversed or normal phase	Astec (-"-)
	Cyclobond I RSN	β-Cyclodextrin-*R,S*-naphthylethyl-isocyanate	-"-	Astec (-"-)
	Cyclobond I RSP	β-Cyclodextrin-*R,S*-hydroxypropyl	-"-	Astec (-"-)
	Cyclobond I SN	β-Cyclodextrin-*S*-naph-thylethyl-isocyanate	-"-	Astec (-"-)
	Cyclobond I SP	β-Cyclodextrin-*S*-hydroxypropyl	-"-	Astec (-"-)
	Cyclobond II	γ-Cyclodextrin	Aqueous buffers/ methanol	Astec (-"-)
	Cyclobond III	α-Cyclodextrin	-"-	Astec (-"-)
	Cyclobond I-Ac	Acetylated β-Cyclo-dextrin	Reversed or normal phase	Astec (-"-)
	Cyclobond III acetylated	α-Cyclodextrin	-"-	Astec (-"-)
	α-Cyclodextrin	α-Cyclodextrin	Reversed phase	Serva (Serva, 69115 Heidelberg)
	β-Cyclodextrin	β-Cyclodextrin	-"-	Serva (-"-)
	Grom-Chiral-Beta-CD	β-Cyclodextrin	-"-	Grom (M. Grom, 71083 Herrenberg)
Protein	Chiral-AGP	α_1-acid glycoprotein	Phosphate buffers/ organic modifier	ICT (65929 Frankfurt); Baker (64521 Groß-Gerau)
	Enantiopac	α_1-acid glycoprotein	-"-	Pharmacia (79111 Freiburg)
	Resolvosil BSA-7	Bovine serum albumin	Phosphate buffers/ propanol	Macherey-Nagel (52348 Düren)
	Ultron OVM	Ovomucoid	Phosphate buffers/	Chrompack

Type	Name	Chiral unit	Typical mobile phase	Supplier
			ionic modifier	(60437 Frankfurt)
Ion exchange	Ceramospher Chiral RU-1	Spherical ceramics ion-exchanged with optical active ruthenium(II)-complex -	Methanól	Shiseido (Stagroma, 8304 Wallisellen, Switzerland)

3.4.3.1 Chiral phases of the 'brush type' (*type 1*) [39]

For enantiomers to be separated on brush-type CSPs the analyte must exhibit the necessary three (or more) interaction sites. Chiral recognition models often allow the order of elution to be related to absolute configuration. Finn [40] has provided a labeling scheme for chiral recognition mechanisms. Major interaction sites are classifiable as π-basic or π-acidic aromatic rings, acidic sites, basic sites, steric interaction sites, or sites for electrostatic interaction. Lipophilic interactions are also possible in reversed mobile phases. Aromatic rings are potential sites for π-π interaction. Acidic sites supply hydrogens for potential intermolecular hydrogen bonds. The hydrogen involved is often an amido proton (NH) from an amide, carbamate, urea, amine or alcohol. Basic sites such as π-electrons, sulphinyl or phosphinyl oxygens, hydroxy or ether oxygens or amino groups may also be involved in hydrogen bond formation. Electrostatic interactions may occur at charged groups or with permanent or induced dipoles. Steric interactions occur between large groups.

The more strongly adsorbed enantiomer is frequently the one that interacts with the higher number of bonding sites or the fewer number of steric repulsion sites. While enthalpic contributions to the binding energy are important, so are the entropic contributions. The latter are more difficult to understand and are only now being studied systematically. Typically, the enantiomer adsorbed with the greatest exothermicity also loses the greatest entropy, a situation which reduces the level of enantioselectivity.

Usually, brush-type CSPs utilize π-π interaction in the recognition process. This type of intermolecular interaction occurs between aromatic ring systems in the enantiomers and those in the CSP. It is analogous to the interaction in aromatic charge transfer complexes. π-π interactions can be described as electron donor-acceptor interactions. The extent of interaction is influenced mainly by the electron affinity of the electron acceptor and the first ionization potential of the electron donor. The schematic representation of a 'brush-type' CSP is given in Figure 3.4.14.

The CSPs π-donor, acidic, and basic binding sites (naphthyl ring, amido hydrogen, carbonyl oxygen, respectively) interact with the complementary sites in the analyte, which is usually an N-(3,5-dinitrobenzoyl) (DNB) derivative. The more strongly retained enantiomer is able to undergo these interactions from a lower energy conformation than the less strongly retained enantiomer.

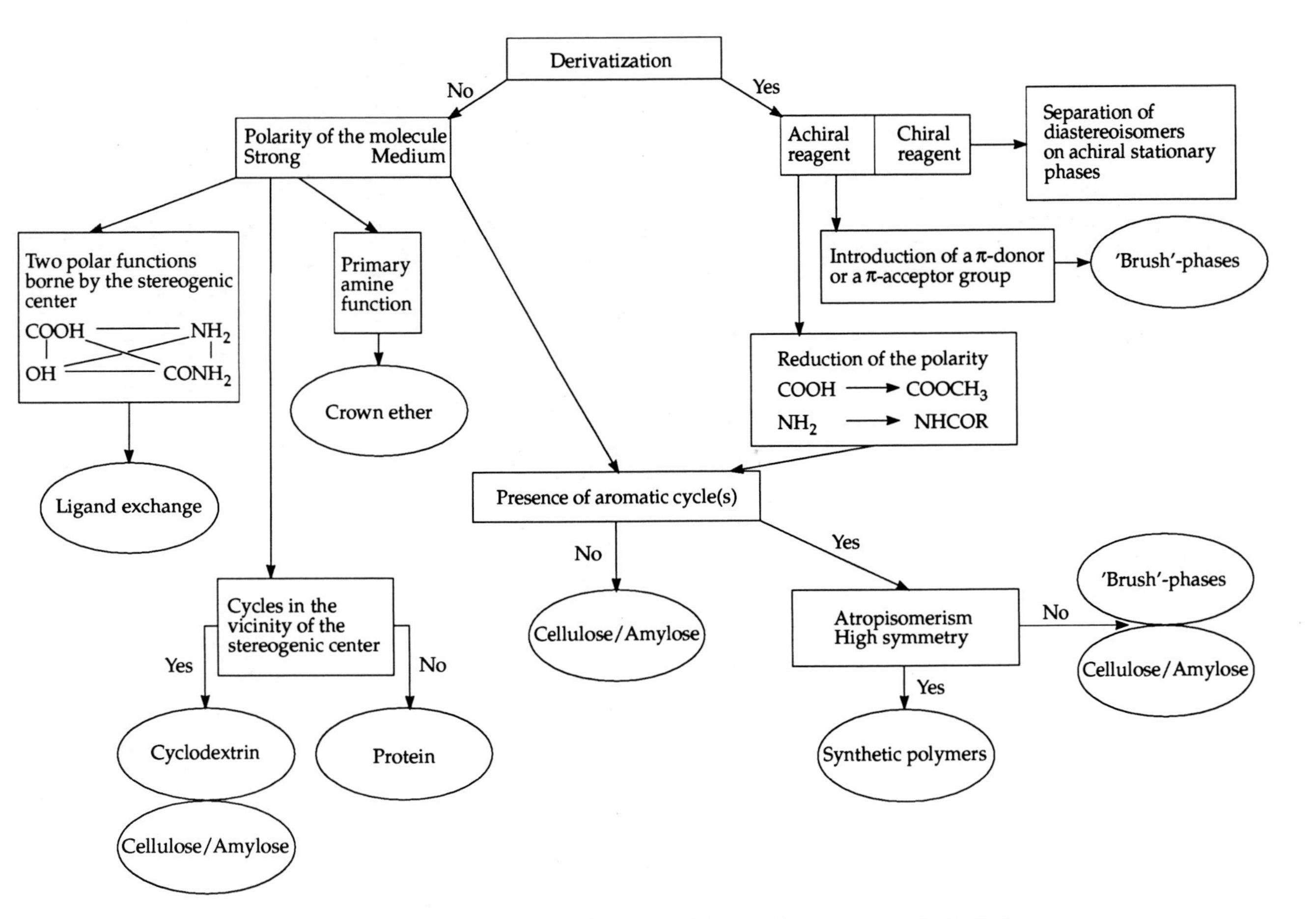

Figure 3.4.13 Strategy for choosing the right chiral stationary phase for HPLC according to solute polarity [38].

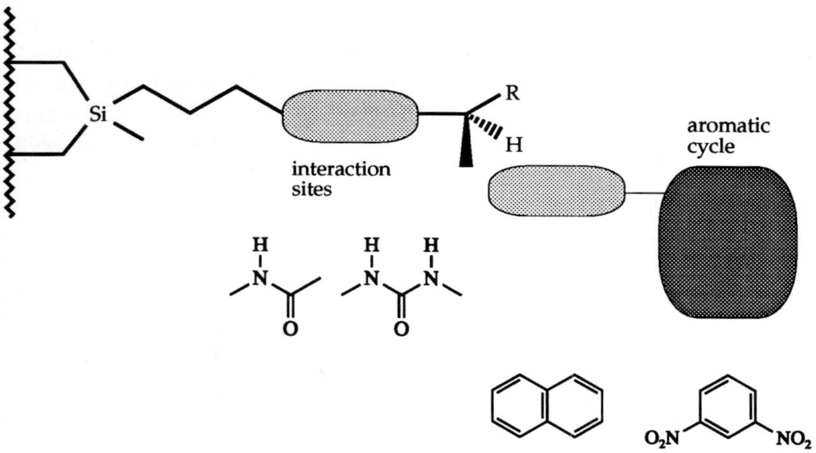

Figure 3.4.14 Schematic representation of a 'brush type' chiral stationary phase [38].

Description of the most frequently used 'brush-type' chiral phases (see Table 3.4.3)

Phenylglycine (CSP1). The π-acceptor phenylglycine CSP is based on the 3,5-dinitrobenzoyl derivative of phenylglycine, bound covalently to aminopropyl silica. CSP1 is also available in the ionic version. Columns in which the phenylglycine has either the *R* or *S* absolute configuration are available, enabling one to invert the elution order by the choice of column. The phase is also available in the racemic form. CSP1 resolves a wide variety of compounds which contain π-basic groups. When such groups are lacking, they can be introduced through derivatization. For example, CSP1 resolves the 1-naphthamides of amines and amino acids, the anilides of carboxylic acids, and the 1-naphthyl carbamates of alcohols.

Leucine (CSP2). The π-acceptor leucine (CSP2) is based on the 3,5-dinitrobenzoyl derivative of leucine, bound covalently to aminopropyl silica. Columns derived from either *R* or *S* leucine are available. CSP2 resolves the same types of compounds as does CSP1 but CSP2 often exhibits an enhanced enantiomeric selectivity for certain classes of compounds, e.g., benzodiazopinones.

Naphthylalanine (CSP3). The π-donor naphthylalanine (CSP3) is based on the N-(2-naphthyl) derivative of alanine, covalently bound to 11-undecanyl silica through an ester linkage. Columns with either the *R* or *S* configurations are available, in addition to the racemic version. For instance, these columns have been used to check racemization during peptide coupling reactions. CSP3 resolves 3,5-dinitrophenylcarbamate or 3,5-dinitrophenylurea derivatives of a high number of amines, alcohols, diols, carboxylic acids, amino acids, thiols, and hydroxy acids. Superior separations of the dinitrobenzoyl derivatives of primary amines have also been reported.

(1-Naphthyl)ethylamine terephthalamide (CSP4). The π-donor CSP4 is based on the terephthalamide derivative of *S*-(1-naphthyl)ethylamine, covalently bound to

aminopropyl silica. CSP4 contains a chiral amide which is able to serve as either a donor or an acceptor in hydrogen bonding. Thus, hydrogen bonding contributes to the separation of amide derivatives. CSP4 exhibits excellent selectivity for the separation of enantiomers of 3,5-dinitrobenzoyl derivatives of amines and amino acids, the esters and amides of amino acids, and other derivatives of carboxylic acids.

Chrysanthemoyl-phenylglycine (CSP5). This phase is based on the *R,R*-1,3-trans-chrysanthemoyl derivative of *R*-phenylglycine; it has shown better selectivity then other phases for chiral amides. CSP5 resolves the 3,5-dinitrobenzoyl derivatives of amines and amino acids, and of the esters and amides of these acids; the 3,5-dinitroanilide derivatives of carboxylic acids; and some underivatized alcohols of interest, such as fungicides.

N-(tert-butylaminocarbonyl) (CSP6). This phase is based on the N-(tert-butylaminocarbonyl)-*S*-valylaminopropyl phase, covalently bound to silica. The enantioselectivity of CSP6 only derives from hydrogen bonding. It is very effective in the separation of N-acetyl-O-alkyl esters of amino acids. CSP6 may also be used in the separation of the 3,5-dinitrophenyl carbamates of hydroxy acids.

R- and S-(1-naphthyl)ethylamine (CSP7). These two diastereomeric π-donor CSPs can reverse the elution order of the optical isomer pairs they separate. They are derived from the *R*- or *S*-(1-naphthyl)ethylamine derivatives of *S*-valine, each of which is covalently bound to aminopropyl silica. These phases exhibit excellent enantioselectivity for the π-acceptor derivatives of amines, carboxylic acids, and amino acids, the esters and amides of amino acids, and of underivatized alcohols.

Naphthylethylurea (CSP8). CSP8 is derived from the reaction between 1-naphthylethylisocyanate and aminopropyl silica to form a urea-based derivative. Columns with either the *R* or *S* configuration are available. They resolve a wide variety of dinitrobenzoyl derivatives of amines and amino acids.

Covalent derivatization with achiral reagents forming enantiomeric derivatives

In general, chiral recognition is pronounced when the three chiral recognition sites are adjacent to the stereogenic center; it becames weaker when these sites are further away. If a molecule contains all of the afore-mentioned properties (π-donor, acidic, and basic binding sites) derivatization is usually not required. Examples include the β-binaphthols, S-aryl-β-hydroxyl sulphoxides, C-aryl hydantoins, C-aryl succinimides, and indoline-2-carboxylates.

In many analytes an adequate number of the necessary interaction sites may not be present. Derivatization with an achiral reagent can provide additional interaction sites and appropriate achiral derivatizing reagents can be applied to enhance detectability as well. Many enantiomers that otherwise cannot be separated by present 'brush-type' CSPs can be separated after suitable achiral derivatization.

For example, derivatization is typically required for analytes lacking aromatic substituents. The presence of aromaticity in the analytes is one of the usual requirements for chiral recognition by 'brush-type' CSPs as well as by most other chiral selectors. While many analytes contain π-basic aryl groups and, therefore, can often be resolved on π-acidic CSPs without derivatization, relatively few analytes originally possess π-acidic groups. Hence, incorporation of such a group by derivatization is a common first step, if a π-donor column is to be used. Reaction with 3,5-dinitrobenzoyl chloride, 3,5-dinitrophenylisocyanate, or 3,5-dinitroaniline is simple, easy, and effective. For both types of columns, derivatization can reduce analysis time and improve band shapes. Thus, amines are often acylated, carboxylic acids esterified or converted to amides or anilides, and alcohols converted to esters or carbamates.

Derivatization of groups remote from the stereogenic center can improve resolution by reducing the extent to which these groups contribute to achiral retention, i.e., both enantiomers are retained equally. If a chiral analyte lacks one or perhaps two of the three sites required for chiral recognition, then derivatization with an achiral reagent may allow separation of the enantiomers of the derivative on the CSP.

Some examples of analytes containing derivatized functional groups (close to or around the stereogenic center) are: *carboxylic groups* (amino-, hydroxy- and carboxylic acids); *amine groups* (amines, amino acids, amino alcohols); *hydroxyl groups* (alcohols and hydroxy acids); and *thiol groups* (thiols and amino thiols).

For the selection of the appropriate achiral derivatization reagent and CSP, Table 3.4.4 summarizes examples of chemicals that have been resolved using commercially available 'brush-type' chiral selectors.

Four examples of possible derivatization schemes are given below (descriptions of the preparations, see Section 3.4.4). In addition, several other methods for derivatization are described in the literature [39].

(i) The analyte contains one derivatizable functional group near the stereogenic center and the aim is to select the most appropriate achiral derivatization reagent. If, for example, the analyte is 2-phenylpropionic acid, the choice of an achiral derivatization reagent is 3,5-dinitroaniline (DNA).

(ii) The analyte contains two similar or identical functional groups near the stereogenic center. The same achiral reagent is used to derivatize either one or both functional groups; an example is 1-phenyl-1,2-ethanediol. The acylation reagent, N-imidazole-N'-carbonic acid-3,5-dinitroanilide (ICDNA), is used for derivatization.

(iii) The analyte contains two dissimilar functional groups which may be treated with the same derivatization reagent. An example of an analyte to be derivatized as follows is propranolol (1-isopropylamino-3-(1-naphthyloxy)-2-propanol). β-Naphthylisocyanate (NIC) derivatizes only the amino group in most cases. This reagent is safe and easy to use.

(iv) The analyte contains two dissimilar functional groups that may be treated with two different reagents. The analyte in this example can be any amino acid that is acylated with 3,5-dinitrobenzoyl chloride (DNBC).

Table 3.4.4 Examples for the resolution of chemicals using 'brush type' phases [39].

Analytes	Reagent[a]	Derivative	CSP	Reference
ALCOHOLS				
Aromatic/aliphatic	ICDNA	Carbamate	3,7	41,42,46a,48,56
Aromatic/aliphatic	NIC-1	Carbamate	1	54
Bi-β-naphthols	N/A		1	49,50
2,2,2-Trifluoro-1-(9-anthryl)-ethanol	N/A		1	52,53
AMINES				
Aromatic/aliphatic	ICDNA	Urea	3	41,42
Aromatic/aliphatic	DNBC	Amide	4,7	46,48
Aromatic/aliphatic α- and β-amino acids	DNBC/ROH	Amide/ester	3,4,5,7	41,42,46,47,48,56
Aromatic/aliphatic α- and β-amino alcohols	ICDNA	Urea	3	41,42
CARBOXYLIC ACIDS				
Aromatic/aliphatic	DNA	Amide	3,4,7	42,46,48,56
DIOLS				
Aromatic/aliphatic	ICDNA	Carbamate	3	42,43
HYDROXY ACIDS				
Aromatic/aliphatic	ICDNA	Carbamate	3	42
PEPTIDES				
Dipeptides	DNBC	Amide	1,3	52
PHARMACEUTICALS				
Propranolol and similar β-blockers	NIC-1	Urea	1	55
Ephedrine	NIC-1	Urea	1	55
Ibuprofen	NMA-1	Amide	1	43a,51a
Amphetamine	NC-1	Amide	1	44,51,57
THIOLS				
Aromatic/aliphatic	ICDNA	Carbamate	3	41,42

[a]ICDNA: N-Imidazole-N′-carbonic acid-3,5-dinitroanilide; NIC-1: 1-Naphthylisocyanate; N/A: Not applicable, no derivatization required; DNBC: 3,5-Dinitrobenzoyl chloride; ROH: Any alcohol; DNA: 3,5-Dinitroaniline; NMA-1: 1-Naphthalenemethylamine; NC-1: 1-Naphthoyl chloride. Examples of derivatization procedures are described in Section 3.4.4.

Advantages and disasvantages of 'brush type' phases

'Brush-type' CSPs exhibit several advantages as compared with other CSPs. They are usually less expensive and much more durable than the biopolymer CSPs (cellulose, cyclodextrine, protein). They have a higher density of binding sites than the biopolymer CSPs and are consequently better suited for preparative separations.

Biopolymer CSPs are capable of separating the enantiomers of some compounds that cannot be resolved on 'brush-type' CSPs. One can, however, look forward to 'brush-type' CSPs of increasing scope and enantioselectivity. Drawbacks are the limitation to compounds of low or medium polarity and the necessity for deriving polar solutes (amines, acids) and solutes with no aromatic systems.

3.4.3.2 Chiral ligand exchange chromatography (CLEC, *type* 2)

CLEC is a particular form of ion exchange. It involves the reversible formation of a metal complex by coordination of substrates that can act as ligands to the metal ion; it has a wider application than the resolution of racemates [59,60]. CLEC uses the ability of Cu^{2+} or other divalent metal ions to form reversibly diastereomeric multiligand complexes, with bidentate chiral molecules, especially amino acids. The separation of two analytes in CLEC is based on the difference in thermodynamic stabilities of adsorbates formed between an immobilized amino acid, a divalent cation, and the analyte enantiomers. The stability of such complexes is highly dependent on the transition metal used; Cu^{2+} complexes are generally the most stable and are preferred for CLEC applications, whereas the less stable complexation with Ni^{2+} or Co^{2+} is used more in chiral gas chromatography (see Section 3.5). Factors affecting the selectivity and efficiency of the separations include pH, ionic strength of the mobile phase, and temperature. The complexes should be kinetically labile, i.e. their formation and dissociation should be fast on the chromatographic time scale.

Chiral resolutions by CLEC were first reported by Davankov [61] in 1971 using resin packings (styrene-divinylbenzene) to which amino acid residues (e.g., proline, hydroxy proline, valine) were bonded. The resin was saturated with Cu^{2+} (2 mEq/g: 1 Cu per 2 proline units) reversibly forming the bis-proline copper complex. Other amino acids passing through the resin then exchange coordination sites with proline residues at pH 10-11. Selectivity as high as four was observed, with *R*-isomers usually being more strongly retained unless they contained a very polar side chain, as in glutamic acid. Unfortunately, efficiencies were very low for resin-packed LC columns, and separation took several hours due to packing weakness at high pressure [62]. Another major disadvantage of this ligand-exchange technique is the complexity of the aqueous mobile phase which contains buffers and the essential copper ions.

Strongly improved efficiency has been achieved by Gübitz et al. [63]. They used silica gel-supported packings, with the amino acid now being chemically bonded to silica via an alkyl chain (about 750 µmol/g) and the copper reversibly attached as before. The procedure works well for direct CLEC resolution of racemic amino acids; very high selectivities (α = 5-10) have been observed. L-(*S*)-isomers are usually more strongly retained on silica packing with a bound L-amino acid unit. Saigo et al. [64] have synthesized more complex CLEC CSPs in order to improve selectivity. The efficiency of such packings has been further improved and their substrate selectivity has been modified by surrounding the ligand-bearing chain with inert spacers such as n-butyl or n-decyl groups [65,66].

Such packings have been applied not only to the resolution of derivatized and free racemic amino acids, but also to the Schiff base derivatives of biologically important amino alcohols, including β-hydroxyphenylethylamines, propanolamines, and catecholamines, as well as to α-hydroxyacids.

The simplest procedure to adopt the ligand-exchange approach would be to purchase a commercial epoxy phase and treat it with an α-amino acid such as *S*-proline. Although commercial packings of this type are available (see Table 3.4.3) they can also be prepared 'on column' [67]. TLC plates of this type are also available commercially ('Chiralplates'; see Section 3.8). However, it has to be noted that CLEC only proves useful if the analyte is a bidentate ligand (amino acids, amino alcohols, hydroxy acids) for Cu^{2+} or other comparable metal ions. Even then, derivatization may be necessary to enable the eluting isomers to be detected. Advantages of this type of CSP are good stability and high selectivity values. With this technique amino acids (as copper complexes) can be detected at 254 nm.

3.4.3.3 Chiral polymer phases with a helical structure (*type 3*)

This useful class is based on polymeric chiral selectors. Some of these polymers are of natural origin, such as those derived from cellulose; others are synthetic and are usually based upon polyacrylates and polyacrylamides.

Natural polymers

Cellulose [68] is an abundant, naturally occurring polymer. It has a highly ordered structure, consisting of hydrogen-bonded chains of 1,4-linked β-D-(+)-glucose units. These asymmetric strands form helical structures which should be able, in principle, to discriminate between enantiomers that interact differently with these strands. Resolution can sometimes be achieved with unsupported natural cellulose [69] as a CSP, but slow mass transfer and slow diffusion through the polymer lead to band broadening and, thus, poor resolution. The highly polar hydroxyl groups often give rise to nonstereoselective binding between the solute and the cellulose. Thus, derivatization of strongly cationic analytes (e.g., amines) and strongly anionic analytes (e.g., carboxylic acids) is used to block these unspecific associations and provide the analyte molecule with groups capable of interacting with the CSP [70].

Attempts to improve the chromatographic and enantioselective properties of cellulose have resulted in derivatization of its hydroxyl groups to decrease bulk polarity and to provide additional bulk for interaction between CSP and analyte. Hesse and Hagel [70] have prepared microcrystalline cellulose triacetate (CTA). They have observed that derivatization of the hydroxyl groups did not destroy the helical structure of cellulose, which retained its potential for chiral recognition as first noted in the early experiments of Dalgliesh [34] and others. Mannschreck and coworkers [71-73] introduced microcrystalline CTA for HPLC separations and resolved, for example, racemic compounds carrying aromatic groups. Microcrystalline CTA can be used as a bulk packing, especially when cross-linked, but its pressure resistance is limited to around 8 MPa. Its usefulness can be improved significantly by coupling a swollen microcrystalline cellulose triacetate column with an achiral alkyl-silica column [74].

Silica-supported cellulose triacetate was the first commercially successful packing of this type. Its versatility led Okamoto's group to develop a range of such phases by changing the derivatization of cellulose [75].

Figure 3.4.15 shows the commonly used CSPs with cellulose derivatives as the chiral selector. They can be used with mobile phases such as pure ethanol and aqueous methanol and have successfully separated a wide range of analytes.

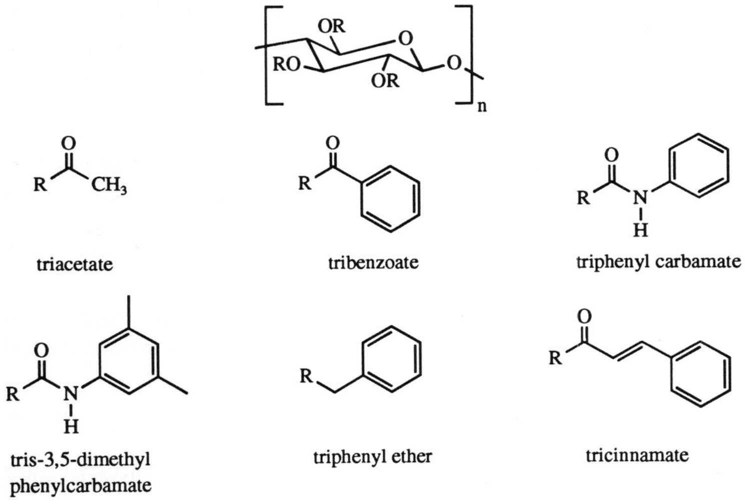

triacetate

tribenzoate

triphenyl carbamate

tris-3,5-dimethyl phenylcarbamate

triphenyl ether

tricinnamate

Figure 3.4.15 Cellulose derivatives used as chiral stationary phases.

Aspects of the mechanism of cellulose-based chiral stationary phases

Attractive interactions play an important part in the retention on *type* 3 CSP. If the solute molecule lacks the required interactive groups, they can be introduced by derivatization. Many racemates have been resolved without derivatization, such as β-amino alcohols [75]. Ichida et al. [76] suggested the following applications for the cellulose based phases:

(i) cellulose triacetate is suitable for many racemates, in particular, for compounds with a phosphorus atom at the asymmetric center;

(ii) cellulose tribenzoate shows good chiral recognition for compounds in which a carbonyl group (or groups) is close to the asymmetric center;

(iii) cellulose triphenylcarbamate has strong affinity for polar racemates; it is sensitive to the molecular geometry of the substrate;
(iv) cellulose tribenzyl ether is effective with protic solvents as mobile phases;
(v) cellulose tricinnamate is suitable for many aromatic racemates and barbiturates;
(vi) cellulose tris(3,5-dimethylphenylcarbamate) is especially suitable for β-adrenergic blocking agents.

Enantioresolution on these phases depends upon inclusion into shape-selective chiral cavities within the polymer network. Hesse and Hagel [70] suggested that the channels in the stationary phase have a high affinity for aromatic groups. Francotte et al. [77] proposed that this portion of the molecule enters the cavity, while at least one of the substituents on the solute chiral center interacts with the steric environment outside the cavity. Wainer [78] studied the separation of a number of analytes on silica-supported cellulose triacetate and proposed an attractive - rather than an inclusive - mechanism of retention with dipole-stacking interactions. Ichida et al. [76] demonstrated the importance of hydrogen bonding between the urethane groups of the phenylcarbamate and the polar hydroxyl or amino groups of suitable solutes. Later, Wainer [79] studied the resolution of a series of enantiomeric aromatic alcohols on the cellulose tribenzoate CSP and proposed a chiral recognition mechanism which involved the following parameters:

(i) the formation of diasteromeric solute-CSP complexes through a hydrogen bonding interaction between the solute's alcoholic hydrogen and an ester carbonyl on the CSP;
(ii) the stabilization of this complex through the insertion of the aromatic portion of the solute into a chiral cavity of the CSP;
(iii) chiral discrimination between enantiomeric solutes due to differences in their steric fit in the chiral cavity.

Aboul-Einen [80] has reviewed the structural factors and functional group selectivity requirements affecting direct HPLC enantiomeric resolution of drug racemates on several derivatized cellulose chiral stationary phases. Polysaccharides other than cellulose have also been used [81], i.e. amylose, chitosin, xylan, curdlan, dextran, and inulin. The latter was the substance favoured by Tswett in his historic early chromatographic experiments. Okamoto [82] evaluated twenty different tris(phenylcarbamate) derivatives of cellulose (bound to silica) and noted that the optical resolving power depended on the substituents introduced on the phenyl groups. Amylose tris(phenylcarbamate/s) and aryl-/alkylcarbamates such as benzyl- and 1-phenylethylcarbamates of cellulose and amylose were also investigated. In most cases, either cellulose tris(3,5-dimethylphenylcarbamate) or amylose tris(3,5-dimethylphenylcarbamate) exhibited the highest resolving power. The 1-phenylethylcarbamate of amylose also performed well. β-Lactams were resolved on tris-(phenylcarbamate) of both cellulose and amylose [83].

Practical considerations: The main influencing factor for chiral recognition in these stationary phases is the choice of the mobile phase. For the complete series of cellulose derivatives, mobile phases may be aqueous or non-aqueous. In general, suitable solvents are limited to hexane, hexane/2-propanol (99 to 50% hexane), and methanol or ethanol or mixtures of these with water. Non-polar mobile phases usually provide greater resolution than methanol/water mixtures, but lower resolution than ethanol. For cellulose tribenzoate and tribenzylether, α-values become larger as the polarity of the mobile phase increases. For a series of alcohol modifiers on the tribenzoate, an increase in the steric bulk around the hydroxyl moiety tend to result in an increased retention and stereoselectivity for enantiomeric amides. This is due to the competition for binding to the chiral sites, or to binding of the alcohol to achiral sites nearby the cavity, or to a combination of these.

With all these phases chlorinated solvents, such as dichloromethane, remove the derivatized cellulose from its silica support. Similarly, chloroform, acetone, tetrahydrofurane, dihydrofurane, dimethylsulphoxide, toluene or acetonitrile (more than 1%) cannot be used. A flow rate of 0.5 to 1.5 ml/min is recommended, although 2 ml/min can be used for column cleaning. Normal back pressures for all columns are generally around 6500 kPa at 1.0 to 1.5 ml/min. The resolution is often improved at temperatures lower than ambient temperature.

Synthetic polymers

Chiral synthetic polymers have been developed in order to improve the natural polymers, such as the cellulose phases. These phases require only simple derivatization procedures to yield selective column materials for enantioseparation. Synthetic chiral polymers, however, cannot be produced without the use of a chiral reagent or a chiral catalyst. In the first case, a chiral derivatization of a suitable monomer is performed and the product is then polymerized to form a polymer network exhibiting chiral substituents. In the second case, the monomer is polymerized under the influence of a chiral catalyst, which produces an optically active polymer since the stereoregulatory influence of the catalyst yields an isotactic polymeric structure of a certain preferred helicity. Here, the chirality of the polymer is inherent, i.e. it is caused only by the helical structure.

Sorbents based on chirally substituted synthetic polymers: These types were developed by Blaschke [84-86]. His work has been emphasized on polyacrylamide and polymethacrylamide derivatives, where the chiral substituents come from an optically active amine or amino acid component. Polymer particles of desired mean diameter and acceptable size homogenity can be obtained. Free-radical initiation is used and the porosity of the gel particles is regulated by the relative amount of cross-linking agent added. The particles swell depending upon the composition of organic solvents. The material is only used in low-pressure LC systems, because it is not stable under the pressure used in HPLC.

The resolving capacity of these sorbents is strongly dependent on a variety of factors. Apart from the substituents on the polyacrylamide backbone, the degrees of cross-linking, the nature of the cross-link and the mobile phase compositions are the most important factors.

These substituted polyacrylamides have been found particularly suitable for polar compounds with functional groups capable of hydrogen bonding. Consequently, relatively non-polar mobile phases have been found most useful. Combinations of hydrocarbons, ethers and, possibly, small amounts of an alcohol are typical; examples include toluene/dioxan, hexane/dioxan and toluene/dioxan/methanol mixtures. Protic co-solvents such as methanol strongly decrease the retention and, therefore, normally constitute less than 10% of the mobile phase.

Although hydrogen bonding is apparently the major binding contribution to retention of the solute, the mechanism of enantiomer discrimination appears to be more complex. As in the case of polysaccharide derived sorbents, enantioselection is considered to be the result of inclusion phenomena, i.e. the binding groups in the asymmetric cavities into which the solute enantiomers are thought to diffuse are more favourably located for one of the enantiomers, which therefore is preferentially retained. Columns packed with sorbents of this kind have mainly been used for semipreparative work (sample amounts between approx. 1 and 250 mg) on racemic pharmaceuticals.

One modification of the sorbent is a silica-bonded non-cross-linked polyacryloyl *S*-phenylalanine ethyl ester, which is used for analytical HPLC due to its considerably improved column efficiency. Polymerization of monomer *S*-N-acryloyl phenylalanine ethylester in the presence of 'Lichrosorb Diol' (Merck, Germany) results, via an unknown mechanism, in an anchoring of the polyamides to the silica gel. Possibly, monomers and radical initiators are adsorbed by the diol phase. Upon heating, polymerization would then occur preferentially on the surface of the diol phase, whereby mechanical tangling of the growing polymer chain in the pores of the silica gel would result in anchoring. The polymer is not washed out by solvents such as n-hexane, toluene, dioxan, or 2-propanol. This phase is commercially available as 'ChiraSpher' from Merck, Germany.

Huffer and Schreier [87] have used this silica-bonded polyacrylamide phase for the HPLC separation of homologous chiral 4(5)-alkylated γ(δ)-lactones and attempted to correlate selectivity and the alkyl chain length of the lactones with the separation mechanism of the system used (see also 3.4.4 'Practical examples'). Table 3.4.5 summarizes several chiral separations performed by 'ChiraSpher'.

Sorbents based on isotactic linear polymethacrylates of helical conformation: A vinyl polymer with chirality resulting from its helicity alone was first prepared in 1979 [88-92]. Optically active poly(triphenylmethyl methacrylate) was obtained by asymmetric anionic polymerization of triphenylmethyl methacrylate under the influence of a chiral initiator ([-]-sparteine) in toluene at low temperature. The success of the reaction is strongly dependent on the initiator used, which is based on a complex formed between an optically active diamine and butyllithium or lithium amide. Both (-)-sparteine-butyllithium and (+)-*S*,*S*-2,3-dimethoxy-1,4-bis(dimethylamino)-butane-lithium amide were found to give a good yield of the (+)-polymer. It is insoluble in most common organic solvents.

Table 3.4.5 Examples of the chiral resolution of racemates using the polyacrylamide phase 'ChiraSpher' (Merck, Germany).

Racemic substance	References
LACTONES	
3-Alkyl/aryl-γ-lactones	87a
4-Alkyl-γ-lactones	87
5-Alkyl-δ-lactones	87
5-Oxo-4-hydroxyhexanoic acid γ-lactone (Solerone)	87b
4,5-Dihydroxyhexanoic acid γ-lactone (Solerole)	87b
PHARMACEUTICALS	
Oxazepam and derivates	86,87c
Baclofen-lactame	86
Lorazepam	86
Chlormezanone	86
Chlorthalidone	86
Thalidomide	86
FLAVANONES	87d
AMINO ACIDS	
N-(3,5-DNB)-amino acids*	87e
MTH-amino acids**	87e
PTH-amino acids***	87e

*DNB: Dinitrobenzoyl; **MTH: Methylthiohydantoin; ***PTH: Phenylthiohydantoin

This CSP has proven to be excellent for optical resolution of racemic aromatic hydrocarbons of linear and planar chirality. It is characterized by high hydrophobicity and a rigid molecular geometry. It can be assumed that there is a difference in the extent to which the enantiomers interact with the helical CSP - in which the triphenylmethyl group is considered to acquire a propeller-like conformation. This is particularly true in the case of hexahelicene, for which the highest α–value (>13) was reported. The enantiomer most strongly retained on the (+)-CSP is the (+)-form, which has P-(right-handed) helicity. Since it was found that the (+)-CSP interacts very strongly with itself, but only weakly with the (-)-polymer, it is very likely that the (+)-CSP also has P-helicity. The same P-helicity of the most strongly retained enantiomer was also found for all other compounds investigated that possessed this type of chirality. This correlation means that chromatographic retention data can be used for determining the absolute configuration of these types of compounds.

Because the structure of this CSP is not cross-linked and the coated silica version of the sorbent is based only on physical adsorption, some limitations exist on the choice of mobile phase, for solubility reasons. Thus, aromatic hydrocarbons, chloroform and tetrahydrofuran (which dissolves the polymer) should be avoided. To date, methanol has been the preferred mobile phase. There is a tendency towards an increase in α with increasing polarity of the solvent, but in many cases retention times may be unacceptably long.

A drawback of these phases is the weak ester bond. The triphenylmethyl ester undergoes solvolysis in methanol to form methyl triphenylmethyl ether. As a result, the separation must carried out at low temperature when an alcoholic eluent is used, and the solvent for storing the column must be nonalcoholic. The solvolysis rates of diphenyl-2-pyridylmethyl methacrylate and diphenyl-4-pyridylmethyl methacrylate are more than 100-fold slower than that of triphenylmethyl methacrylate. This means that the polymethacrylates containing a pyridyl group are more resistant to solvolysis in methanol. Although the optical resolution capacity of these polymers is low compared with that of the triphenylmethyl methacrylates when polar eluents are used, they show attractive chiral recognition towards polar compounds when a hexane/2-propanol mixture is used as the mobile phase. Hydrogen bonding between the pyridyl groups of the polymer and enantiomers may contribute to the resolution.

Helical synthetic polymer phases are commercially available (see Table 3.4.3). Particularly promising are methods of generating such CSPs by applying short-chain material to silica and then cross-linking.

Saotome et al. [93] prepared a number of new vinyl polymers carrying the penicillin sulphoxide nucleus, and investigated the influence of the substituents on the optical resolution capabilities. A less hydrogen-bonding solvent in the polymer preparation step was preferred for effective resolution.

There is a virtually infinite number of possibilities for preparing synthetic polymer CSPs from chiral monomeric units. The advances in coating silica with polymers [94] will certainly result in much progress in this area. Such polymeric CSPs exhibit broad applicability, high capacity, good efficiency, and can be manipulated by changes in the mobile phase modifier. However, the solute must often be derivatized and the phases often show strict steric requirements for resolution [95].

3.4.3.4 Chiral phases with inclusion effects (*type* 4)

Native and functionalized cyclodextrin-bonded phases

Cyclodextrins (CD) are chiral, toroidal-shaped molecules composed of six or more D(+)-glucose residues bonded through α-(1,4)-glycosidic linkages in which all of the glucose units are held in chair conformation. Cyclodextrins are referred to by the number of glucose residues they contain; α-CD (cyclohexaamylose) contains six, β-CD (cycloheptaamylose) contains seven, and γ-CD (cyclooctaamylose) contains eight glucose residues. Cyclodextrins with fewer than six glucose residues have not been found; this is probably due to an excessive ring strain that would result from a hypothetical five glucose-residue torus. Cyclodextrins with more than eight glucose residues have been identified, and some with branched structures have been reported.

The mouth of the cyclodextrin molecule has a larger circumference than its base and is lined with secondary hydroxyl groups of the C-2 and C-3 atoms of each glucose unit. The primary C-6 hydroxyl groups are located at the base of the cyclodextrin torus. While the primary hydroxyl groups (C-6) are free to rotate, partially blocking the opening at the base, the secondary hydroxyl groups (C-2 and C-3) are fixed in space with all of the C-2 hydroxyl groups pointed in a clockwise direction and all of the C-3 hydroxyl groups pointed in a counter-clockwise direction (when looking down into the mouth of the CD cavity). Each glucose residue contains five stereogenic centers; thus, there are 30, 35, and 40 stereogenic centers in α-CD, β-CD, and γ-CD, respectively. The interior of the cyclodextrin cavity has two rings of C-H (C-5 and C-3) and one ring of glycosidic oxygens. A cyclodextrin molecule can be described as a truncated cone with a partially blocked base, a hydrophobic cavity, hydrophilic edges, and multiple stereogenic centers (Figure 3.4.16).

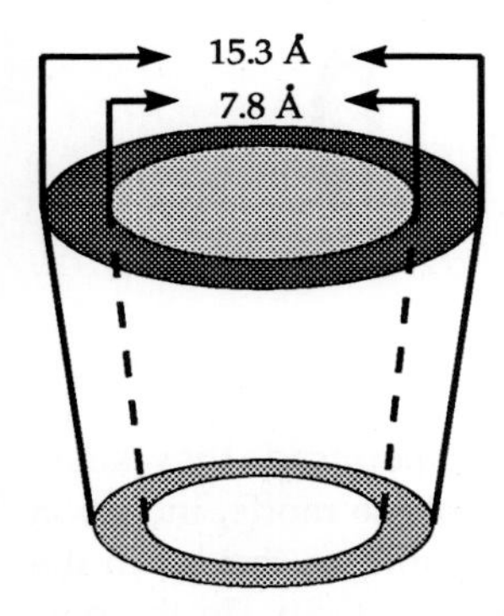

Figure 3.4.16 Schematical structure of a cyclodextrin molecule [35].

Essential physical properties of cyclodextrins are: (i) they are non-reducing; (ii) glucose is the only product of acid hydrolysis; (iii) their molecular masses are always integral units of 162.1 (glucose); and (iv) they do not appreciably absorb in the 200 to 800 nm wavelength range [97]. Several physical properties of α-CD, β-CD, and γ-CD are listed in Table 3.4.6.

Table 3.4.6 Physical properties of cyclodextrins (CD).

CD	Glucose units	Molecular mass	External diameter (A°)	Internal diameter (A°)	Depth (A°)	Water solubility, M
α-CD	6	973	13.7	5.7	7.8	0.114
β-CD	7	1135	15.3	7.8	7.8	0.016
γ-CD	8	1297	16.9	9.5	7.8	0.179

With the growing interest in using cyclodextrins in chiral separations, CD-bonded stationary phases with the mechanical strength necessary for HPLC became necessary. One of the earliest attempts to develop such a stationary phase was to bond CD to silica gel via nitrogen-containing linkage arms [98]. All of these CD-based stationary phases suffered from low coverage and instability under typical reversed-phase chromatographic conditions.

In 1983, Armstrong developed a hydrolytically stable linkage [99]. The spacer arm is attached to the silica gel through a silane linkage. Either an epoxy silane, an organohalosilane or a vinyl alkyl silane may be used. The reaction was performed in the absence of water. The CD is then attached through its hydroxyls to the epoxide or the organohalide functionality. This non-nitrogen-containing, bonded CD silica gel stationary phase has proven to be relatively stable under standard HPLC conditions. A schematic of this linkage is shown in Figure 3.4.17.

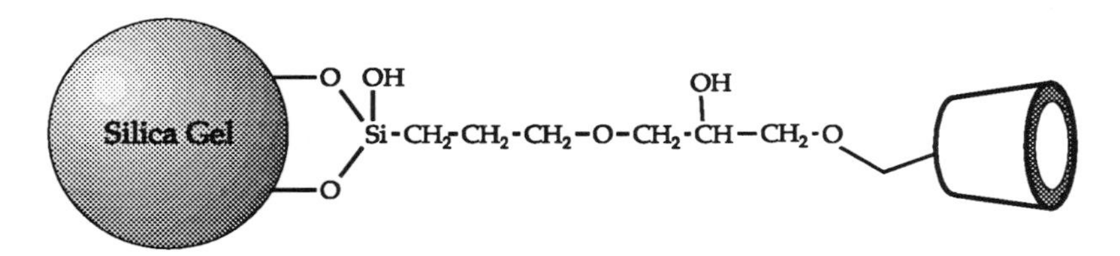

Figure 3.4.17 Diagram of a bonded cyclodextrin stationary phase.

The impact of CD phases for chiral separations has been pronounced; they show unusual selectivity and stability. When used in the reversed phase mode, inclusion complex formation - in which the hydrophobic portion of the analyte resides in the cyclodextrin cavity - is largely responsible for retention and selectivity. In the normal phase mode, the hydrophobic mobile phase occupies the cavity of the cyclodextrin. Analyte interactions are primarily with the hydroxyl groups at the mouth and base of the cyclodextrin. Thus, in the normal phase mode, selectivity is similar to that of a diol type column. Chiral selectivity has been achieved only in the reversed phase mode. Separations of geometric position isomers have been obtained in both the reversed and normal phase modes using α-, β- and γ-cyclodextrins.

In recent years, the useful range of the CD phases has been extended by derivatizing the CD-bonded silica gel. Five different functional pendant groups have been placed on the CD, i.e. (i) acetyl ; (ii) 2-hydroxypropyl; (iii) naphthylethyl carbamate; (iv) dimethylphenyl carbamate; (v) para-toluoyl ester. These derivatized CD columns have been used in both the reversed and normal phase modes making them the first successful 'multimodal' chiral stationary phases [100].

Basic aspects of the retention mechanism on native cyclodextrin phases

In the reversed-phase mode, the feature of the native CD-bonded stationary phases which gives them their unique selectivity and influences retention is inclusion complexation. This complexation involves interaction of a nonpolar portion of the analyte molecule with the relatively nonpolar cyclodextrin cavity. This type of interaction is shown in Figure 3.4.18.

The inclusion complex gains stability from Van der Waals' forces and hydrogen bonding. The size, shape, and polarity of the analyte are the most critical factors influencing the stability of the inclusion complex. If an analyte is too large to fit into the hydrophobic cavity, then inclusion is impossible. Hydrogen bonding of polar molecules to the C-2 and C-3 hydroxyl groups generally occurs preferentially to inclusion complexation. Aromatic groups have the ability to share electrons with the glycosidic oxygens of the cyclodextrin cavity, given the proper orientation. Saturated rings can also be included, but a more random fit and less selectivity is observed. Amines and carboxylate groups interact strongly with the hydroxyl groups of cyclodextrin through hydrogen bonding if the proper geometric and pH conditions are used. These can be the important second and third interactions necessary for chiral selectivity.

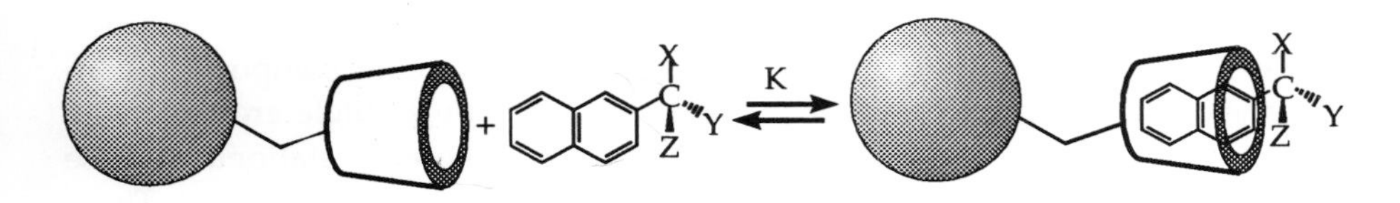

Figure 3.4.18 Diagram of the reversible inclusion complex formation of cyclodextrin.

The polarity of the CD-bonded stationary phase contributes to retention. Many other interactions are possible and may also contribute to retention. For example, interactions with residual silanols on the surface of the silica gel and with the linkage arm are two reactions that have been discussed. Retention has been observed in both the reversed and normal phase modes on native CD-bonded stationary phases. To date, the greatest success for enantioselective separation has been in the reversed phase mode with aqueous/organic mobile phases.

Using native cyclodextrin-bonded stationary phases in the reversed-phase mode, at least two conditions must be satisfied in order to achieve enantioselectivity. First, a relatively tight fit is needed between the analyte and the cavity of the cyclodextrin. Second, there must be some interaction between the analyte and the C-2 and C-3 hydroxyls at the mouth of the cyclodextrin cavity. The better the molecular fit, the greater the potential for chiral recognition. It has been observed that, in general, at least one or more rings must be present in an analyte to achieve this fit; in most cases, an aromatic ring is necessary.

Of the three commercially available native cyclodextrin bonded stationary phases, β-CD has been the most successful in chiral separations. β-CD-bonded stationary phases generally require a structure with at least two aromatic rings, but some success has also been achieved with single ring structures [101-103]. Structures with naphthyl or biphenyl moieties, in general, fit tightly in β-CD-bonded stationary phases. The large γ-CD-bonded stationary phase has been shown to separate successfully in the reversed phase mode enantiomers containing fused rings, such as naphthalene or multiple fused ring moieties [104].

Chromatographic studies as well as computer modeling [105,106] relating analyte structures to enantioselectivity have shown that definite structure/ selectivity relationships exist and that they may be used to predict what types of compounds can be separated by CD-bonded stationary phases. The α-CD bonded stationary phase can be used to separate compounds containing one or two rings. Examples are the enantiomers of aromatic amino acids, e.g., phenylalanine, tryptophan, tyrosine, and analogues, which were separated on an α-CD-bonded stationary phase [107]. All structures had a single aromatic ring or the aromatic indole substituent near the stereogenic center. Opposite the stereogenic center to the aromatic ring were the amine and acid functionalities, both of which effectively create hydrogen bonds. Substitutions on the aromatic ring tended to decrease selectivity with the exception of para-NO_2, -OH or -halogens, which are able to enter the CD cavity and enhance complexation. Nonpolar substituents that forced the analyte to tilt inside the cavity (steric interactions) increased selectivity.

The β-CD-bonded stationary phase can be used to evaluate compounds containing one or more rings. Chiral compounds which contain a single aromatic ring and whose enantiomers could be separated on a β-CD-bonded stationary phase showed similarities to those separated on α-CD-bonded stationary phases. These racemates have an aromatic ring α to the stereogenic center except for the amino acids. They were either β- or γ-substituted. All the compounds also have a carbonyl substituent off the stereogenic center. All of the racemates are believed to have at least one hydrogen bonding group in close proximity to the hydroxyls at the mouth of the cyclodextrin when the inclusion complex is formed. The stereogenic center of all racemates appears to be the main point of rotation between two π systems. This 'sandwich' of two π systems would impart a degree of rigidity to the structure. This rigidity is believed to be responsible, in part, for enhanced enantioselectivity. Structures containing two or more rings can be further subdivided into groups that contain either a single or double ring aromatic substituent off the stereogenic center, two aromatic ring substituents off the stereogenic center, and structures that have the stereogenic center as part of a ring.

For solutes, all of the previous requirements for single ring compounds apply as well. Aromatic substituents in α-position to the stereogenic center showed better selectivity than β– and γ–substituted aromatic compounds. Compounds with two or more aromatic ring systems generally showed greater retention than single aromatic ring structures. If the stereogenic center was between two aromatic rings or an aromatic ring system and another π system, then enantioselectivity was higher.

The separation of racemates with the stereogenic center as part of a ring system has also been studied, e.g., 5-phenylhydantoin analogues. When two aromatic ring systems were attached to the stereogenic center, enantioselectivity was higher than when only one aromatic ring was attached. If the stereogenic center is in α-position to two aromatic rings and contains a fused ring system such as chlorthalidone, the chiral selectivity is very high. This can be attributed to the rigidity and the optimum spatial arrangement of the analyte. In Table 3.4.7 several examples of enantioresolutions performed by using native cyclodextrin phases are summarized.

Table 3.4.7 Examples of separations using native cyclodextrin phases [96].

Compound	Separation factor [α]	Stationary phase	Mobile phase
AROMATIC AMINO ACIDS			
Phenylalanine	1.09	α-CD	100:0[a]
p-Chlorophenylalanine	1.05	α-CD	90:10[a]
Tryptophan	1.20	α-CD	100:0[a]
6-Methyltryptophan	1.48	α-CD	100:0[a]
Tyrosine	1.40	α-CD	100:0[a]
COMPOUNDS WITH ONE AROMATIC RING			
α-Methylbenzyl acetate	1.03	β-CD	60:40[b]
Ethyl-3-phenylglycidate	3.32	β-CD	50:50[c]
2-Phenylpropanal	1.22	β-CD	90:10[b]
Methyl mandelate	1.05	β-CD	98:2[b]
Ephedrine	1.10	β-CD	98:2[b]
Tyrosine	1.06	β-CD	90:10[c]

[a]Mobile phase reported as (v/v) buffer (1% triethylammonium acetate TEAA, pH 5.1): organic modifier (methanol); [b]same mobile phase, pH 7.1; [c]same mobile phase, pH 4.1.

Mechanism of separation of functionalized cyclodextrins

Acetyl- and hydroxy-derivatized β-CD-bonded stationary phases have extended the selectivity of cyclodextrin phases in the reversed phase mode. The secondary hydroxyl groups are relatively fixed in space on the native β-CD. Acetylated and hydroxylpropyl derivatives allow the rim of the cyclodextrin to be 'extended' and somehow flexible. The 2-hydroxylpropyl β-CD-bonded stationary phase has been found to be useful in the separation of enantiomers with stereogenic centers incorporated into ring systems (indanol), external or β to an aromatic ring, and for atropic compounds such as (+/-)-1,1'-di-β-naphthol [108]. The acetylated β-CD-bonded stationary phase has been shown to be useful for the separation of the enantiomers of norphenylephrine and multi-ring structures such as (+/-)-norgestrel [109] and scopolamine [110].

In the normal phase mode, nonpolar organic phases occupy the cyclodextrin cavity. Without inclusion, enantiomeric selectivity does often not occur. To overcome this problem, a variety of different substituents were attached to the hydroxyl groups of native cyclodextrin. To date, several derivatized β-CD stationary phases have been used in performing chiral separation in the normal phase mode [111]. Acetylated (e.g., 'Cyclobond I-Ac', Astec), dimethylphenyl carbamate (e.g., 'Cyclobond I-DMP', Astec), naphthylethyl carbamate (e.g., 'Cyclobond I-N', Astec), and para-toluyl ester (e.g., 'Cyclobond I-PT', Astec) β-CD bonded stationary phases have been used successfully in the normal phase mode for the separation of enantiomers. With the addition of aromatic and carbonyl functionalities, opportunities for π-π interactions arise. Interactions with the residual hydroxyl groups of β-CD also exist.

The isocyanate-derivatized β-CD phases seem to have the best enantioselectivity. This may be attributable to the additional hydrogen bonding and stronger dipole-dipole interactions available through the carbamate linkage relative to the ester linkage. In addition to similarities to derivatized cellulose phases, the naphthylethyl isocyanate-derivatized β-CD shows enantioselectivity similar to the naphthylvaline reciprocal chiral stationary phase developed by Pirkle and Pochapsky [112].

Some of the racemates separated on derivatized β-CD-bonded stationary phases in the normal phase mode have been reported to be resolved on native cyclodextrin in the reversed phase mode. However, these normal phase separations were achieved with shorter retention times and increased resolution [111]. Mechanistic studies concerning requirements for the enantioselectivity of the derivatized cyclodextrin chiral stationary phases have recently been investigated. General retention may be related to the degree of substitution of each type of pendant on the β-CD.

The acetylated β-CD stationary phase appears to be approximately 90% derivatized, while the aromatic pendant groups show an approximate average degree of substitution of 6 for naphthylethyl isocyanate, 10 for dimethylphenyl isocyanate, and 13 for para-toluyl ester. This corresponds to 33, 50, and 66%, respectively. The decreased percentage of coverage for the larger pendant groups is probably due to increased steric hindrance.

Acetyl- and naphthylethyl isocyanate-derivatized β-CD-bonded phases have shown enantioselectivity in both the reversed and normal phase modes [113]. With the increasing number of commercially available chiral stationary phases, the need for consolidation and elimination of different chiral stationary phases is apparent. The derivatized cyclodextrin stationary phases have been effective in separations analogous to derivatized cellulosic and 'Pirkle' type chiral stationary phases. Derivatized cellulosic phases consist of wide range molecular weight polymers coated onto large-pore silica gels. Aqueous and some organic (CH_2Cl_2, $CHCl_3$, THF, DMF, etc.) mobile phases would destroy this coating. Even less polar alcoholic mobile phases can destroy the secondary structure often thought to be responsible for some of their chiral selectivity. Polar mobile phases are also not recommended for 'Pirkle'-type chiral stationary phases.

Bonded CD stationary phases have shown the ability to operate in both the normal and reversed phase modes. However, for a chiral stationary phase to be considered 'multimodal', it needs to be able to separate different classes of compounds in both modes. The naphthylethyl carbamate β-CD-bonded stationary phase is the first useful 'multimodal' column. In the reversed phase mode, a completely different set of racemates was separated than was separated in the normal phase mode

[111,113]. Successful separations of enantiomers of the pesticides dyfonate, ruelene, ancymidol, and coumachlor, as well as a variety of pharmaceuticals such as tropicamide, indapamide, althiazide, tolperisone, and a sulphonamide have been reported [113].

Cyclodextrin columns have successfully been applied in the separation of many drug stereoisomers [105,106,114-116]. Table 3.4.8 lists some examples.

Table 3.4.8 Examples of the separation of drugs using cyclodextrin phases [96].

Drug	Separation factor [α]	Column[a]	Mobile phase
Ibuprofen	1.10	CBI	70:30[b]
Ketoprofen	1.06	2 CBI	73:27
Hexobarbital	1.14	CBI	85:15
Phensuximide	1.15	CBI	90:10[c]
Propranolol	1.04	2 CBI	75:25
Metaprolol	1.03	2 CBI	68:32
Norgestrel	1.24	2 CBII	70:30[c,d]
Chlorthalidone	1.44	CBI	70:30
Nomifensine	2.50	CBI	80:20
Methadone	1.04	CBI	gradient[c]
Atropine	1.04	CBI	96:4[c]
Cocaine	1.04	CBI	96:4[c]

[a]Column is 250 x 5mm Cyclobond I (CBI), Cyclobond II (CBII) or 2 Cyclobond columns in series; [b]Mobile phase reported as (v/v) aqueous buffer (1% TEAA pH 4.1): organic modifier (methanol); [c]Acetonitrile was used as modifier; [d]The mobile phase was unbuffered.

Applications

Many clinical applications have been reported for cyclodextrin-bonded stationary phases, including separation of chiral drugs and chiral metabolites. The analysis and quantitation of the enantiomers of ibuprofen and its metabolites [117] and the structural isomers of suprofen [118] in biological fluids have been reported. The determination of the enantiomeric composition of urinary phenolic metabolites of phenytoin [119], hydantoin and their analogues [120] have also been shown. The separation of the enantiomers of hexobarbital and chiral anti-inflammatory agents and metabolites in rat blood has also been achieved by means of a β-CD-bonded stationary phase [121,122].

In the past decade, the separation of carbohydrates by liquid chromatography has received increased attention [123-125]. CD-bonded stationary phases have shown a high degree of selectivity for sugars and other carbohydrates. Mono-, di-, tri-, tetra-, and desoxysaccharides and sugar alcohols have been separated using either α-CD- or β-CD-bonded stationary phases. Retention of saccharides on CD-bonded stationary phases was related to size, with the smaller saccharides usually eluting before the larger saccharides. There was a close relationship between the number of hydroxyl groups on the sugar available for hydrogen bonding and retention, with the desoxy sugars eluting first. Further, α-CD- and β-CD-bonded stationary phases successfully separated cyclodextrins from each other as well [124].

The separation of sugar anomers has long been a difficult problem. The rapid mutarotation of anomers leads to peak doubling and band broadening. The separation of different pairs of anomers is possible using either α-CD- or β-CD-bonded stationary phases [125]. A high percentage of organic modifier was used in the majority of sugar separations employing these phases. Thus, hydrogen bonding to the surface of cyclodextrin, and not inclusion complexation, is believed to be the dominant retention mechanism. This high percentage of organic modifier also slowed down mutarotation.

Practical considerations

Mobile phase: Methanol and acetonitrile have been the most frequently used organic modifiers for chiral separations with cyclodextrin-bonded phases in the reversed phase mode. This does not mean that other organic modifiers such as ethanol, propanol, dimethylformamide, and dioxane could not be used. Organic modifiers compete with the solute molecule for residence in the cyclodextrin cavity. The nature of the cyclodextrin inclusion complex with organic molecules is such that it is usually strongest in water and decreases upon the addition of organic modifiers.

Methanol is the weakest displacer of the alcohols, i.e. the poorest competitor for residence in the cyclodextrin cavity; acetonitrile is a stronger displacer than methanol or ethanol. Enantiomeric selectivity is generally independent of the organic modifier used. However, sometimes enantioselectivity is observed when acetonitrile is used but not methanol. It is not possible to predict which modifier will give the best separation, but many of the published separations have been achieved using methanol/water mixtures and the choice is usually to try methanol first. To decrease retention or increase efficiency or selectivity acetonitrile may be added to the methanol/water mixture or used as a substitute for methanol completely.

It is recommended that, initially, a mobile phase containing methanol/water (1+1) be used. The retention time of the solute can be adjusted by changing this ratio. The percent changes should be made in smaller steps than are made with reversed-phase columns as the cyclodextrin column is more sensitive to these changes.

Normal phase eluants for CD-bonded stationary phases are, in general, hexane/isopropanol mixtures. The fact that the CD are covalently bonded to silica gel allows a wider range of solvents to be used in comparison to other normal phase chiral stationary phases [111].

Salt concentration and pH: Because the ionization of the solute molecule is an important factor in inclusion complexing, it is often advantageous to change the pH or to vary the ionic strength to effect or enhance a separation. The addition of salts to the mobile phase, in the form of buffers, has a greater effect on retention and efficiency than it does on selectivity.

Buffers have been known to compete with the analyte for the cyclodextrin cavity. With increasing buffer concentration, retention is decreased and efficiency is increased; however, these effects can be pH- and concentration-dependent. The usual range of salt concentration is from 0.005 to 0.1 mol. Ionizable solutes tend to show the greatest sensitivity to changes in pH and there can be a tremendous variation in retention with changes in pH. Columns with cyclodextrin covalently bound to a silica support should be operated below pH 7.5. The cyclodextrin groups can be cleaved at pH < 3.0. Buffers that can be used with these columns are trifluoroacetic acid, ammonium acetate and phosphate.

Triethylammonium acetate (TEAA) is more useful because it is noncorrosive and the pH can be easily adjusted. The column manufacturers recommend that the buffers should be filtered and passed through a C_{18}-column to remove organic impurities that would collect on the column and impair its performance.

Temperature: The binding constant of a solute to cyclodextrin is strongly temperature-dependent, i.e., there is an increase in the binding constant at lower temperatures. Chromatographic studies using CD-bonded stationary phases have shown that decreasing the temperature almost always increases selectivity, but the same is not valid for resolution. Increasing the binding constant of the solute in the cyclodextrin cavity by lowering the column temperature may result in reduced resolution. This is due to the lowering of efficiency that is a consequence of poorer mass transfer. The increasing selectivity and band broadening effects may work antagonistically and are difficult to predict.

Increasing the column temperature may even improve the separation of strongly retained solutes. In the normal phase mode, since inclusion is not the only mechanism responsible for retention and enantioselectivity, band broadening at reduced temperatures may not be a problem, but this effect has not been studied specifically. If there are significant changes in the ambient temperature it may be recommended to use a column oven to ensure reproducible separations and retention times.

Flow rate: Typical flow rates for a 25 cm analytical cyclodextrin column range from 0.5 to 2.0 ml/min. At high flow rates, i.e. 1.0 to 2.0 ml/min., band broadening from mass transfer effects tends to decrease efficiency and leads to poorer resolution in the reversed-phase mode. Lowering the flow rate can improve efficiency by a factor of 1.5 to 4; the greatest changes are to be seen between 1.0 and 0.4 ml/min.

Advantages and disadvantages of cyclodextrin phases

Cyclodextrin columns offer various advantages because they are operated in the reversed-phase mode and solvents containing water and organic modifiers can be employed. The columns are stable and if cared for properly will last a long time. Unlike, e.g., protein-based columns, they are also available in preparative sizes. Actual column technology works with more cyclodextrin bonded to the silica gel, which leads to an increase of the retention time and, in most cases, the resolution.

Crown-ether phases

Macrocyclic polyethers are known as crown ethers, because molecular models of them often resemble a crown in shape. Their ability to form strong complexes with metal cations, as well as with substituted ammonium ions, has led to extensive research. One interesting use is as phase-transfer agents, where the formation of lipophilic alkali metal ion inclusion complexes is utilized for the transfer of alkali metal salts into organic solvents. This field is known as a branch of 'host-guest' complexation chemistry.

The first successful synthesis of optically active crown ethers, obtained by incorporating atropisomers into the 'mouth' of the crown ether, led Cram and his group to investigate their use for optical resolution purposes [126]. This resolution principle was then transformed into an LC separation technique by the use of a chiral crown ether in the mobile phase or covalently bound to a silica support [127].

Such a chiral host is able to discriminate between enantiomeric ammonium compounds, such as D,L-amino acids, and many compounds bearing a primary amino group near the chiral center. Due to the multiple hydrogen bonds formed between the ammonium group and the ether oxygen, for steric reasons a less stable complex with one of the enantiomers will be formed (Figure 3.4.19). In addition, the absolute configuration of the sample can be inferred from the order of elution on the crown ether phase (commercially available as 'Crownpak CR' by Daicel). In fact, for amino acids, D-isomers are eluted faster than the L-isomers.

Practical considerations

This phase should be operated with an aqueous acidic mobile phase to form the ammonium ion. Various kinds of acids can be used, however, the use of perchloric acid is recommended because of the better resolution and low UV absorption. Low UV absorption is necessary, because this technique requires UV detection at very low wavelengths.

The capacity factors on 'Crownpak CR' depend on the degree of hydrophobicity of the samples. Hydrophobic compounds are retained strongly, hydrophilic ones to a lesser extent. When the sample shows only little retention and poor resolution because of its hydrophilic nature, the separation may be improved by decreasing analysis temperature (in the range of 0-10°C) and pH (in the range of 1.5 - 2.0).

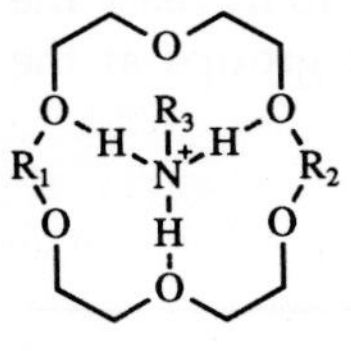

Figure 3.4.19 Simplified diagram of the interactions between an ammonium cation and a crown ether.

3.4.3.5 Protein phases (*type 5*) [128]

Proteins are high-molecular mass polymers composed of L-amino acids as chiral subunits. As proteins are chiral polymers, it is logical that the binding of small enantiomeric molecules is often stereospecific. This property is especially pronounced for the serum proteins α_1-acid glycoprotein (AGP), serum albumin (SA) and ovomucoid (OV) leading to the commercial development of three chiral stationary phases for HPLC. The AGP-CSP was synthesized by Hermansson [129]; Allenmark et al. [130] used bovine serum albumin (BSA) to create the BSA-CSP and Miwa et al. [131,132] have immobilized ovomucoid, resulting in the OV-phase. These phases can be used to resolve a wide variety of enantiomeric compounds; they will now be briefly discussed.

α_1-acid glycoprotein (AGP) phase

Human AGP has a molecular mass of about 41 kDa and is composed of a single 181-unit peptide chain and five carbohydrate units. The protein has an isoelectric point of 2.7 and is one of the key sites for the binding of cationic molecules [133]. The commercially available form of AGP-CSP is prepared by binding the AGP to diethylaminoethyl silica and then crosslinking the proteins using a process which involves the oxidation of the terminal alcohol groups to aldehydes, Schiff base formation and reduction of the resulting enamines to secondary amines. This process yields a stable immobilized protein coating of approximately 180 mg of protein per gram of silica or 4.4×10^{-6} moles of protein per gram of silica. The resulting column is a useful CSP with a wide range of applications, although it exhibits a very low capacity.

Solute selectivity: The ability of human AGP to bind cationic molecules stereoselectively is the basis of chiral resolutions achieved with the AGP-CSP. Many enantiomeric cationic molecules have been resolved on this CSP. Table 3.4.9 lists some of the reported resolutions.

The magnitude of the stereochemical resolution of a solute appears to be highly dependent upon the molecular structure of the solute. The results of initial studies of the AGP-CSP indicate that the solute should contain at least two bonding groups, e.g., an ammonium ion and a hydrogen bonding moiety, and a bulky or rigid structure at or near one of the binding sites. Structural features which tend to increase the magnitude of the stereoselectivity include steric bulk and aromatic groups at the chiral carbon and the chiral centre as part of a ring system.

Table 3.4.9 Examples of chiral cationic drugs resolved on AGP-phases [133a].

Solute	Solute
Atropine	Promethazine
Brompheniramine	Pronethalol
Butorphanol	Propranolol (as oxazolidone derivative)
Cocaine	Pseudoephedrine
Disopyramide	Salbutamol
Doxyamine	Sotalol
Ephedrine	Terbutaline
Homatropine	Tetrahydrazoline
Methadone	Verapamil
Methylatropine	
Metoprolol	
Phenoxybenzamine	

The effect on the stereoselectivity of the steric bulk at the ammonium ion is illustrated by the chromatographic results obtained with terbutaline (**5**) and metaproterenol (**6**) [134]. Compounds **5** and **6** differ only in the size of the alkyl moiety (tert-butyl and 2-propyl) on the amine function, respectively. When these compounds are chromatographed on the AGP-CSP, terbutaline is resolved with a stereochemical selectivity α of 1.22, while metaproterenol remains unresolved.

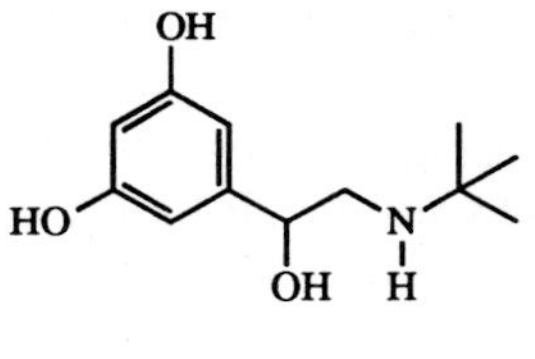

OH HO N OH H

5

OH HO N OH H

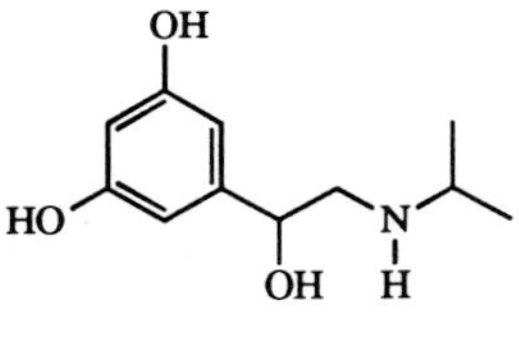

6

The distance between the ammomium and hydrogen bonding moieties also effects the stereoselectivity. This is illustrated by the chromatographic results of a series of compounds related to metoprolol (**7**) [134]. When the number of methylene groups in the chain between the amine and the alcohol groups is increased from 1 to 3, α is reduced from 1.49 to 1.00.

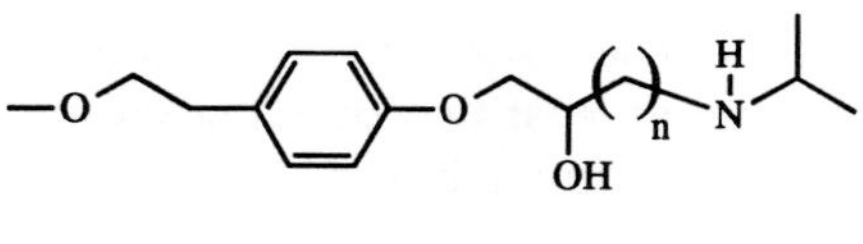

—O O n H N OH

7

Chiral molecules with an amine function can be resolved on the AGP-CSP without precolumn derivatization. Unlike a number of other CSPs, the amine function does not have to be at the chiral centre. Enantiomeric α,β-amino alcohols such as ephedrine, propranolol and labetalol can also be resolved without precolumn derivatization [134,135]. The efficiency and stereoselectivity of the resolution can be improved by converting the compounds to oxazolidone derivatives [136].

Although AGP primarily binds cationic compounds, a number of anionic chiral molecules can also be resolved on the AGP-CSP. These resolutions are obtained when the mobile phase contains an ion pairing agent such as N,N-dimethyloctylamine or tetrabutylammonium bromide [134,135,137]. The acidic compounds resolved on the AGP-CSP include α-methylarylacetic acids such as ibuprofen and naproxen and barbiturates such as hexobarbital. Some of the anionic compounds which have been resolved on the CSP are listed in Table 3.4.10.

Table 3.4.10 Examples of chiral anionic drugs resolved on AGP-phases

Solute	Reference
Ethotoin	137
Fenoprofen	135
Hexobarbital	137
Ibuprofen	135,137
Ketoproxen	137
Naproxen	135,137
2-Phenylbutanoic acid	135
3-Phenylbutanoic acid	135
2-Phenoxypropanoic acid	135,137
2-Phenylpropanoic acid	135

Chiral carboxylic acid derivatives such as amides and esters can also be resolved on the AGP-CSP, often with an increase in stereoselectivity. For example, mandelic acid is not resolved on this CSP, but its methyl and ethyl esters are resolved with α-values = 1.27 and 1.93, respectively [136].

Practical considerations

Mobile phase: Most mobile phases used with the AGP-CSP are composed of phosphate buffer and an organic or inorganic modifier. The partition ratio k and stereoselectivity α of a solute can often be drastically altered by changing the pH and/or the composition of the mobile phase.

Effect of pH: The pH of the mobile phase should be kept between 3.0 and 7.5, as the silica-based column is unstable outside this range [135]. Within this range, altering the pH of the mobile phase is an excellent means of adjusting k. For cationic solutes, a decrease in pH usually results in a drop in k for anionic solutes a decrease in pH often results in an increase in k. The stereoselectivity of the AGP-CSP can also be changed by modifying the pH of the mobile phase. This effect is dependent on the composition of the mobile phase. For example, when cyclopentolate and doxylamine are chromatographed using a mobile phase composed of 0.02 mol phosphate buffer modified with 0.003 mol tetrabutylammonium bromide, a decrease in the pH of the mobile phase from 7.0 to 6.0 results in an increase in α-value. However, if the mobile phase is changed to 0.02 mol phosphate buffer modified with 0.1 mol sodium chloride and 0.33 mol 2-propanol, a decrease in the pH from 7.5 to 6.5 reduces the α value for cyclopentolate from 1.96 to 1.79, whereas the α-value for doxylamine is virtually unchanged (1.15 as against 1.16) [135].

Table 3.4.11 Various mobile phase modifiers used with the AGP-phases [138].

Uncharged modifiers	Cationic modifiers	Anionic modifiers
MONOVALENT ALCOHOLS	TERTIARY AMINES	CARBOXYLIC ACIDS
1-Propanol	N,N-Dimethyloctylamine	Octanoic acid
2-Propanol	N,N-Dimethylethylamine	Butanoic acid
Ethanol		
	QUARTERNARY AMINES	AMINO ACID
DIOLS	Tetrapropylammonium	Aspartic acid
1,2-Ethanediol	bromide	
1,2-Propanediol		
1,2-Butanediol	AMINO ACID	SULPHAMILIC ACID
	1,2-Diaminobutanoic acid	Cyclohexylsulphamic acid
AMINO ACIDS		
6-Aminohexanoic acid		
ß-Alanine		
Leucine		

Mobile phase modifiers: While altering the pH of the mobile phase can be used to change k and α, the most dramatic changes can be provoked by the use of mobile phase modifiers. Both uncharged and charged modifiers have been studied and their effects often differ [134-137]. The compounds which have been used as mobile phase modifiers for the AGP-CSP are presented in Table 3.4.11.Uncharged modifiers such as monovalent alcohols are primarily used to reduce k. For example, the k for *R*- and *S*-disopyramide are 8.51 and 31.48, respectively, when the mobile phase is composed of 0.02 mol phosphate buffer modified with 4% 2-propanol. When the amount of 2-propanol is increased to 8%, the k-values are reduced by about 80% to 1.77 and 5.66, respectively. However, these modifiers also tend to reduce α. In this case, the stereoselectivity for disopyramide was reduced from α = 3.70 to 3.20 [136].

A variety of cationic and anionic compounds have been used as mobile phase modifiers for AGP-CSP (Table 3.4.11). The effect of these modifiers on k and α varies according to the structures of the solute and the modifier and the pH of the mobile phase. There is, in fact, a complex relationship between these variables, which is still not fully understood. In general, the addition of a cationic mobile phase modifier decreases the retention of a cationic solute and increases the k for anionic solutes, whereas an anionic modifier shows the opposite effect. Cationic mobile phase modifiers usually increase α for anionic solutes, whereas very low values for α or no chiral resolution at all is observed with anionic modifiers. The effect on the stereoselectivity of cationic solutes is more variable.

The bovine serum albumin (BSA) phase

Bovine serum albumin is a globular, hydrophobic protein with a molecular mass of about 66 kDa. The protein contains a single 581 amino acid chain with 17 intra-chain disulphide bridges forming nine double loops. BSA has an isoelectric point of 4.7 and a net charge of -18 at pH 7.0. The protein is able to bind uncharged hydrophobic molecules as well as anionic compounds.

In 1973, Stewart and Doherty [139] immobilized BSA on succinoylaminoethyl-Sepharose and used the resulting CSP to resolve D,L-tryptophan. Allenmark et al. [130] reported the covalent immobilization of BSA on 10-µm silica. This work was used as the basis for the development of the commercially available BSA-CSP.

Solute selectivity: The binding properties and the observed stereoselectivity of the native BSA are the basis for the enantioselectivity of the BSA-CSP. A high variety of anionic and neutral enantiomeric molecules can be resolved on the BSA-CSP, but not as cationic compounds. Some representative classes of solutes are given in Table 3.4.12.

Table 3.4.12 Examples of chiral compounds resolved on BSA-phases.

Class	Solute	Reference
Aromatic amino acids	Kynurenine	140
Amino acid derivatives	N-benzoyl-	142
	Dansyl-	141
	N-(2,4-dinitrophenyl)-	142
		141
Aromatic hydroxy ketones	Benzoin	140
		142
Benzodiazepinones	Oxazepam	140
Coumarin derivatives	Warfarin	140
Lactams	4-Amino-3-(p-chlorophenyl)butyric acid lactam	140
Reduced folates	Leucovorin	143
Sulphoxides	Omeprazole	140

From initial studies, it appears that enantioselectivity on the BSA-CSP requires aromatic and polar moieties in the solute. Steric effects also seem to be important. Some aromatic amino acids, such as kynurenine, can be resolved stereochemically on the BSA-CSP without derivatization of the amine moiety. The N-derivatives already used include: acetyl, benzenesulphonyl, phthalimido, dansyl, 2,4-dinitrophenyl and 2,4,6-trinitrophenyl [140-142]. The retention of the N-nitrobenzoyl- and N-nitrophenyl amino acids tends to increase with the degree of nitro substitution. This may be due to π-π interaction between the aromatic moiety on the amino acids and two tryptophanyl residues in the protein [140].

The BSA-CSP has also been used to resolve stereochemically enantiomeric molecules containing asymmetric sulphoxide and sulphoximine moieties and an aromatic group near or at the chiral centre [140,141]. The coumarin derivatives phenprocoumon and warfarin, as well as a number of benzodiazepine derivatives, can be resolved without pre-column derivatization [140]. In addition, the BSA-CSP has been used to resolve 2-hydroxy-2-phenylacetophenone (benzoin) and the reduced folates leucovorin and 5-methyltetrahydrofolate [143].

Practical considerations

Mobile phase: Several modifiers have been investigated, but the standard mobile phase used with the BSA-CSP is composed of phosphate buffer modified with 1-propanol. Mobile phases containing methanol and acetonitrile should not be used with this CSP. The buffer concentration can range from 0.01 to 0.2 mol and the pH from 4.5 to 8.0. The resolution and retention of solutes can be regulated by changing these parameters.

In general, for N-derivatized amino acids an increase in pH results in a decrease in retention and stereochemical resolution. This effect is believed to be due to the fact that a decrease in the pH results in a corresponding decrease in the net negative charge of the BSA. This means that the Coulomb interaction between the BSA and the N-derivatized amino acids or other uncharged carboxylic acids increases, resulting in an increase in k.

For solutes containing a carboxylic acid moiety, the general rule appears to be that the lower the buffer concentration the higher the retention. This is valid for buffer concentrations ranging from 0 to 100 mmol. However, when N-benzoyl-amino acids were chromatographed with buffer concentrations of 0.2 mol and above, an increase in the buffer concentration resulted in an increase in k. These results have been interpreted as a reflection of the two types of binding forces between the solute and the CSP, i.e. Coulomb attraction and hydrophobic interaction. The rapid decrease in k with increasing buffer strength up to 0.1 mol probably mainly reflects the decrease in Coulomb attraction, whereas the increase above 0.2 mol is probably due to the increasing role of binding by hydrophobic interaction.

Mobile phase modifiers: 1-Propanol has been the most commonly used mobile phase modifier with the BSA-CSP. As with the AGP-CSP, the addition of an alcohol to the mobile phase seems to reduce the hydrophobic interactions between the solute and the CSP, resulting in lower k-values and a reduction in α. This effect has been demonstrated for N-benzoyl amino acids [142], N-(2,4-dinitrophenyl)- and dansyl--amino acids [141]. In the latter case, an increase in 1-propanol concentration from 1% to 5% resulted in a decrease in k for *S*-warfarin from 41.1 to 12.0 and for *R*-warfarin from 54.6 to 13.8. However, there was also a decrease in α from 1.33 to 1.15.

The utility of this approach is demonstrated by the effect of the mobile phase additive trichloroacetic acid. When *S*- and *R*-warfarin are chromatographed using a mobile phase composed of phosphate buffer (0.2 mol, pH 7.5): 1-propanol (97:3, v/v) the k for *S*- and *R*-warfarin are 18.08 and 21.83, respectively, and α = 1.21. The addition of trichloroacetic acid (5 mmol) reduces the k-values by almost 50% (9.42, *S*-warfarin; 10,96, *R*-warfarin), whereas the stereoselectivity falls by only 5% to α = 1.16.

The ovomucoid (OV) phase

Ovomucoid is a protein which can be easily purified from chicken egg whites. Miwa et al. [131,132] have immobilized this protein on aminopropyl silica gel. The resulting chiral stationary phase (OV-CSP) was capable of resolving a number of different chiral compounds including amines and carboxylic acids.

The cationic compounds resolved on the OV-CSP include chloropheniramine (α = 2.28) and pindolol (α = 1.48); the anionic compounds include flurbiprofen (α = 1.28) and ketoprofen (α = 1.21). The effect on k and α of the pH of the mobile phase has been investigated [132].

For anionic compounds, an increase in pH results in a decrease in k with very little change in α. For example, when the pH of the mobile phase (0.2 mol phosphate buffer) was raised from 5.0 to 6.0, the k of the first eluted enantiomer of flurbiprofen fell from 13.12 to 7.28, whereas the stereoselectivity was only slightly reduced from α = 1.28 to 1.24. An increase in pH has the opposite effect on the cationic solute chloropheniramine.

When the pH of the mobile phase (0.2 mol phosphate buffer) was raised from 5.5 to 6.0, the k of the first eluted enantiomer rose from 3.20 to 9.50 and α also increased from 2.28 to 3.14. The effect is not, however, consistent. When pindolol was studied under the same conditions, the k of the first eluted enantiomer increased from 0.46 (pH 5.5) to 1.31 (pH 6.0), but α decreased from 1.48 to 1.44.

Cationic and anionic mobile phase modifiers can also be used to alter k and α [132]. For example, when a cationic molecule such as tetra-n-butylammonium hydrogen sulphate (TBS-HSO_4) was added to the mobile phase there was a moderate effect on the k and α of the anionic solutes and a dramatic effect on the chromatography of the cationic solutes.

For example, with 0.5 M TBS-HSO_4 in the mobile phase (0.5 mol phosphate buffer, pH 5.0) the k of the first eluted enantiomer of fluriprofen fell from 13.12 to 11.32 and α rose from 1.28 to 1.30. For chloropheniramine the addition of 0.5 mol TBS-HSO_4 to the mobile phase (0.5 mol phosphate buffer, pH 5.5) reduced the k of the first eluted enantiomer from 3.20 to 0.46 and the stereoselectivity was lost. In addition to the composition of the mobile phase, k and α also depend on the silica support.

When the ovomucoid was immobilized on LiChrosorb NH_2, the k-values for the enantiomers of flurbiprofen were 52.26 and 79.31, with $\alpha = 1.52$ (mobile phase: 0.2 mol phosphate buffer, pH 5.0). Using the same mobile phase on an OV-CSP based upon Unisil Q-NH_2, the k-values for these solutes were 13.12 and 16.73, with α = 1.28.

3.4.3.6 Chiral ion-exchange phase (*type 6*)

For the phase Ceramospher Chiral RU-1 (Shiseido), synthetic hectorite (sodium magnesium silicate) was used as packing material [144]. A natural clay is composed of microcrystalline particles of diameter 0.1-1 µm. These particles are too small to allow a natural clay to be used as a packing material. In addition, the particles may collapse to smaller particles under high pressure. The synthetic ceramic exists of spherically shaped particles with a size of 5 µm and a high durability under pressure. The chirality is introduced by ion exchanging the optically active metal complex tris(1,10-phenanthroline)-ruthenium(II) into the hectorite.

The influence of different separation parameters

When methanol or ethanol were used as the eluting solvent, many compounds were resolved completely into their enantiomers. If a compound is not separated, one should add 1% isopropylamine or dimethylamine in methanol for basic compounds, or 1% acetic acid for acidic compounds. If the sample elutes too fast, an ethanol/n-hexane system should be tried. The addition of acetonitrile resulted in a decrease of the separation factor from $\alpha = 2.09$ to 1.00 until, at 18% acetonitrile, the column no longer resolved the enantiomers of 1,1'-binaphthol at all. When the concentration of acetonitrile exceeded 18%, the order of elution of the enantiomers reversed from *R,S* to *S,R* and the separation factor α increased from 1.0 to 1.3. The reversal of selectivity with the addition of acetonitrile implies that solvent molecules affect the binding states of the enantiomers in different ways. One possible explanation for the occurrence of this effect is that the enantiomers are adsorbed at different binding sites on the column and that acetonitrile molecules interact specifically with either one of the sites [144].

The effect of the flow-rate on the separation factor was studied by varying it from 0.3 to 1.0 ml/min using ethanol as mobile phase and 1,1'-binaphthol as solute. As a result, the compound was resolved completely with a slight decrease in α (from 2.02 to 1.98). In the same study, varying the column temperature in the range 30 to 50°C led to a decrease in α. But the peaks of the separated enantiomers were sharper at higher temperatures, leading to an increase in the theoretical plate number. It is, therefore, recommended that one work at higher temperatures (50°C) with this phase. A disadvantage of this chiral phase is that one cannot use an aqueous solution for the mobile phase and the sample solution.

3.4.4 *Practical examples*

3.4.4.1 Description of the preparation of derivatives of chiral compounds using 'brush type' CSP

Preparation of the DNA-derivatives: A mixture of 0.02 mmol of the chiral compound and 0.12 mmol of oxalyl chloride (in a 4-ml vial) is refluxed for 15 minutes. After evaporation of the solvent to dryness under a stream of N_2 1.0 ml of $CHCl_3$ and 0.04 mmol of DNA are added. After stirring for 1 hour the organic layer is washed with 3 x 1 ml of a 0.1 mol $NaHCO_3$ solution, 3 x 1 ml of a 0.1 mol HCl solution, and 3 x 1 ml of distilled water. The organic layer is then filtered through anhydrous Na_2SO_4 and the solvent evaporated to dryness under a stream of nitrogen. The residue is redissolved using 1-2 ml of the mobile phase; from this solution 10 µl are injected.

Preparation of the ICDNA-derivatives: A mixture of 0.02 mmol of the chiral compound, 0.05 mmol of ICDNA, and 2 ml of DMF (in a 5-ml vial) is stirred under heating at 60-70°C for 1-2 hours. After cooling the solvent is evaporated to dryness under a stream of nitrogen. The residue is redissolved in 3 ml of CH_2Cl_2 and washed with 2 x 1 ml of distilled water. The organic layer is filtered through anhydrous Na_2SO_4 and the solvent evaporated to dryness under a stream of nitrogen. The residue is redissolved in 1-2 ml of the mobile phase; from this solution 10 µl are injected.

Preparation of the NIC-derivatives: A mixture of 0.02 mmol of the chiral compound, 0.05 mmol of NIC, and 2 ml of toluene (in a 5 ml-vial) is stirred under heating at 60-70°C for 30 minutes. After cooling the solvent is evaporated to dryness under a stream of nitrogen. The residue is redissolved in 3 ml of CH_2Cl_2 and washed with 2 x 1 ml of H_2O. The organic layer is filtered through anhydrous Na_2SO_4 and the solvent evaporated to dryness under a stream of nitrogen. The residue is redissolved in 1-2 ml of the mobile phase; from this solution 10 µl are injected.

Preparation of the DNBC-derivatives: A mixture of 0.02 mmol of the chiral compound, 2 ml of dichloromethane and 0.03 mmol of DNBC (in a 5-ml vial) is stirred (or shaked) for 2 minutes. Then 2 ml of a 0.1 N NaOH solution is added and the mixture shaked for 1 minute. The upper aqueous layer is discarded and again 2 ml of 0.1 N NaOH solution is added. After shaking for 2 minutes the upper aqueous layer is discarded and the organic layer washed with 2 x 1 ml of distilled water. After discarding the upper layer the organic layer is filtered through anhydrous Na_2SO_4 and the solvent is evaporated to dryness under a stream of nitrogen. The residue is redissolved in 1-2 ml of the mobile phase; from this solution 10 µl are injected.

3.4.4.2 Enantiodifferentiation of chiral hydroperoxides and their corresponding alcohols using a Chiracel OD column

The enzyme-catalyzed enantioselective synthesis of organic substances has become increasingly important (see Section 1). For example, heme-containing peroxidases, i.e. haloperoxidases, lactoperoxidases, cytochrome P peroxidases, and horseradish peroxidase (HRP) can be used as suitable catalysts in bioorganic synthesis.

Table 3.4.13 · Kinetic parameters and enantiomeric excess (ee) of HRP-catalyzed kinetic resolution of racemic hydroperoxides (reaction rate 50%) [146].

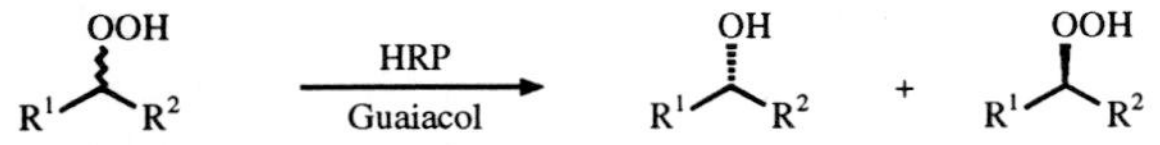

OOH / R^1 R^2 —HRP, Guaiacol→ OH / R^1 R^2 + OOH / R^1 R^2

	k_{cat} [a] [min^{-1}]	k_{cat} / K_m [b] [mM^{-1} min^{-1}]	ee [%] (+)-***R***-ROH	ee [%] (-)-*S*-ROOH
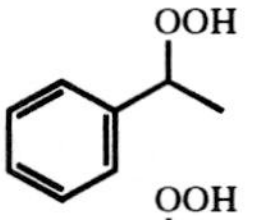 OOH	568	811	>99	>99
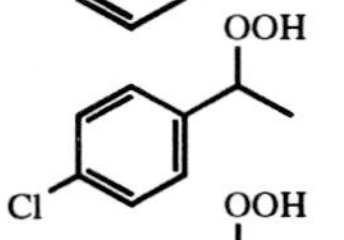 OOH, Cl	376	110	>95	>95
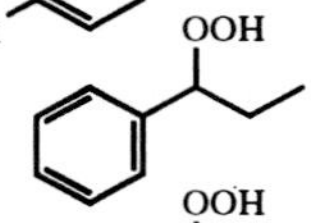 OOH	92	3.4	95	93
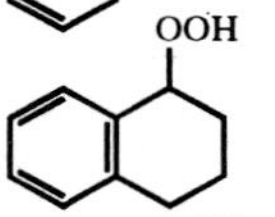 OOH	n.d.[d]	n.d.	>95	97
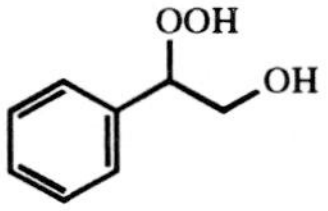 OOH, OH	256	39.4	95	95
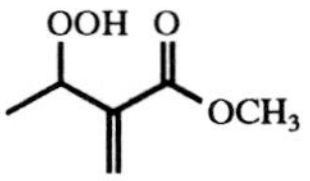 OOH O, OCH_3	720	19	82	75
			(-)-ROH	(+)-ROOH
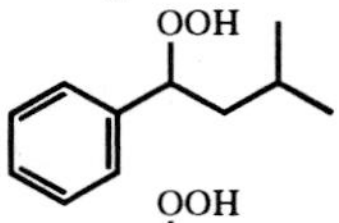 OOH	n.d.	n.d.	48	36
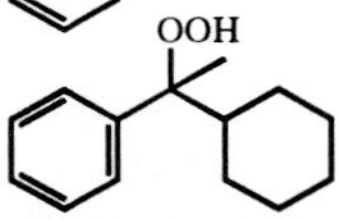 OOH		no reaction		

[a] Turnover number. [b] K_m: Michaelis constant. [c] Not determined.

The studies of Adam et al. [146] revealed that HRP catalyzes very efficiently the kinetic resolution of racemic hydroperoxides (cf. Figure 1.7). Selected results of the HRP-catalyzed kinetic resolution of a number of aromatic and functionalized aliphatic hydroperoxides are summarized in Table 3.4.13.

This novel HRP-catalyzed oxidoreduction can be performed on preparative scale and provides in many cases optically pure hydroperoxides and alcohols conveniently. The absolute configuration of the *S*-hydroperoxide and the *R*-alcohol was determined by comparison of the HPLC data with those of the authentic substances. The enantiomeric excesses of the hydroperoxides and their alcohols was established from the ratios of areas corresponding to the two enantiomers in the HPLC chromatogram.

Description of the experimental procedure: In the preparative-scale reaction, 1.50 mmol rac. hydroperoxide and 1.50 mmol guaiacol are dissolved in 10 ml 0.1 mol phosphate buffer (pH 6) and subsequently 0.25-1.25 x 10^{-4} mmol HRP is added. The reaction mixture is allowed to stand at 20°C for 2 hours and is then extracted with CH_2Cl_2 (3 x 20 ml). After drying of the combined organic phases over anhydrous Na_2SO_4 and removing the solvent (rotavapor) the hydroperoxide and the alcohol are separated by column chromatography on silica gel eluting with petroleum ether/diethyl ether (4+1). The enantioresolution is performed by using a Chiracel OD column (250 x 4 mm, Daicel) with n-hexane/isopropanol (9+1) as eluent (0.6 ml/min) and combined UV (220 nm) and on-line polarimetric (ChiraLyser, Knauer, Germany) detection.

References

[1] Ahnhoff, M., Einarsson, S. In *Chiral Liquid Chromatography,* Lough, W.J., Ed., Blackie, Glasgow and London, 1989, Chapter 4, p. 39

[2] Scott, C.G., Petrin, M.J., McCorkle, T. *J. Chromatogr.* (1976), *125,* 157

[3] Snyder, L.R. *J. Chromatogr.* (1982), *245,* 165

[4] Snyder, L.R., Palamareva, M.D., Kurtev, B.J., Viteva, L.Z., Stefanovsky, J.N. *J. Chromatogr.* (1986), *354,* 107

[5] Helmchen, G., Ott, R., Sauber, K. *Tetrahedron Lett.* (1972), 3873

[6] Helmchen, G., Völter, H., Schühle, W. *Tetrahedron Lett.* (1977), 1417

[7] Helmchen, G., Nill, G., Flockerzi, D., Schühle, W., Youssef, M.S.K. *Angew. Chem. Int. Ed. Engl.* (1979), *18,* 62

[8] Pirkle, W.H., Hauske, J.R. *J. Org. Chem.* (1977), *42,* 1839

[9] Pirkle, W.H., Simmons, K.A. *J. Org. Chem.* (1983), *48,* 2520

[10] Barksdale, J.M., Clark, C.R. *J. Chromatogr. Sci.* (1985), *23,* 176

[11] Wikby, A., Thalen, A., Oresten, G. *J. Chromatogr.* (1978), *157,* 65

[12] Nimura, N., Ogura, H., Konoshita, T. *J. Chromatogr.* (1980), *202,* 375

[13] Goto, J., Ito, M., Katsuki, S., Saito, N., Nambara, T. *J. Liq. Chromatogr.* (1986), *9,* 683

[14] Einarsson, S., Josefsson, B., Möller, P., Sanchez, D. *Anal. Chem.* (1987), *59,* 1191

[15] Shimada, K., Oe, T., Nambara, T. *J. Chromatogr.* (1987), *419,* 17

[16] Shimada, K., Hainuda, E., Oe, T., Nambara, T. *J. Liq. Chromatogr.* (1987), *10*, 3161

[17] Björkman, S. *J. Chromatogr.* (1985), *339*, 339

[18] Allenmark, S.G. *Chromatographic Enantioseparation: Methods and Applications*, Ellis Horwood: Chichester, 1988, Chapter 7.3, p. 132

[19] LePage, J., Lindner, G., Davies, G., Seitz, D., Karger, B.L. *Anal. Chem.* (1979), *51*, 433

[20] Lindner, G., LePage, J., Davies, G., Seitz, D., Karger, B.L. *J. Chromatogr.* (1979), *185*, 323

[21] Taphui, Y., Miller, N., Karger, B.L. *J. Chromatogr.* (1981), *205*, 325

[22] Davankov, V.A., Bochkov, A.S., Kurganov, A., Roumelioties, P., Unger, K.K. *Chromatographia* (1980), *13*, 677

[23] Davankov, V.A., Bochkov, A.S., Belov, Y.P. *J. Chromatogr.* (1981), *218*, 547

[24] Dobashi, A., Dobashi, Y., Hara, S. *J. Liq. Chromatogr.* (1986), 9, 243

[25] Dobashi, Y., Hara, S. *J. Am. Chem. Soc.* (1985), *107*, 3406

[26] Debowski, J., Jurczak, J., Sybilska, D. *J. Chromatogr.* (1983), *282*, 83

[27] Cabrera, K., Schwinn, G. *Intern. Laboratory* (1990), *7/8*, 28

[28] Knox, J.H., Jurand, J. *J. Chromatogr.* (1982), *234*, 222

[29] Pettersson, C., Schill, G. *J. Chromatogr.* (1981), *204*, 179

[30] Pettersson, C., Josefsson, M. *Chromatographia* (1986), *21*, 321

[31] Pettersson, C., Stuurman, H.W. *J. Chromatogr. Sci.* (1984), *22*, 441

[32] Prelog, V., Stojanac, Z., Konacevic, K. *Helv. Chim. Acta* (1982), *65*, 377

[33] Wainer, I.W. *Trends Anal. Chem.* (1987), *6*, 125

[34] Dalgliesh, C.E. *J. Chem. Soc.* (1952), *137*, 3940

[35] Taylor, D.R., Maher, K. *J. Chromatogr. Sci.* (1992), *30*, 67

[36] Pirkle, W.H., Hyun, M., Tsipouras, A., Hamper, B.C., Banks, B. *J. Pharm. Biomed. Anal.* (1984), *2*, 173

[37] Topiol, S. *Chirality* (1989), *1*, 69

[38] Siret, L., Bargmann-Leyder, N., Tambuté, A., Caude, M. *Analusis* (1992), *20*, 427

[39] Perrin, S.R., Pirkle, W.H. In *Chiral Separations by Liquid Chromatography*, Ahuja, S., Ed., ACS Symposium Series 471, American Chemical Society: Washington, DC, 1991, Chapter 3, p. 43

[40] Finn, J.M. In *Chromatographic Chiral Separations*, Zief, M., Crane, L.J., Eds., Dekker: New York, 1987, Chapter 3, p. 53

[41] Pirkle, W.H., Pochapsky, T.C., *J. Am. Chem. Soc.* (1986), *108*, 352

[42] Pirkle, W.H., Pochapsky, T.C., Mahler, G.S., Corey, D.E., Reno, D.S., Alessi, D.M. *J. Org. Chem.* (1986), *51*, 4991

[43] Pirkle, W.H. Mahler, G.S., Pochapsky, T.C., Hyun, M.H. *J. Chromatogr.* (1987), *388*, 307; (a) Wainer, I.W., Doyle, T.D. *J. Chromatogr.* (1984), *284*, 117

[44] Wainer, I.W., Doyle, T.D. *J. Chromatogr.* (1984), *306*, 405

[45] Pirkle, W.H., Dappen, R., Reno, D.S. *J. Chromatogr.* (1987), *407*, 211

[46] Oi, N., Nagase, M., Doi, T. *J. Chromatogr.* (1983), *257*, 111; (a) Oi, N., Kitahara, H. *J. Chromatogr.* (1983), 265, 117

[47] Oi, N., Nagase, M., Inda, Y., Doi, T. *J. Chromatogr.* (1983), *259*, 487

[48] Oi, N., Kitahara, H. *J. Liq. Chromatogr.* (1986), *9*, 511

[49] Pirkle, W.H., Finn, J.M., Hamper, B.C., Schreiner, J. L., Pribish, J.R. In *A Useful and Conveniently Accessible Chiral Stationary Phase for the Liquid Chromatographic Separation of*

Enantiomers, Eliel and Otsuka, Eds., ACS Symposium Series 185, American Chemical Society: Washington, D.C., 1982, chapter 18, p. 245
[50] Pirkle, W.H., Finn, J.M., Schreiner, J.L., Hamper, B.C. *J. Am. Chem. Soc.* (1981), *103,* 3964
[51] Crowther, J.B., Covey, T.R., Dewey, E.A., Henion, J.D. *Anal. Chem.* (1984), *56,* 2921; (a) McDaniel, D., Snider, B.G. *J. Chromatogr.* (1987), *404,* 123
[52] Pirkle, W.H., McCune, J.E. *J. Chromatogr.* (1988), *441,* 311
[53] Perry, J.A., Rateike, J.D., Szczerba, T.J. *J. Chromatogr.* (1987), *389,* 57
[54] Pirkle, W.H., McCune, J.E. *J. Liq. Chromatogr.* (1988), *11(9&10),* 2165
[55] Yang, Q., Sun, Z.P., Ling, D.K. *J. Chromatogr.* (1988), *477,* 208
[56] Pirkle, W.H., Deming, K. C., Burke, J.A. *J. Chirality* (1991)
[57] Doyle, T.D., Wainer, I.W. *High Resolut. Chromatogr. Chromatogr. Comm.* (1984), *7,* 38
[58] Pirkle, W.H., McCune, J.E. *J. Chromatogr.* (1989), *471,* 271
[59] Doury-Berthod, M., Poitrenaud, C., Tremilon, B. *J. Chromatogr.* (1977), *131,* 73
[60] Davankov, V.A., Semechkin, A.V. *J. Chromatogr.* (1977), *141,* 313
[61] Davankov, V.A., Rogozhin, S.V. *J. Chromatogr.* (1971), *60,* 280
[62] Davankov, V.A., Rogozhin, S.V., Semechkin, A.V., Sachkova, T.P. *J. Chromatogr.* (1973), *82,* 359
[63] Gübitz, G., Jellenz, W., Santi, W *J. Chromatogr.* (1981), *203,* 377
[64] Saigo, K., Yuki, Y., Kimoto, H., Nishida, T., Hasegawa M. *Bull. Chem. Soc.* (1988), *61,* 322
[65] Feibush, B., Cohen, M.J., Karger, B.L. *J. Chromatogr.* (1983), *282,* 3
[66] Gelber, L.R., Karger, B.L., Neumeyer, J.L., Feibush, B. *J. Am. Chem. Soc.* (1984), *106,* 7729
[67] Grierson, J.R., Adam, M.J. *J. Chromatogr.* (1985), *325,* 103
[68] Taylor, L.T., Busch, D.H. *J. Am. Chem. Soc.* (1967), *89,* 5372
[69] Gübitz, G., Jellenz, W., Schönleber, J. *J. High Res. Chromatogr. Commun.* (1980), *3,* 31
[70] Heese, G., Hagel, R. *Chromatographia* (1973), *6,* 277
[71] Lindner, K.R., Mannschreck, A. *J. Chromatogr.* (1980), *193,* 308
[72] Köller, H., Rimbock, K.-H., Mannschreck, A. *J. Chromatogr.* (1983), *282,* 89
[73] Mintas, M., Mannschreck, A., Schneider, M.P. *J. Chem. Soc. Commun.* (1979), 602
[74] Rizzi, A.M. *J. Chromatogr.* (1990), *513,* 192
[75] Okamoto, Y., Kawashima, M., Aburatani, R., Hatada, K., Nishiyama, T., Masuda, M. *Chem. Lett.* (1986), 1237
[76] Ichida, A., Shibata, T., Okamoto, Y., Yuki, Y., Namikashi, H., Toga, Y. *Chromatographia* (1984), *19,* 280
[77] Francotte, E., Wolf, R.M., Lohmann, D., Mueller, R. *J. Chromatogr.* (1985), *347,* 25
[78] Wainer, I.W., Alembik, M.C. *J. Chromatogr.* (1986), *358,* 85
[79] Wainer, I.W., Stiffin, R.M., Shibata, T. *J. Chromatogr.* (1988), *411,* 139
[80] Aboul-Einen, H.Y., Islam, M.R. *J. Liq. Chromatogr.* (1990), *13,* 485
[81] Okamoto, Y., Kawashima, M., Hatada, K. *J. Am. Chem. Soc.* (1984), *106,* 5357
[82] Okamoto, Y., Kaida, Y. *J. High Res. Chromatogr.* (1990), *13,* 708
[83] Okamoto, Y., Senoh, T., Nakane, H., Hatada, K. *Chirality* (1989), *1,* 216
[84] Blaschke, G. In *Chromatographic Chiral separations,* Zief, M., Crane, L.J., Eds., Dekker: New York, 1988, Chapter 7, p. 179
[85] Blaschke, G., Broker, W., Fraenkel, W. *Angew. Chem. Int. Ed.* (1986), *25,* 830
[86] Kinkel, J.N., Fraenkel, W., Blaschke, G. *Kontakte* (1987), *1,* 3

[87] Huffer, M., Schreier, P. *J. Chromatogr.* (1989), *469*, 137; (a) Hünig, S., Klaunzer, N., Günther, K. *J. Chromatogr.* (1989), *481*, 387; (b) Mosandl, A., Hollnagel, A. *Chirality* (1989), *1*, 293; (c) Yang, S.K., Lu, X.L. *J. Pharm. Sci.* (1989), *78*, 789; (d) Krause, M., Galensa, R. *Lebensmittelchem. Gerichtl. Chem.* (1989), 43, 1; (e) Kinkel, J.N. *GIT Supplement Chromatographie* (1988), *3*, 29

[88] Okamoto, Y., Honda, S., Hatada, K., Yuki H. *J. Chromatogr.* (1985), *350*, 127

[89] Okamoto, Y., Hatada, K. *J. Liq. Chromatogr.* (1986), *9*, 369

[90] Yuki, H., Okamoto, Y., Okamoto, I. *J. Am. Chem. Soc.* (1980), *102*, 6356

[91] Okamoto, Y., Okamoto, T., Yuki, A., Murata, S., Noyori, R., Takaya, H. *J. Am. Chem. Soc.* (1981), *103*, 6971

[92] Okamoto, Y. *Bull. Chem. Soc. Japan* (1988), *61*, 255

[93] Saotome, Y., Miyazawa, T., Endo, T. *Chromatographia* (1989), *28*, 505, Parts II-IV, 509

[94] Schomburg, G. *LC-GC* (1988), *6*, 37

[95] Wainer, I.W., Alembik, M.C., Smith, E. *J. Chromatogr.* (1987), *388*, 65

[96] Menges, R.A., Armstrong, D.W. In *Chiral Separations by Liquid Chromatography*, Ahuja, S., Ed., ACS Symposium Series 471, American Chemical Society: Washington, DC, 1991, Chapter 4, p. 67

[97] Bender, M.L., Komiyama, M. *Cyclodextrin Chemistry*, Springer: Berlin, 1978, p. 1

[98] Fujimura, K., Veda, T., Ando, T. *Anal. Chem.* (1983), *55*, 446

[99] Armstrong, D.W., DeMond, W. *J. Chromatogr. Sci.* (1984), *22*, 411

[100] Armstrong, D.W., Chang, C.D., Lee, S.H. *J. Chromatogr.* (1991), *539*, 83

[101] Armstrong, D.W., DeMond, W., Czech, B.P. *Anal. Chem.* (1985), *57*, 481

[102] Han, S.H., Han, Y.I., Armstrong, D.W. *J. Chromatogr.* (1988), *441*, 376

[103] Armstrong, D.W., Han, Y.I., Han, S.M. *Anal. Chim. Acta* (1988), *208*, 275

[104] Stalcup, A.M. Jin, H.L., Armstrong, D.W. *J. Liq. Chromatogr.* (1990), *13*, 473

[105] Armstrong, D.W., Ward, T.J., Armstrong, R.D., Beesley, T.E. *Science* (1986), *232*, 1132

[106] Armstrong, R.D., Ward, T.J., Pattabiraman, N:, Benz, C., Armstrong, D.W. *J. Chromatogr.* (1987), *414*, 192

[107] Armstrong, D.W., Xang, X., Han, S.M., Menges, R.A. *Anal. Chem.* (1987), *59*, 2594

[108] Stalcup, A.M., Chang, S.C., Armstrong, D.W., Pitha, J. *J. Chromatogr.* (1990), *513*, 181

[109] Beesley, T.E. *American Laboratory* (1985), *5*, 78

[110] Stalcup, A.M., Faulkner, J.R., Tang, Y., Armstrong, D.W., Levy, L.W., Regaldo, E. *Biomedical Chromatogr.* (1991), *5*, 3

[111] Armstrong, D.W., Stalcup, A.M., Hilton, M.L., Duncan, J.D., Faulkner, J.R., Chang, S.C. *Anal. Chem.* (1990), *62*, 1610

[112] Pirkle, W.H., Pochapsky, T.C. *J. Am. Chem. Soc.* (1986), *108*, 352

[113] Armstrong, D.W., Chang,. C.D., Lee, S.H. *J. Chromatogr.* (1991), *539*, 83

[114] Berthod, A., Jin, H.L., Beesley, T.E., Duncan, J.D., Armstrong, D.W. *J. Pharm. Biomed. Anal.* (1990), *8(2)*, 123

[115] Armstrong, D.W., Han, S.M., Han, Y.I. *Anal. Biochem.* (1987), *167*, 261

[116] Aboul-Eneim, H.Y., Islam, M.R., Bakr, S.A. *J. Liq. Chromatogr.* (1988), *11(7)*, 1485

[117] Geisslinger, G., Dietzel, K., Leow, P., Schuster, O., Rau, G., Lachmann, G., Brune, K. *J. Chromatogr.* (1989), *491*, 139

[118] Marziani, F.C., Sisco, W.r. *J. Chromatogr.* (1989), *465*, 422

[119] McClanahan, J.S., Maguire, J.H. *J. Chromatogr.* (1986), *381*, 438
[120] Maguire, J.H. *J. Chromatogr.* (1987), *387*, 453
[121] Chandler, M.H.H., Guttendorf, R.J. Blouin, R.A., Wedlend, P.J. *J. Chromatogr.* (1988), *426*, 417
[122] Krstulovic, A.M., Gianviti, J.M., Burke, J.T., Mompon, B. *J. Chromatogr.* (1988), *426*, 417
[123] Armstrong, D.W., Jin, H.L. *J. Chromatogr.* (1989), *462*, 219
[124] Jin, H.L., Stalcup, A.M., Armstrong, D.W. *J. Liq. Chromatogr.* (1988), *11(6)*, 3295
[125] Armstrong, D.W., Jin, H.L. *Chirality* (1989), *1*, 27
[126] Kyba, E.P., Timko, J.M., Kaplan, L., de Jong, F., Gokel, G.W., Cram, D.J., *J. Am. Chem. Soc.* (1978), *100*, 4555
[127] Sousa, L.R., Sogah, G.D.Y., Hoffmann, D.H., Cram, D.J., *J. Am. Chem. Soc.* (1978), *100*, 4569
[128] Wainer, I.W. In *Chiral Liquid Chromatography*, Lough, W.J., Ed., Blackie: Glasgow, 1989, Chapter 7, p. 129
[129] Hermansson, J. *J. Chromatogr.* (1983), *269*, 71
[130] Allenmark, S. and B. Bomgren, *J. Chromatogr.* (1983), *365*, 63
[131] Miwa, T., M. Ichikawa, M. Tsuno, T. Hattori, T. Miyakawa, M. Kayano and Y. Miyake, *Chem. Pharm. Bull.* (1987), *35*, 682
[132] Miwa, T., T. Miyakawa, M. Kayano and Y. Miyake, *J. Chromatogr.* (1987), *408*, 316
[133] Foster, J.F. In *The Plasma Proteins*, Ed. K. Schmid, Academic Press: New York, 1975; (a) Wainer, I.W., Alembik, M.C. In *Chromatographic Chiral Separations*, Zief, M., Crane, L.J., Eds., Dekker: New York, 1988
[134] Schill, G., I.W. Wainer and S.A. Barkan, *J. Liq. Chromatogr.* (1986), *9*, 641
[135] Schill, G., I.W. Wainer and S.A. Barkan, *J. Chromatogr.* (1986), *365*, 73
[136] Hermansson, J. *J. Chromatogr.* (1985), *325*, 379
[137] Hermansson, J. and M. Eriksson *J. Liq. Chromatogr.* (1986), *9*, 621
[138] Wainer, I.W., Barkan, S.A., Schill, G. *L.C./G.C.* (1986), *4*, 422
[139] Stewart, K.K. and R.F. Doherty, *Proc. Natl. Acad. Sci. USA*, (1973), *70*, 2850
[140] Allenmark, S., *J. Liq. Chromatogr.* (1986), *9*, 425
[141] Allenmark, S. and B. Bomgren, *J. Chromatogr.* (1982), *252*, 297
[142] Allenmark, S., B. Bomgren and H. Boren, *J. Chromatogr.* (1984), *316*, 617
[143] Wainer, I.W., Stiffin, R.M. *J. Chromatogr.* (1988), *424*, 158
[144] Nakamura, Y., Yamagishi, A., Matumoto, S., Tohkubo, K., Ohtu, Y., Yamaguchi, M. *J. Chromatogr.* (1989), *482*, 165
[145] Information sheet of Shiseido
[146] Adam, W., Hoch, U., Saha-Möller, C.R., Schreier, P. *Angew. Chem. Int. Ed. Engl.* (1993), 32, 1737

3.5 Gas chromatography

A number of monographs and reviews including detailed introductions and fundamental descriptions have been published on the gas chromatographic separation of enantiomers (Table 3.5.1). However, these publications are more or less specialized presentations of groups of applications or types of stationary phases. A particular overview about chiral stationary phases (CSP) for, e.g., flavour and pheromone analysis, does not exist; only a few groups of stationary phases have been reported on [1-3]. Therefore, besides the historical development of gas chromatographic enantiomer separation, in the following classification particular applications in the field of flavour and pheromone analysis are documented. In addition, information on many other important chiral compounds of pharmaceutical, pharmacological, or toxicological interest, as well as pesticides or auxiliaries for asymmetric syntheses, is given. Finally, the formation of diastereomeric derivatives with chiral derivatization agents (CDA) for enantiomeric analysis on achiral stationary phases will be discussed.

Table 3.5.1 The gas chromatographic separation of enantiomers (literature overview).

A. Monographs	Author(s)	Year	Ref.
'Asymmetric Synthesis'	Morrison	1983	[4]
'Chromatographic Separations of Stereoisomers'	Souter	1985	[5]
'The Science of Chromatography'	Bruner	1985	[6]
'The Practice of Enantiomer Separation by Capillary Gas Chromatography'	König	1987	[7]
'Chromatographic Chiral Separations'	Zief/Crane	1988	[8]
'Chiral Separations'	Stevenson/Wilson	1988	[9]
'Recent Advances in Chiral Separations'	Stevenson/Wilson	1990	[10]
'Chromatographic Enantioseparation'	Allenmark	1991	[11]
'Gas Chromatographic Enantiomer Separation with Modified Cyclodextrins'	König	1992	[12]

B. Reviews	Year	Ref.	Reviews	Year	Ref.
Gil-Av	1975	[13]	Lochmüller/Souter	1975	[14]
Smolková-Keule.	1982	[15]	König	1982	[16]
Liu/Ku	1983	[17]	Schurig	1984	[18]
Schurig	1986	[19]	Szejtli	1987	[20]
Armstrong/Han	1988	[21]	König	1989	[22]
Schurig/Nowotny	1990	[23]	König	1990	[24]
Brandt	1991	[25]	Werkhoff et al.	1991	[1]
König	1992	[26]	Werkhoff et al.	1993	[3]

Thermodynamic parameters of gas chromatography

The gas chromatographic separation of enantiomers is a thermodynamically controlled process. The discrimination of enantiomers by a racemic selectand and an optically active selector is described by the Gibbs-Helmholtz relation. This relation is equivalent to the sum of the differences of the enthalpy and the entropy during the association of the enantiomers with the chiral phase [23]:

$$-\Delta_{R,S}(\Delta G^0) = -\Delta_{R,S}(\Delta H^0) + T\,\Delta_{R,S}(\Delta S^0)$$

$-\Delta_{R,S}(\Delta G^0)$ = difference in the free association enthalpy [J mol^{-1}]
$-\Delta_{R,S}(\Delta H^0)$ = difference in the association enthalpy [J mol^{-1}] (formally, the *S*-enantiomer elutes faster than the *R*-enantiomer)
$\Delta_{R,S}(\Delta S^0)$ = difference in the association entropy [J mol^{-1} K^{-1}]

The thermodynamic parameters can be determined from the retention parameters [28]. The difference in the free association enthalpy $-\Delta_{R,S}(\Delta G^0)$ (enantiodiscrimination) of a pair of enantiomers can be calculated by using the difference in the association constants for the combination of both enantiomers of the selectand to the optically active selector:

$$-\Delta_{R,S}(\Delta G^0) = R\,T \ln \alpha_{max}$$

T = retention temperature [K]
R = gas constant [8.314510 J mol^{-1} K^{-1}]
α_{max} = maximal separation factor valid for quantitative enantiomeric purity (calculation of α, cf. Section 3.3).

The Gibbs-Helmholtz parameters $-\Delta_{R,S}(\Delta H^0)$ and $\Delta_{R,S}(\Delta S^0)$ of the chiral recognition are determinable due to the temperature dependence of the separation factor α_{max} :

$$R \ln \alpha_{max} = [-\Delta_{R,S}(\Delta H^0) \,/\, T] + \Delta_{R,S}(\Delta S^0).$$

A plot of R ln α_{max} against 1/T gives a straight line ('Van't Hoff plot'), the slope of which shows the difference in the association enthalpies; the axial section is equal to the difference in the association entropies of the enantiomers, which are highly diluted in the stationary phase. A low value for $-\Delta_{R,S}(\Delta H^0)$ will lead to a low temperature dependance of the enantiomeric separation.

Watabe et al. [29] as well as Schurig et al. [30] observed for the first time the phenomenon of temperature-dependent inversion of the elution order. The temperature at which a peak coalescence of the third kind (see Section 3.3) occurs is called the isoenantioselective temperature T_{iso} (coalescence temperature), i.e. at this temperature there is no separation of enantiomers possible. Therefore, α_{max} = 1 and consequently ln α_{max} = 0 and $T_{iso} = \Delta_{R,S}(\Delta H^0) \,/\, \Delta_{R,S}(\Delta S^0)$.

For a better understanding of the mechanisms of enantiomeric discrimination, the above-mentioned Gibbs-Helmholtz relation is important, especially for the explanation of phenomena such as the inversion of elution orders. Since in the practice mainly gas chromatographic (retention) parameters are used, in the following, the use of thermodynamic parameters will be omitted. Moreover, the separation factor α is directly proportional to $-\Delta_{R,S}(\Delta G^0)$. Using α in connection with the elution temperature, the data obtained from enantiomeric separations can be compared with thermodynamic data reported in the literature.

To compare unknown substances with authentic references the linear retention index I is very useful under the (mostly used) programmed temperature GC conditions. The index I (or R_i) gives the retention of a compound in relation to that of n-alkanes (the number of carbon atoms is multiplied by 100) [31]:

$$I = 100\,[(t'_{Rx} - t'_{Rz}) / (t'_{Rz+1} - t'_{Rz})] + 100\,z$$

t'_{Rx} = net retention time of the compound of interest [min]

t'_{Rz} = net retention time of the alkane which elutes before the compound of interest [min]

t'_{Rz+1} = net retention time of the alkane which elutes after the compound of interest [min]

z = number of carbon atoms of the alkane which elutes before the compound of interest.

References

[1] Werkhoff P., Brennecke S., Bretschneider W. *Chem. Mikrobiol. Technol. Lebensm.* (1991), *13*, 129

[2] Werkhoff P., Brennecke S., Bretschneider W., Güntert M., Hopp R., Surburg H. *Z. Lebensm. Unters. Forsch.* (1993), *196*, 307

[3] Mosandl A. *J. Chromatogr.* (1992), *624*, 267

[4] Morrison J.D. (Ed.) *Asymmetric Synthesis. Volume 1. Analytical Methods*, Academic Press: New York, 1983

[5] Souter R.W. *Chromatographic Separations of Stereoisomers*, CRC Press: Boca Raton, 1985

[6] Bruner F. (Eds.) *The Science of Chromatography* J.Chromatogr. Libr. Vol. 32, Elsevier: Amsterdam, Oxford, New York, Tokyo, 1985

[7] König W.A. *The Practice of Enantiomer Separation by Capillary Gas Chromatography*, Hüthig: Heidelberg, 1987

[8] Zief M., L.J. Crane (Eds.) *Chromatographic Chiral Separations*, Dekker: New York, Basel, 1988

[9] Stevenson D., Wilson I.D. (Eds.) *Chiral Separations*, Plenum Press: New York, London, 1988

[10] Stevenson D., Wilson I.D. (Eds.) *Recent Advances in Chiral Separations*, Plenum Press: New York, London, 1990

[11] Allenmark S. *Chromatographic Enantioseparation: Methods and Applications*, 2nd Ed., Horwood: Chichester, 1991

[12] König W.A. *Gas Chromatographic Enantiomer Separation with Modified Cyclodextrins,* Hüthig: Heidelberg, 1992
[13] Gil-Av E. *J. Mol. Evol.* (1975), *6*, 131
[14] Lochmüller C.H., Souter R.W. *J. Chromatogr.* (1975), *113*, 283
[15] Smolková-Keulemansová E. *J. Chromatogr.* (1982), *251*, 17
[16] König W.A. *J. High Resol. Chromatogr. Chromatogr. Commun.* (1982), *5*, 588
[17] Liu R.H., Ku W.W. *J. Chromatogr.* (1983), *271*, 309
[18] Schurig V. *Angew. Chem.* (1984), *96*, 733
[19] Schurig V. *Kontakte (Darmstadt)* (1986), *1*, 3
[20] Szejtli J. *Stärke* (1987), *39*, 357
[21] Armstrong D.W., Han S.M. *Crit. Rev. Anal. Chem.* (1988), *19*, 175
[22] König W.A. *Nachr. Chem. Tech. Lab.* (1989), *37*, 471
[23] Schurig V., Nowotny H.-P. *Angew. Chem.* 1990, *102*, 969
[24] König W.A. *Kontakte (Darmstadt)* (1990), 2, 3
[25] Brandt K. *Nachr. Chem. Tech. Lab.* (1991), *39*, M1-M14
[26] König W.A. in: *Drug Stereochemistry: Analytical Methods and Pharmacology,* Wainer I.W. (Ed.), 2nd Ed., Dekker: New York, 1992, p 107
[27] König W.A. *Trends Anal.Chem.* (1993), *12*,130
[28] Schurig V., Chang R.C., Zlatkis A., Feibush B. *J. Chromatogr.* (1974), *99*, 147
[29] Watabe K., Charles R., Gil-Av E. *Angew. Chem.* (1989), *101*, 195
[30] Schurig V., Ossig J., Link R. *Angew. Chem.* (1989), *101*, 197
[31] Günther W., Schlegelmilch F. (Eds.) *Gaschromatographie mit Kapillar-Trennsäulen. Grundlagen* Vol.1, Vogel: Würzburg, 1984

3.5.1 *Enantiomeric separation via diastereomeric derivatives*

Racemic mixtures can be resolved into their enantiomers using two different ways [1,2]: (i) *Indirect method:* Conversion of the enantiomers into diastereomeric derivatives by reaction with an enantiomerically pure chiral auxiliary, with subsequent gas chromatographic separation of diastereomers on an achiral stationary phase. (ii) *Direct method:* Gas chromatographic separation of enantiomers on a chiral stationary phase that contains an optical auxiliary in high (but not necessarily absolute) enantiomeric purity.

For a long time derivatization to diastereomers (indirect method) was the only means of achieving gas chromatographic separation of enantiomers. This method was originally developed by Bailey and Hass [3]. Even at that time agents like 2-acetyl lactic acid or (-)-menthol were used for derivatization of volatile compounds. The diastereomers formed were partially separated via rectification. Weygand et al. [4] reported the gas chromatographic diastereomer separation of a synthesized N-TFA-dipeptide methyl ester (made with optically pure *S*-alanine and *R*-containing *S*-phenylalanine). Later, Casanova and Corey [5] described the first specific use of the indirect method. They separated racemic camphor into its enantiomers as cyclic acetals after derivatization with *R,R*-2,3-butanediol.

Virtually every separation problem involving enantiomeric compounds can be solved with CDAs. Compounds like alcohols, polyols, amines, aldehydes, ketones, carbohydrates, alkyl acids, hydroxy acids, amino acids, esters, and even hydrocarbons are accessible to enantiomeric separation. Further, some CDAs were developed to improve detection limits or to allow detection with an electron capture detector (ECD) such as *S*-N-heptafluorobutanoylprolyl chloride [6].

More than 100 different CDAs have been used so far in gas chromatography. But nowadays the indirect method is being increasingly replaced by the direct method. There are, however, still certain problems that cannot be solved by using chiral stationary phases. This means that there is still need for the development of new chiral derivatization agents [7-10].

Insufficient optical purity of the derivatization agent or incomplete conversion reaction with the sample, racemization, random fractionation during sample preparation, as well as longer sample preparing times, are the main disadvantages of the indirect method. On the other hand, the freedom to use conventional columns can be regarded as an advantage. Most of the diasteriomeric pairs can be resolved on common achiral high resolution capillaries. In Table 3.5.2 the most important derivatization agents for chiral compounds are listed and examples for their use are given.

Table 3.5.2 Chiral derivatization agents (CDA) for gas chromatographic analysis (* = commercially available; ** = both enantiomers commercially available; Suppliers: Aldrich, Fluka, Sigma).

CDA	Derivative	Examples
Alkanols		
R-(-)-2-Butanol [11] [CDA-1]** (Aldrich)	Esters	2-Hydroxycarboxylic acids (ac.) [12]; aldonic acids (ac.) [12]; hexuronic acids (TMS) [13]; 2-hydroxydicarboxylic acids (ac.) [12]; α-amino acids (N-TFA) [11]; α-hydroxymethyl-α-amino acids (N,O-PFP) [14]
	Glycosides	Monosaccharides (pentoses and hexoses) (TMS) [15]; 2-acetamido-2-deoxymonosaccharides (hexoses) (TMS) [13]
2*R*,3*R*-(-)-2,3-Butanediol [5] [CDA-2]** (Aldrich)	Cyclic acetals	Cycloalkanones [16]; terpene ketones (e.g., menthone, camphor [5,16]; terpene hydrocarbons (e.g. α-pinene, limonene) (after oxidation to ketones) [17]
	Orthoesters	5-Alkanolides [18]
S-(+)-3-Methyl-2-butanol [12] [CDA-3]* (Aldrich)	Esters	2-Alkylcarboxylic acids [19]; 2-hydroxycarboxylic acids (ac.); 2-hydroxycarboxylic acids (TFA or TMS) [19]; aldonic acids and 2-hydroxydicarboxylic acids (ac.) [12]; α-amino acids (N-PFP) [20]; N-methyl-α-amino acids [21]
1*R*,3*R*,4*S*-(-)-Menthol [22] [CDA-4]** (Aldrich)	Esters	Sec. alkylcarboxylic acids (e.g. phytanic acid [23] or chrysanthemic acid [24]); 2-arylcarboxylic acids [25]; sec. hydroxycarboxylic acids [26]; *E*,*Z*-3-,ethyl-phenyl glycidic acid [27]; α-amino acids (N-TFA) [22]; insecticides (e.g., fenvalerate] (as 2-chlorophenyl-3-methylbutanoic acid after saponification [28]); herbicides (e.g., mecoprop) [29]

CDA	Derivative	Examples
Amines		
R-(+)-1-Phenylethyl-amine (PEA) [30] [CDA-5]** (Aldrich)	Amides	Sec. alkylcarboxylic acids (e.g. chrysanthemic acid) [31]; 2-alkyl-1-alkanols(e. g. lavandulol, citronellol) (after oxidation to carboxylic acids) [32]; 2-hydroxycarboxylic acids (methylether ot TFA or TMS) [30]; pharmaceuticals (e. g. 2-[4-(1-oxo-2-isoindolinyl)phenyl] propanoic acid [33] or ibuprofen [34] or benoxaprofen [35])
	Amino-alditols	Monosaccharides (aldoses, ketoses, deoxyhexoses, N-acetyl-hexosamines (after reduction; TMS or ac.) [36]
S-(+)-Amphetamine [CDA-6]* [37] (Sigma)	Amides	Sec. arylcarboxylic acids (e. g,. ibuprofen) [37]; sec. alkyl-carboxylic acids [37a]
	Pyrrol der.	Polygodial [38]
Acids		
S-Acetyl lactic acid [CDA-7] [39]	Esters	2-Alkanols [39]
S-(-)-Trolox™ methyl ether [8] [CDA-8]* (Fluka)	Esters	Sec. Alkanols [8]; 2-alkyl-1-alkanols (e.g., citronellol) [8]
Acid anhydrides		
R-2-Phenylbutanoic acid anhydride [40] [CDA-9] (**) (Fluka; acid only)	Amides	Sec. amines (e.g., amphetamine and related compounds [40]); aminoalkanols (e.g., ephedrine (O-TMS) [41]; α-amino acids (ME) [40]
Acid chlorides		
S-(-)-Acetyl lactyl chloride (ALC) [42] [CDA-10]* (Fluka)	Esters	2-Alkanols [42]; 3-alkanols [43]; 1-alken-3-ols [44]; terpeneoic alkanols (e.g., borneol) [45]; cyanhydrines [42]; 4-alkanolides (after ring opening as ME) [47]
	Diesters	1,3-Alkanediols [46]; 1-thio-3-alkanols [46]; 4- and 5-alkanolides (after reduction to 1,4- and 1,5-alkanediols) [46]; 3-methyl-4-alkanolides (after reduction to 3-methyl-1,4-alkane-diols) [48]
1*S*,3*S*-*E*-Chrysanthem-oyl chloride [49] [CDA-11]	Esters	Sec. alkanols (e.g., menthol, pantolactone) [49]; carboxylic acids (ME) [49]; fungicides (e.g., bitertanol, diclobutrazol, triadimenol) [50]
	Amides	Sec. amines (e.g., 1-PEA [51] and amphetamine [49]); α-amino acids (ME) [49]
1*S*,4*R*-(-)Camphanoyl chloride [52] [CDA-12]** (Aldrich)	Esters	2-Alkanols [52]; 3-hydroxycarboxylic acid esters [52]
	Amides	Sec. arylamines (e.g., 1-(1,3-benzodioxol-5-yl)-2-aminoalkanes) [53]; nicotine and related compounds (after demethylation) [54]; alkylpyrrolidines and alkylpiperidines [55]
(-)-α-Teresantalinoyl chloride [56] [CDA-13]	Amides	α-Amino acid esters [56]

CDA	Derivative	Examples
1*R*,3*R*,4*S*-(-)-Menthyl chloroformate (MCF) [CDA-14]** [57] (Aldrich)	Carbonates	Sec. alkanols [57]; 2-hydroxycarboxylic acids (ME) [57]; 2-hydroxydicarboxylic acids (ME) [58]; pantolactone [59]; 1,4,5,6-tetrachlorobicyclo[2.2.1]hept-5-en-2-ol and related compounds [60]
	Carbamates	Sec. amines (e.g., 1-PEA [57] and methadone [61]); α-amino acids (ME) [57]
Drimanoyl chloride [CDA-15] [49]	Esters	Sec. alkanols (e.g., menthol) [49], hydroxycarboxylic acids (ME) [49]; pantolactone
	Amides	Sec. amines (e.g., amphetamine) [49]; α-amino acids (ME) [49]
3β-Acetoxy-Δ^5-etiocholenic acid chloride [62] [CDA-16] (*) (Fluka; acid only)	Esters	Sec. alkanols (e.g., 1-phenyl-2,2,2-trinonylfluoroethanol [62]
S-2-Phenylpropanoyl chloride [63] [CDA-17] (**) (Aldrich; acid only)	Esters	Sec. alkanols [63]; 2- and 3-hydroxycarboxylic acids (ME) [63]; 4-hydroxycarboxylic acids (iPE) [64]; 5-alkanolides (after ring opening as iPE) [65]
R-(-)-2-Methyoxy-2-tri-fluoromethyl-2-phenyl acetyl chloride [66] (MTPA) [CDA-18]** (Aldrich)	Esters	Sec. alkanols [67]; terpenoic alkanols (e.g., menthol) [68]; 1,2-ketols (e.g., acetoin) [69]; methyl-2-hydroxyheptanoate; dimethyl-2-hydroxyheptandioate [70]; 3-hydroxycarboxyl acid esters [71]; 2- and 4-hydroxycarboxylic acid esters [72]; 4-alkanolides (after ring opening as EE) [72]-
	Amides	Sec. amines (e.g., amphetamine [73] and tocainidine [74])
	Diamides	Sec. diaminoalkanes [75]
S-Naproxen chloride [CDA-19] (*) [76] (Aldrich; acid only)	Amides	Sec. amines (e.g., tocainide) [76]; α-amino acids (ME) [77]
S-Tetrahydro-5-oxo-2-furancarboxyl acid chloride (TOF) [78] [CDA-20] (**) (Aldrich; acid only)	Esters	Sec. alkanols and alkenols [78]; 2-, 3-. 5-hydroxycrboxylic acid esters [79]; 5-alkanolides (after ring opening as iPE) [80]
	Diesters	1,4-Alkanediols [79]
S-(-)-N-TFA-Prolyl chloride (TPC) [81] [CDA-21]* (Aldrich) (also PFP and HFB; better for ECD) [6;82]	Esters	Sec. alkanols (e.g., menthol) [83]; N-hydroxy-1-arylisopropylamines (e.g., N-hydroxyamphetamine) [84]
	Amides	Sec. amines (e.g., 2-amino-3-methylpentane [85] and amphetamine [86]); cyclic amines (e.g., 3-methylpyrrolidine [55], 2-methylpiperidine [87] and 2-methylindoline [88]), octahydro-1-(4-methoxybenzyl)isoquinoline [89]; pharamceuticals (e.g., propranolol and related compounds) (O-TMS) [82], tranylcypromine [90]
	Peptides	α-Amino acids (ME) [81]
1*R*,4*R*-(+)-Camphor-10-sulfonyl chloride [91] [CDA-22]** (Aldrich)	Amides	Sec. arylalkylamines (e.g., 1-PEA) [91]

CDA	Derivative	Examples
Esters		
2*R*,3*R*-(+)-Dimethyl tartrate [92] [CDA-23]** (Aldrich)	Cyclic acetals	Ketones (e.g., 3,3,5-trimethylcyclohexanone [92] and menthone [93]
2*R*,3*R*-(+)-Diisopropyl tartrate [7] [CDA-24]** (Aldrich)	Cyclic acetals	Acyclic terpenoic aldehydes (e.g., 3,7-dimethyloctanal) [7]
R-(-)-Cysteine methyl ester [94] [CDA-25]* (Fluka)	Thiazolidine	Monosaccharides (aldoses) (TMS) [94]
S-(+)-Methyl3-iodo-2-methylpropanoate [10] [CDA-26]	Ether	Sec. alkanols [10]
Isocyanates		
R-(+)-1-Phenylethyl-isocyanate (PEIC) [95] [CDA-27]** (Aldrich)	Carbamates	Sec. alkanols nd alkenols (e.g., menthol) [95]; tert. alkanols (e.g., linalool) [96]; 2-, 3-, 4- and 5-hydroxycarboxylic acid esters [97]; ω-2-hydroxycarboxylic acids (ME) [98]; 5-hydroxy-4-alkanolides [99]; 4- and 5-alkanolides (after ring opening as EE) [97]; 4- and 5-alkanolides (after ring opening as N-bütyl amides) [100]
	Dicarbamates	Alkanediols (e.g., *E*-2,3-butanediol) [69]; 4- and 5-alkanolides (after reduction to 1,4- and 1,5-alkanediols) [79]
	Amides	2-Alkylcarboxylic acids [101]
	Ureides	Propranolol (O-TMS) [102]
R-(-)-1,1'-Naphthyl-ethyl isocyanate (NEIC) [103] [CDA-28]** (Aldrich)	Carbamates	Sec. alkanols [103]; tert. alkanols (e.g., linalool) [104]
Miscellaneous		
R-(-)Pantolactone [CDA-29] [105]* (Aldrich)	Esters	2-Alkyl- and 2-arylcarboxylic acids [106]; α-amino acids (N-isobutyloxy-carbonyl) [105]
1,2,3,4,6-Pentaacetyl β-D-(+)-glucopyranoside [CDA-30] [107]* (Aldrich)	Glucosides	Sec. Alkanols (e.g., menthol and borneol) [107]
1*R*,3*R*,4*S*-(-)-Menthyl hydroxylammonium chloride [CDA-31] [108] (**) (Aldrich; menthol only)	Oximes	Monosaccharides (aldoses) (TFA) [108] , ketoses (TFA) [109]
R-(+)-2,2,2-Trifluoro-1-phenylethyl hydrazine [CDA-32] [110]	Hydrazones	Terpene ketones (e.g., camphor and menthone) [110]; androstanone der. [110]

CDA	Derivative	Examples
S-(-)-N-Pentafluoro-benzoylprolyl-1-imidazolide (for ECD) [CDA-33] [111]	Amides	Sec. amines (e.g., 1-PEA and amphetamine) [111]
(+)-1-Phenyl-ethanethiol [CDA-34] [112]	Dithio-acetals	Monosaccharides (ac.) or TMS [112]

ac. = acetyl, CDA = chiral derivatization agent, der. = derivative, EE = ethyl ester, HFB = heptafluorobutanoyl, iPE = isopropyl ester, ME = methyl ester, PEA = phenylethylamine, PFP = pentafluoropropanoyl, TFA = trifluoroacetyl, TMS = trimethylsilyl

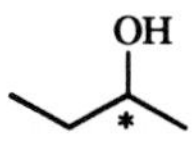

CDA-1 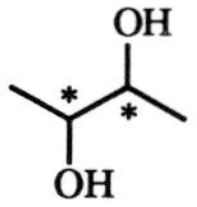 CDA-2 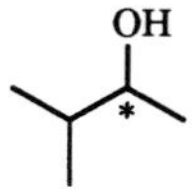 CDA-3 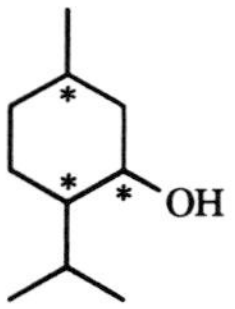 CDA-4 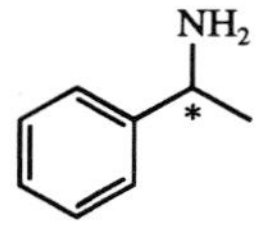 CDA-5 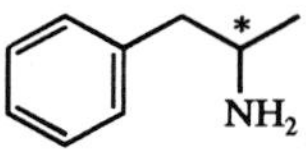 CDA-6

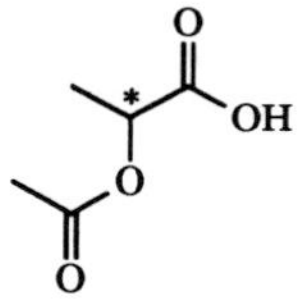

CDA-7 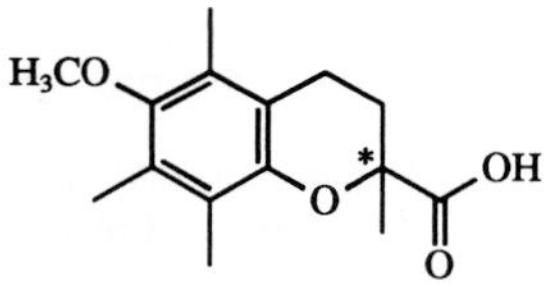 CDA-8 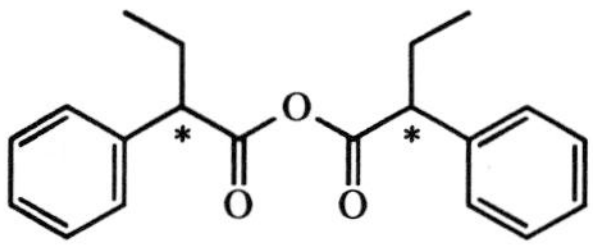 CDA-9 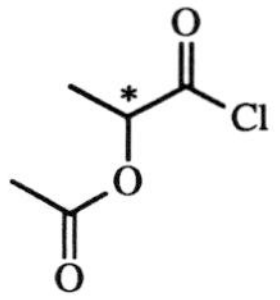 CDA-10

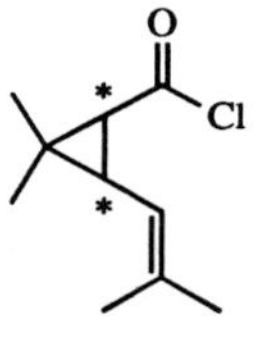

CDA-11 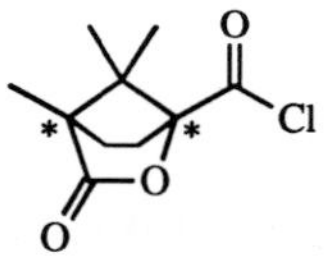 CDA-12 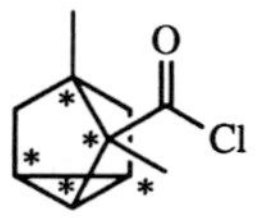 CDA-13 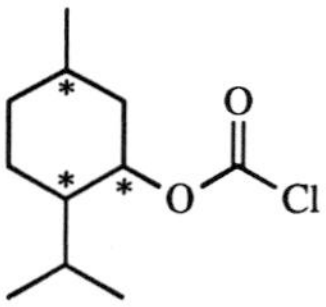 CDA-14 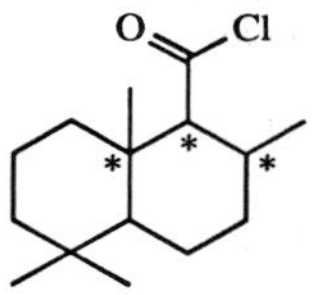 CDA-15

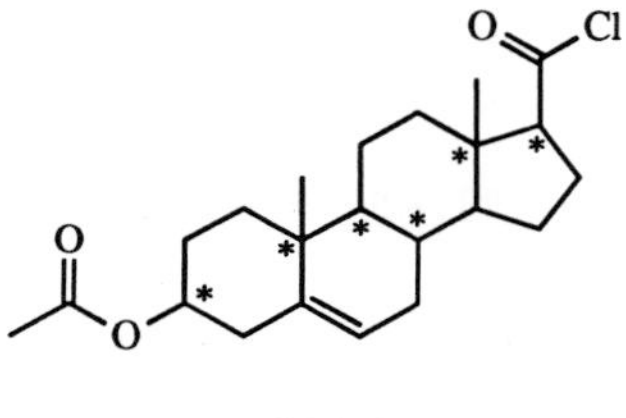

CDA-16 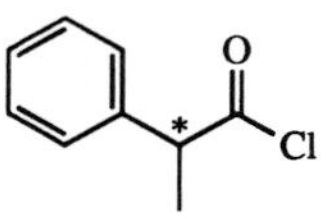 CDA-17 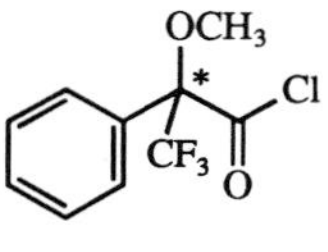 CDA-18 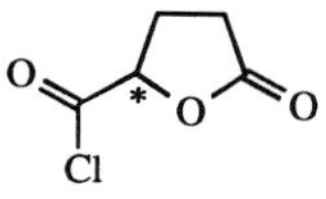 CDA-20

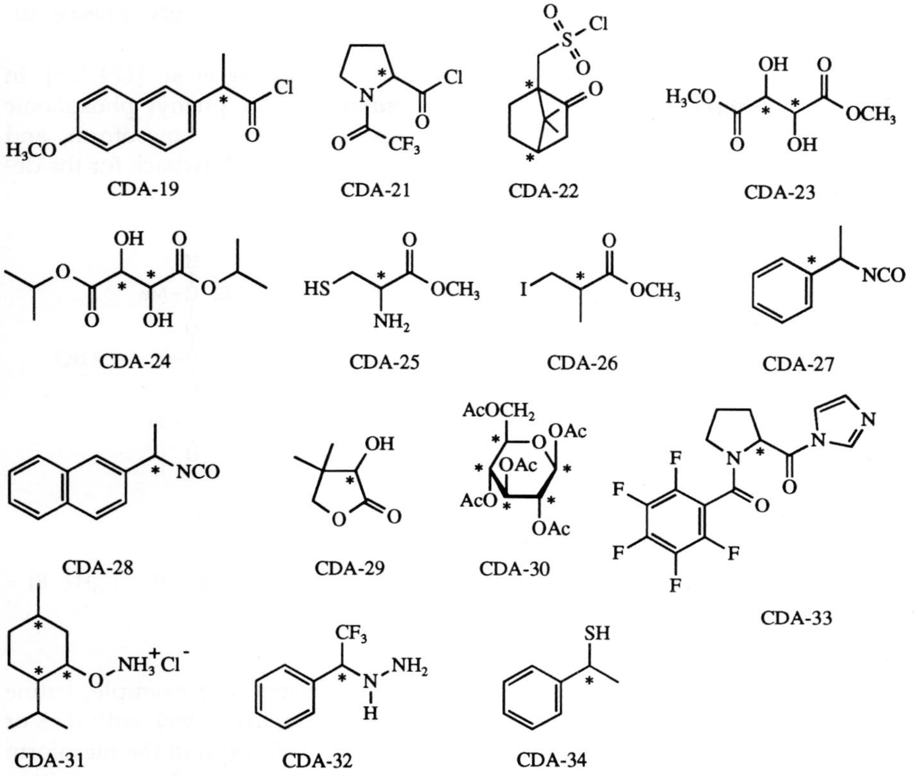

CDA-19 CDA-21 CDA-22 CDA-23

CDA-24 CDA-25 CDA-26 CDA-27

CDA-28 CDA-29 CDA-30 CDA-33

CDA-31 CDA-32 CDA-34

Figure 3.5.1 Structures of CDAs used for gas chromatographic analysis (cf. Table 3.5.2).

Detailed information is provided in several reviews [113,114] and in various monographs [115-117]. Finally, it should be mentioned that diastereomeric derivatives have been used together with chiral stationary phases in order to improve separation efficiency [21,55,118-121].

To avoid the problem of optical impurity of derivatization agents, Blink et al. [122] have introduced the achiral bifunctional agent phenyl phosphonic dichloride for esterification of 2-butanol. The resulting phosphonate contains three chiral centres (two centres due to the two 2-butyl groups and one due to the pseudo asymmetric phosphorous [123]). This leads to three peaks in a chromatogram. Two of them indicate optically inactive mesoforms and one an optically active form (Figure 3.5.2). Since during the esterification with optically pure 2-butanol no mesoforms are produced, the enantiomeric excess of 2-butanol with a lower optical purity can easily be calculated via the ratio of both mesoform peaks to the peak of the optically pure form. But it must be clear which enantiomer is predominant. For example, a ratio of 60% *R*- to 40% *S*-2-butanol produces the same peak shapes and peak ratios as a ratio of 40% *R*- to 60% *S*-2-butanol. Blink et al. [122] have used also other aryl substituents and thio homologues (e.g., methyl thiophosphonic dichloride).

A simplier achiral bifunctional agent was used by Stan'kov et al. [124,125]. In principle, the method is the same as Blink's but instead of phenyl phosphonic dichloride they used phosgene ($COCl_2$). This leads only to one mesoform, and therefore to two peaks in the chromatogram, but with the same drawback for the determination of the enantiomeric composition.

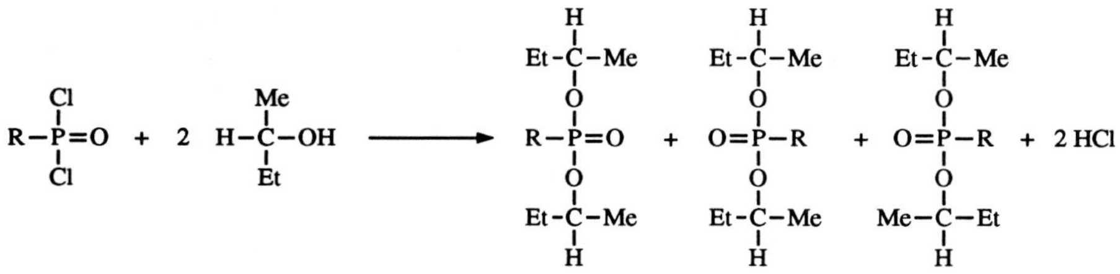

Phosphonic dichloride (*R,S*)-2-Butanol (*R,S*)-meso[1] (*R,S*)-meso[2] (*RR,SS*)

Figure 3.5.2 Reaction scheme of phenylphosphonic dichloride with 2-butanol (R = C_6H_5; Et = Ethyl; Me = Methyl) [122].

The compounds analyzed were α-amino acid methyl esters. For example, valine methyl ester reacts to the ureide N,N'-carbonyl-bis(valine methyl ester), with *R*+*R* or *S*+*S* resulting in the optically active forms, while *R*+*S* or *S*+*R* result in the mesoform (Figure 3.5.3). Alternatively, carbondisulphide (CS_2) was also used (the resulting products are a thioureide and H_2S). According to Stan'kov et al. [124,125] the calculation of the enantiomeric ratios is not linear due to the reaction kinetics. They therefore used, in contrast to Blink et al. [122], a complex equation (inverted quadratic function) corresponding to a mathematical probability model (Figure 3.5.4).

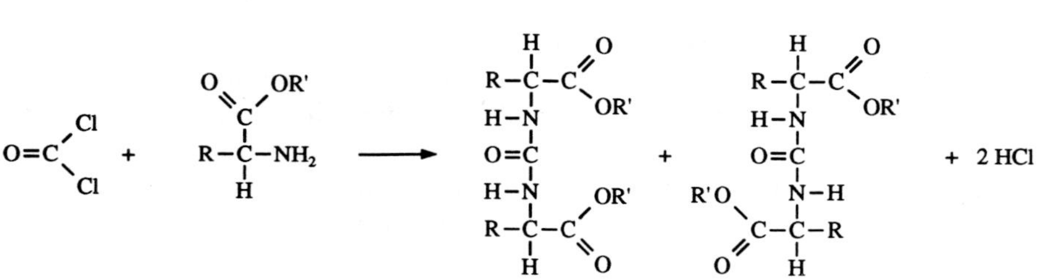

Phosgene α-Amino acid ester (*R,S*)-meso (*RR,SS*)

Figure 3.5.3 Reaction scheme of phosgene with an α-amino acid ester (R = rest of amino acid; R' = rest of ester) [124,125].

Within a variety of (ii) (direct method) an optically active auxiliary is added to the mobile phase; the stationary phase is achiral. For example, the separation of racemic camphor was obtained by coinjection of *S*-borneol; three peaks appeared in the chromatogram [126]. This phenomenon is based on the so-called 'ee effect' (ee = enantiomeric excess). It can occur during achiral chromatography of optically enriched compounds even without addition of chiral auxiliaries and can be explained by formation of self-association complexes [127]. Jung and Schurig [128] used computer simulation to investigate the theoretical background of self association. For the first time this phenomenon was observed in HPLC by Cundy and Crook [129], as well as by Charles and Gil-Av [130]. For detailed informations (review) see also de Min et al. [131].

$$X\,[\%] = \frac{(1+Z) \pm \sqrt{1-Z^2}}{2\,(1+Z)} \times 100$$

Figure 3.5.4 Equation for calculating the content X of enantiomers in a sample mixture (Z = ratio of areas of the meso peak and the peak of the racemic forms) [124,125].

Gaffney et al. [132] worked without any chiral auxiliaries, i.e. without derivatization and chiral stationary phases. To detect optically active substances, e.g., camphor and fenchone, they used a gas chromatographic circular dichroism system with packed columns. It has to be stressed that, according to current scientific knowledge, chromatographic separation of racemic mixtures without the use of an optically active auxiliary are impossible [133].

References

[1] Gil-Av E. *J. Mol. Evol.* (1975), *6*, 131

[2] Gal J. in: *Drug Stereochemistry. Analytical Methods and Pharmacology,* Wainer I.W. (Ed.), 2nd Ed., Dekker: New York, 1992, p. 65

[3] Bailey M.E., Hass H.B. *J. Am. Chem. Soc.* (1941), *63*, 1969

[4] Weygand F., Kolb B., Prox A., Tilak M.A., Tomida I. *Hoppe Seyler's Z. Physiol.Chem.* (1960), *322*, 38

[5] Casanova Jr. J., Corey E.J. *Chem. Ind.(London)* (1961), 1664

[6] Srinivas N.R., Hubbard J.W., Hawes E.M., McKay G., Midha K.K. *J. Chromatogr.* (1989), *487*, 61

[7] Knierzinger A., Walther W., Weber B., Netscher T. *Chimia* (1989), *43*, 163

[8] Walther W., Vetter W., Vecchi M., Schneider H., Müller R.K., Netscher T. *Chimia* (1991), *45*, 121

[9] Yasuhara F., Takeda M., Ochiai Y., Miyano S., Yamaguchi S. *Chem. Lett.* (1992), 251

[10] Rimmer D.A., Rose M.E. *J. Chromatogr.* (1992), *598*, 251

[11] Gil-Av E., Charles R., Fischer G. *J. Chromatogr.* (1965), *17*, 408
[12] Pollock G.E., Jermany D.A. *J. Gas Chromatogr.* (1968), *6*, 412
[13] Gerwig G.J., Kamerling J.P., Vliegenthart J.F.G. *Carbohydr. Res.* (1979), *77*, 1
[14] Kamiski Z.J., Leplawy M.T., Esna-Ashari A., Kühne S, Zivny S., Langer M., Brückner H. in: *Peptides 1988, Proc. 20th Eur. Peptide Symp., Tübingen,* Jung G., Bayer E. (Eds.), de Gruyter: Berlin, New York, 1989, p. 298
[15] Gerwig G.J., Kamerling J.P., Vliegenthart J.F.G. *Carbohydr. Res.* (1978), *62*, 349
[16] Sanz-Burata M., Irurre-Pérez J., Juliá-Arechaga S. *Afinidad* (1970), *27*, 698
[17] Satterwhite D.M., Croteau R.B. *J. Chromatogr.* (1987), *407*, 243
[18] Saucy G., Borer R., Trullinger D.P. *J. Org. Chem.* (1977), *42*, 3206
[19] König W.A., Benecke I. *J. Chromatogr.* (1980), *195*, 292
[20] König W.A. *Chromatographia* (1976), *9*, 72
[21] König W.A., Benecke I., Schulze J. *J. Chromatogr.* (1982), *238*, 237
[22] Vitt S.V., Saporovskaya M.B., Gudkova I.P., Belikov V.M. *Tetrahedron Lett.* (1965), *6*, 2575
[23] MacLean I., Eglington G., Douraghi-Zadeh K., Ackman R.G., Hooper S.N. *Nature* (1968), *218*, 1019
[24] Murano A. *Agr. Biol. Chem.* (1972), *36*, 917
[25] Guetté J.P., Horeau A. *Tetrahedron Lett.* (1965), 3049
[26] Kamerling J.P., Gerwig G.J., Vliegenthart J.F.G. *J. Chromatogr.* (1977), *143*, 117
[27] Mosandl A., Heusinger G. in: *Analysis of Volatiles,* Schreier P. (Ed.), de Gruyter: Berlin, New York, 1984, p. 343
[28] Horiba M., Takahashi K.-I., Yamamoto S., Murano A. *Agr. Biol. Chem.* (1980), *44*, 463
[29] Müller M.D., Bosshardt H.-P. *J. Assoc. Off. Anal. Chem.* (1988), 71, 614
[30] Karlsson K.-A., Pascher I. *Chem. Phys. Lipids* (1974), *12*, 65
[31] Rickett F.E. *The Analyst* (1973), *98*, 687
[32] Huffer M., Schreier P. *J. Chromatogr.* (1990), *519*, 263
[33] Tosolini G.P., Moro E., Forgione A., Ranghieri M., Mandelli V. *J. Pharmaceut. Sci.* (1974), *63*, 1072
[34] Kaiser D.G., VanGiessen G.J. *The Pharmacologist* (1974), *16*, 221
[35] Bopp R.J., Nash J.F., Ridolfo A.S., Shepard E.R. *Drug Metab.Dispos.* (1979), *7*, 356
[36] Oshima R., Kumanotani J., Watanabe C. *J. Chromatogr.* (1983), *259*, 159
[37] Singh N.N., Pasutto F.M., Coutts R.T., Jamali F. *J. Chromatogr.* (1986), *378*, 125; (a) Taylor W.G., Vedres D.D., Elder J.L. *J. Chromatogr.* (1993), *645*, 303
[38] Brooks C.J.W., Watson D.G., Cole W.J. *J. Chromatogr.* (1985), *347*, 455
[39] Gil-Av E., Nurok D. *Proc. Chem. Soc.* (1962), 146
[40] Gilbert M.T., Gilbert J.D., Brooks C.J.W. *Biomed. Mass Spectrom.* (1974), *1*, 274
[41] Gilbert M.T., Brooks C.J.W. *Biomed. Mass Spectrom.* (1977), *4*, 226
[42] Juliá S., Sans J.M. *J. Chromatogr. Sci.* (1979), *17*, 651
[43] Mosandl A., Deger W. *Z. Lebensm. Unters. Forsch.* (1989), *188*, 333
[44] Mosandl A., Deger W., Gessner M., Günther C., Singer G., Kustermann A., Schubert V. in: *Chirality and Biological Activity, Proc. Int. Symp., Tübingen 1988,* Liss: New York, 1990, p. 119
[45] Juliá S., Zurbano M. *Afinidad* (1978), *35*, 150
[46] Mosandl A., Gessner M., Günther C., Deger W., Singer G. *J. High Resol. Chromatogr. Chromatogr. Commun.* (1987), *10*, 67

[47] Oehlschlager A.C., King G.G.S., Pierce Jr. H.D., Pierce A.M., Slessor K.N., Millar J.G., Borden J.H. *J. Chem. Ecol.* (1987), *13*, 1543
[48] Günther C., Mosandl A. *Z. Lebenm. Unters. Forsch.* (1987), *185*, 1
[49] Brooks C.J.W., Gilbert M.T., Gilbert J.D. *Anal. Chem.* (1973), *45*, 896
[50] Burden R.S., Deas A.H.B., Clark T. *J. Chromatogr.* (1987), *391*, 273
[51] Murano A., Fujiwara S. *Agr. Biol. Chem.* (1973), *37*, 1977
[52] Wipf B., Kupfer E., Bertazzi R., Leuenberger H.G.W. *Helv. Chim. Acta* (1983), *66*, 485
[53] Nichols D.E., Hoffman A.J., Oberlender R.A., Jacob P. III, Shulgin A.T. *J. Med. Chem.* (1986), *29*, 2009
[54] Jacob P. III, Benowith N.L., Copeland J.R., Risner M.E., Cone E.J. *J. Pharmaceut. Sci.* (1988), *77*, 396
[55] Büyüktimkin N., Keller F., Schunack W. *J. Chromatogr.* (1989), *467*, 402
[56] Nambara T., Goto J., Taguchi K., Iwata T. *J. Chromatogr.* (1974), *100*, 180
[57] Westley J.W., Halpern B. *J. Org. Chem.* (1968), *33*, 3978
[58] Hamberg M. *Anal. Biochem.* (1971), *43*, 515
[59] Wilken D.R., Dyar R.E. *Anal. Biochem.* (1981), *112*, 9
[60] Berger B., Rabiller C.G., Königsberger K., Faber K., Griengl H. *Tetrahedron Assymmetry* (1990), *1*, 541
[61] Kristensen K., Angelo H.R. *Chirality* (1992), *4*, 263
[62] Feigl D.M., Mosher H.S. *J. Org. Chem.* (1968), *33*, 4242
[63] Hammarstrøm S., Hamberg M. *Anal. Biochem.* (1973), *52*, 169
[64] Mosandl A., Günther C. *J. Agric. Food Chem.* (1989), *37*, 413
[65] Mosandl A., Gessner M. *Z. Lebensm. Unters. Forsch.* (1988), *187*, 40
[66] Dale J.A., Mosher H.S. *J. Am. Chem. Soc.* (1968), *90*, 3732
[67] Dale J.A., Dull D.L., Mosher H.S. *J. Org. Chem.* (1969), *34*, 2543
[68] Tressl R., Engel K.-H. in: *Progress in Flavour Research 1984,* Adda J. (Ed.), Elsevier: Amsterdam, 1985, p. 441
[69] Heidlas J., Tressl R. *Eur. J. Biochem.* (1990), *188*, 165
[70] Kajiwara T., Nagata N., Hatanaka A., Naoshima Y. *Agr. Biol. Chem.* (1980), *44*, 437
[71] Hochuli E., Taylor K.E., Dutler H. *Eur. J. Biochem.* (1977), *75*, 433
[72] Tressl R., Engel K.-H. in: *Analysis of Volatiles,* Schreier P. (Ed.), de Gruyter: Berlin, New York, 1984, p. 323
[73] Nichols, D.E., Barfknecht C.F., Rusterholz D.B. *J. Med. Chem.* (1973), *16*, 480
[74] Gal J., French T.A., Zysset T., Haroldsen P.E. *Drug Metab. Dispos.* (1982), *10*, 399
[75] Gaget C., Wolf E., Heintzelmann B., Wagner J. *J. Chromatogr.* (1987), *395*, 597
[76] Spahn H., Henke W. *Naunyn-Schmiedeberg's Arch. Pharmacol.* (1985), *329*, R11-R11, Abstr.44
[77] Büyüktimkin N., Büyüktimkin S., Grunow D., Elz S. *Chromatographia* (1988), *25*, 925
[78] Doolittle R.E., Heath R.R. *J. Org. Chem.* (1984), *49*, 5041
[79] Engel K.-H. in: *Bioflavour '87,* Schreier P. (Ed.), de Gruyter: Berlin, New York, 1988, p. 75
[80] Mosandl A., Günther C., Gessner M., Deger W., Singer G., Heusinger G. in: *Bioflavour '87,* Schreier P. (Ed.), de Gruyter: Berlin, New York, 1988, p. 55
[81] Halpern B., Westley J.W. *Biochem. Biophys. Res. Commun.* (1965), *19*, 361
[82] Caccia S., Chiabrando C., De Ponte P., Fanelli R. *J. Chromatogr. Sci.* (1978), *16*, 543

[83] Horiba M., Kitahara H., Murano A. Japan. Kokai 77 86, 394 (Cl.G01N31/08), 18 Jul 1977 (C.A. 1978, 88: 15596q)
[84] Beckett A.H., Haya K., Jones G.R., Morgan P.H. *Tetrahedron* (1975), *31*, 1531
[85] Pereira Jr. W.E., Halpern B. *Aust. J. Chem.* (1972), *25*, 667
[86] Halpern B., Westley J.W. *J. Chem. Soc. Chem. Commun.* (1966), 34
[87] Darbre A., Blau K. *J. Chromatogr.* (1967), *29*, 49
[88] Karger B.L., Stern R.L., Rose H.C., Keane W. in: *6th Int. Symp. on Gas Chromatography and Associated Techniques, Rome 1966*, Littlewood A.B. (Ed.), Inst. of Petroleum: London ,1967, p. 240
[89] Manius G., Tscherne R. *J. Chromatogr.Sci.* (1979), *17*, 322
[90] Aspeslet L.J., Baker G.B., Coutts R.T., Mousseau D.D. *Biochem. Pharmacol.* (1992), *44*, 1894
[91] Hoyer G.-A., Rosenberg D., Rufer C., Seeger A. *Tetrahedron Lett.* (1972), *13*, 985
[92] Sanz-Burrata M., Irurre-Pérez J., Juliá-Arechaga S. *Afinidad* (1970), *27*, 705
[93] Abe I., Musha S. Bunseki Kagaku (1974), 23, 755-759 (C.A. 1974, 81: 136310f)
[94] Hara S., Okabe H., Mihashi K. *Chem. Pharm. Bull.* (1986), *34*, 1843
[95] Pereira W., Bacon V.A., Patton W., Halpern B. *Anal. Lett.* (1970), *3*, 23
[96] Gaydou E.M., Randriamiharisoa R.P. *J. Chromatogr.* (1987), *396*, 378
[97] Tressl R., Engel K.-H., Albrecht W., Bille-Abdullah H. in: *Characterization and Measurement of Flavor Compounds*, Bills D.D., Mussinan C.J. (Eds.), ACS Symposium Series 289, American Chemical Society: Washington, D.C., 1985, p. 43
[98] Hamberg M. *Chem. Phys. Lipids* (1971), *6*, 152
[99] Albrecht W., Tressl R. *Tetrahedron Asymmetry* (1993), *4*, 1391
[100] Engel K.-H., Albrecht W., Heidlas J. *J. Agric. Food Chem.* (1990), *38*, 244
[101] Engel K.-H. *Tetrahedron Asymmetry* (1991), *2*, 165
[102] Thompson J.A., Holtzman J.L., Tsuru M., Lerman C.L., Holtzman J.L. *J. Chromatogr.* (1982), *238*, 470
[103] Deger W., Gessner M., Heusinger G., Singer G., Mosandl A. *J. Chromatogr.* (1986), *366*, 385
[104] Rudmann A.A., Aldrich J.R. *J. Chromatogr.* (1987), *407*, 324
[105] Makita M., Ohkaru Y., Yamamoto S. *J. Chromatogr.* (1980), *188*, 408
[106] Barbeni M., Guarda P.A., Cabella P., Allegrone G., Cisero M. in: *Flavour Science and Technology*, Bessière Y., Thomas A.F. (Eds.), Wiley: Chichester, 1990, p. 37
[107] Koshimizu K., Sakata I. *Agr. Biol. Chem.* (1979), *43*, 411
[108] Schweer H. *J. Chromatogr.* (1982), *243*, 149
[109] Schweer H. *Carbohydr. Res.* (1983), *116*, 139
[110] Pereira Jr. W.E., Solomon M., Halpern B. *Aust. J. Chem.* (1971), *24*, 1103
[111] Matin S.B., Rowland M., Castagnoli Jr. N. *J. Pharmaceut. Sci.* (1973), *62*, 821
[112] Little M.R. *Carbohydr. Res.* (1982), *105*, 1
[113] Gil-Av E., Nurok D. in: *Advances in Chromatography*, Vol. 10, Giddings J.C., Keller R.A. (Eds.), Dekker: New York, 1974, p. 99
[114] Drozd J. *J. Chromatogr.* (1975), *113*, 303
[115] Halpern B. in: *Handbook of Derivatives for Chromatography*, Blau K., King G.S. (Eds.), Heyden: London, 1977, p. 457
[116] Knapp D.R. *Handbook of Analytical Derivatization Reactions*, Wiley: Chichester, 1979
[117] Souter R.W. *Chromatographic Separations of Stereoisomers*, CRC Press: Boca Raton, 1985

[118] König W.A., Stölting K., Kruse K. *Chromatographia* (1977), *10*, 444
[119] Liu D.-M. *J. Chromatogr.* (1992), *589*, 249
[120] Harden R.C., Rackham D.M. *J. High Resol. Chromatogr. Chromatogr. Commun.* (1992), *15*, 407
[121] Sjo P., Aasen A.J. *Acta Chem. Scand.* (1993), *47*, 486
[122] Blink A., Suijkerbuijk M.L., Ishiwata T., Feringa B.L. *J. Chromatogr.* (1989), *467*, 285
[123] Testa B. *Grundlagen der Organischen Stereochemie*, Verlag Chemie: Weinheim 1983
[124] Stan'kov I.N., Tarasov S.N., Lysenko V.V., Beresnev A.N., Sakodynskii K.I., Shatilo F.E. *Zh. Fiz. Khim.* (1990), *64*, 2467 [(C.A. 1991,114:114404n)]
[125] Stan'kov I.N., Tarasov S.N., Agureyev V.G., Beresnev A.N., Lysenko V.V. *Anal. Chim. Acta* (1991), *251*, 223
[126] Maestas P.D., Morrow C.J. *Tetrahedron Lett.* (1976), *17*, 1047
[127] Tsai W.-L., Hermann K., Hug E., Rohde B., Dreiding A.S. *Helv. Chim. Acta* (1985), *68*, 2238
[128] Jung M., Schurig V. *J. Chromatogr.* (1992), *605*, 161
[129] Cundy K.C., Crooks P.A. *J. Chromatogr.* (1983), *281*, 17
[130] Charles R., Gil-Av E. *J. Chromatogr.* (1984), *298*, 516
[131] De Min M., Levy G., Micheau J.C. *J. Chim. Phys.* (1988), *85*, 603
[132] Gaffney J.S., Premuzic E.T., Orlando T., Ellis S., Snyder P. *J. Chromatogr.* (1983), *262*, 321
[133] Schurig V. *Angew. Chem.* (1984), *96*, 733

3.5.2 *Separation of enantiomers on chiral stationary phases*

The first attempts to separate racemic mixtures by gas chromatography on chiral stationary phase were carried out by Karagounis and Lippold [1]. They moistened Al_2O_3 with a thin layer of (-)-diethyl tartrate or starch. Besides partial separation of 2-bromobutane and other brominated compounds, they also achieved a partial resolution of 2-butanol. These results could not, however, be reproduced [2]. They were explained as the result of dehydrohalogenation of the brominated compounds in the inlet system [3,4]. A similar experiment using D-tartaric acid was performed by Lynch [5]. A partial resolution of 2-hexanol was achieved.

Goeckner [6] used as stationary phases several sucrose derivatives, menthyl octadecanoate, as well as an optically active polymer of 1-propylene oxide, but no resolution of any of the investigated compounds, e.g., 2-methyl-1-butanol, 2-octanol, 2-butyl acetate or 4-aminodecane, was noted. Karagounis and Lemperle [7] tested several chiral phases successfully. With the optically active metal complex $[Co(en)_3]Br_3$ on Al_2O_3 as stationary phase, these authors achieved a partial separation of acetoin and 2-methoxybutane [8].

The partial separation of the racemic metal complex Cr(III)-tris(hexafluoroacetyl acetonate) was reported by Sievers et al. [9]. They used (+)-quartz powder as the stationary phase. The use of monosodium L-glutamate as the stationary phase in a packed column revealed no enantiomer separation of racemic N-acetyl-α-amino acid pentyl esters [10].

Summarizing, in all cases only partial separations were obtained with poorly reproducible results, and only packed separation columns were used. It was only with the indroduction of high resolution capillary columns into gas chromatography (HRCG) that conditions for a successful separation of enantiomers were obtained.

In Table 3.5.3 actual manufacturers and suppliers of chiral stationary phases for gas chromatography are listed. The properties and applications of the most important phases are discussed in the following sections.

Table 3.5.3 Manufacturers and suppliers of chiral stationary phases for gas chromatography (cyclodextrin concentrations: * = 10 %, ** = 30 %, # = 0.07 molar, dissolved in OV-1701).

CSP	Name	Company
68	Cyclodextrin Chiral (bonded)	Advanced Separation
93	Chiraldex A-PH	Technologies Inc. (Astec)
63	Chiraldex B-PH	37 Leslie Ct., Whippany,
94	Chiraldex G-PH	NJ 07981
90	Chiraldex A-DA	USA
91	Chiraldex B-DA	
92	Chiraldex G-DA	
95	Chiraldex A-TA	
96	Chiraldex B-TA	
65	Chiraldex G-TA	
11	Chirasil-Val (Heliflex)	Alltech Associates Inc.
12	RSL-007	2051 Waukegan Rd., Deerfield
93	Chiraldex A-PH	IL 60015, USA
63	Chiraldex B-PH	
94	Chiraldex G-PH	
90	Chiraldex A-DA	
91	Chiraldex B-DA	
92	Chiraldex G-DA	
95	Chiraldex A-TA	
96	Chiraldex B-TA	
65	Chiraldex G-TA	
11	Chirasil-Val III	Applied Science Laboratories
		P.O. Box 440, State College
		PA. 16801, USA
68	**SCHA	Carlo Erba Reagenti
71	**SCHB	V. le Liguria, 7
69	**SCHC	I-20093 Cologno M. (MI)
75	**SCHD (Nr. s. Megadex 4)	Italy
33	Chirasil-Metal	CC & CC
		PF 14, D-72138
		Kirchenstellinsfurt, FRG
18	Chirasil-D-Val	Chrompack International B.V.
11	Chirasil-L-Val	Kuipersweg 6
13	XE-60-S-Valine-S-α-phenylethylamide	NL-8033-Middelburg
14	XE-60-S-Valine-R-α-phenylethylamide	The Netherlands
68	*CP-Cyclodextrin β-2,3,6-M-19	
93	Chiraldex A-PH	Chrom Tech Inc.
63	Chiraldex B-PH	Apple Valley, MN 55124, USA

CSP	Name	Company
94	Chiraldex G-PH	
91	Chiraldex B-DA	
92	Chiraldex G-DA	
95	Chiraldex A-TA	
96	Chiraldex B-TA	
65	Chiraldex G-TA	
11	Chiral-C	CS-Chromatographie Service
71	#FS-Cyclodex alpha-I/P	Am Wehebach 26
68	#FS-Cyclodex beta-I/P	D-52379 Langerwehe
69	#FS-Cyclodex gamma-I/P	FRG
51	FS-Cyclodex alpha-II	
53	FS-Cyclodex beta-II	
89	FS-Cyclodex gamma-II	
97	#FS-Cyclodex alpha-II/P	
98	#FS-Cyclodex beta-II/P	
109	#FS-Cyclodex gamma-II/P	
99	#FS-Cyclodex alpha-III/P	
100	#FS-Cyclodex beta-III/P	
101	#FS-Cyclodex gamma-III/P	
75	#FS-Cyclodex (2,6-dimethyl-3-TFA-γ-CD)	
78	#FS-Cyclodex (6-tBDMS-2,3-dimethyl-α-CD)	
79	#FS-Cyclodex (6-tBDMS-2,3-dimethyl-β-CD)	
53	Perpent-β	DvK Laborbedarf
89	Perpent-γ	D-52379 Langerwehe, FRG
47	Peraldex	FZB Biotechnik GmbH, Alt-
		Stralau 62, D-10245 Berlin, FRG
18	Chirasil-D-Val	Gsukuro Kogyo, Tokyo, Japan
93	Chiraldex A-PH	ict Handelsgesellschaft mbH
63	Chiraldex B-PH	Antoniterstr. 27
94	Chiraldex G-PH	D-65929 Frankfurt a.M.
90	Chiraldex A-DA	FRG
91	Chiraldex B-DA	
92	Chiraldex G-DA	
95	Chiraldex A-TA	
96	Chiraldex B-TA	
65	Chiraldex G-TA	
68	*C-Dex B (Cyclodex B)	J & W Scientific/Fisons
		Folsom, CA 95630, USA
68	β-Cyclodextrin (diss. in OV-1701)	Kupper + Co., PF 55
		CH-7402 Bonaduz, Switzerland
11	Permabond L-Chirasil-Val	Macherey-Nagel GmbH & Co.
51	FS-Lipodex A	Neumann-Neander-Str.
61	FS-Lipodex B	D-52335 Düren
53	FS-Lipodex C	FRG
58	FS-Lipodex D	
57	FS-Lipodex E	
68	FS-Hydrodex-β-PM (diss. in OV-1701)	
74	FS-Hydrodex β-TA (diss. in OV-1701)	

CSP	Name	Company
68	**Megadex-1	Mega
71	**Megadex-2	Dr. Galli
69	**Megadex-3	Via Plinio, 29
75	*Megadex-4	I-20025 Legnano (MI)
88	**Megadex-5[x]	Italy
110	**Megadex-6[x]	([x]also available in OV-1 and
102	**Megadex-7[x]	OV-225)
68	*C-Dex B (Cyclodex B)	Promochem,D-46485 Wesel,FRG
68	*Cydex-B	SGE International Pty. Ltd.
		7 Argent Place, Ringwood
		Victoria, 3134, Australia
10	OA-200	Sumitomo Chemical Co.
23	OA-300	Sumitomo Bld. 15, 5-chome
24	OA-310	Kitahama, Higashi-ku
25	OA-400	Osaka 541
26	OA-500	Japan
9	OA-510	
27	OA-520	
6	SP-300	Supelco Inc.
71	α-Dex (10 % diss. in SPB-35)	Supelco Park, Bellefonte
68	β-Dex 110 (10 % diss. in SPB-35)	PA. 16823
68	β-Dex 120 (10 % diss. in SBP-35)	USA
69	γ-Dex (10 % diss. in SBP-35)	
93	Chiraldex A-PH	Technicol Ltd.
63	Chiraldex B-PH	Brook St., Higher Hillgate
94	Chiraldex G-PH	Stockport, Cheshire
90	Chiraldex A-DA	SK1 3HS
91	Chiraldex B-DA	UK
92	Chiraldex G-DA	
95	Chiraldex A-TA	
96	Chiraldex B-TA	
65	Chiraldex G-TA	
71	#FS-Cyclodex alpha-I/P	TPC Ziemer
68	#FS-Cyclodex beta-I/P	Mannheim
69	#FS-Cyclodex gamma-I/P	FRG
51	FS-Cyclodex alpha-II	
53	FS-Cyclodex beta-II	
89	FS-Cyclodex gamma-II	
97	#FS-Cyclodex alpha-II/P	
98	#FS-Cyclodex beta-II/P	
109	#FS-Cyclodex gamma-II/P	
99	#FS-Cyclodex alpha-III/P	
100	#FS-Cyclodex beta-III/P	
101	#FS-Cyclodex gamma-III/P	

3.5.2.1 Amide phases

Lower molecular weight amide phases

The first reproducible separations of racemic mixtures were described by Gil-Av et al. in 1966 and 1967 [11,12]. Eighteen racemic proteinogenic amino acids were separated as their N-TFA alkyl esters; in most cases baseline separation was achieved. The authors used glass capillaries coated with N-TFA-*S*-isoleucine dodecylester (CSP-1) or N-TFA-*S*-phenylalanine cyclohexylester (CSP-3). The enantiomer separation was checked by using a column coated with the mirror image stationary phase of CSP-1, N-TFA-*R*-isoleucine dodecylester (CSP-2), since the order of elution (order of emergence) of N-TFA-amino acid isopropyl esters was reversed. The aim of the development of the phases was the imitation of the enantioselectivity of enzymes. With the success of these attempts the foundation for a rapid further development in this field was established.

With the development of dipeptide phases such as N-TFA-*S*-valyl-*S*-valine isopropylester (CSP-4), in comparison to CSP-1 and CSP-3, an additional amide function has been introduced, which is able to associate via hydrogen bonds [12]. Consequently, the enantioselectivity has been improved [13]. The influence of N- and C-terminal amino acids, in which the N-terminal amino acid determines the enantioselectivity of the dipeptide phases [14], was systematically investigated. Subsequently, several different dipeptide phases were developed [15-21]. With tripeptide phases like N-TFA-*S*-valyl-*S*-valyl-*S*-valine isopropylester (CSP-5) no improvement of the enantioselectivity was obtained, but a reduction in the volatility of the phase [22].

Due to the fact that the C-terminal amino acid contributes less to the enantiomeric discrimination, diamide phases have been developed in which this amino acid was dropped and the N-acyl group was extended; one example is N-dodecanoyl-*S*-valine tBA (CSP-6). This stationary phase is commercially available under the name 'SP-300' (see Figure 3.5.5). Although one of the two chiral centres was dropped, these diamide phases showed a better enantioselectivity as compared with the dipeptide phases [23].

Several diamide phases were synthesized and used in gas chromatography for enantiomer separations [24-32]. Their temperature stability (until 190°C) is remarkable, especially, in contrast with so-called ureide phases like N,N'-carbonyl-bis-[*S*-valine isopropyl ester] (CSP-7), which can be used only at temperatures < 80°C. The ureide phases are especially suited for the analysis of N-TFA-aminoalkanes [33]. While proteinogenic amino acids show uniform orders of elution on the above-mentioned ester and dipeptide phases, Feibush et al. [34] found a change in the order of elution in a homologous series of N-TFA-2-amino carboxylic acids methyl esters using CSP-7. Several other ureide-type phases were successfully used as chiral stationary phases [35-39].

In addition to the above-mentioned diamide phases, a series of simple amides of carboxylic acids [29; 40-47] were developed as stationary phases, for example *S*-

mandelic acid *S*-1-phenylethylamide (CSP-8) [41] and *R,R*-1,3-*E*-chrysanthemic acid *R*-1,1'-naphthylethylamide (CSP-9) [43], which show only a low temperature stability. The latter phase is purchasable as 'OA-510' (see Figure 3.5.5).

Due to their structure, triazine phases are assumed to retain an exceptional position [48-52]. To increase the temperature stability, the 1,3,5-triazine ring is connected, e.g., by two tripeptide chains: N,N'-[2,4-(6-ethoxy-1,3,5-triazine)diyl]-bis-[*S*-valyl-*S*-valyl-*S*-valine isopropylester] (CSP-10). Using these phases even underivatized alcohols can be separated into their enantiomers [53].

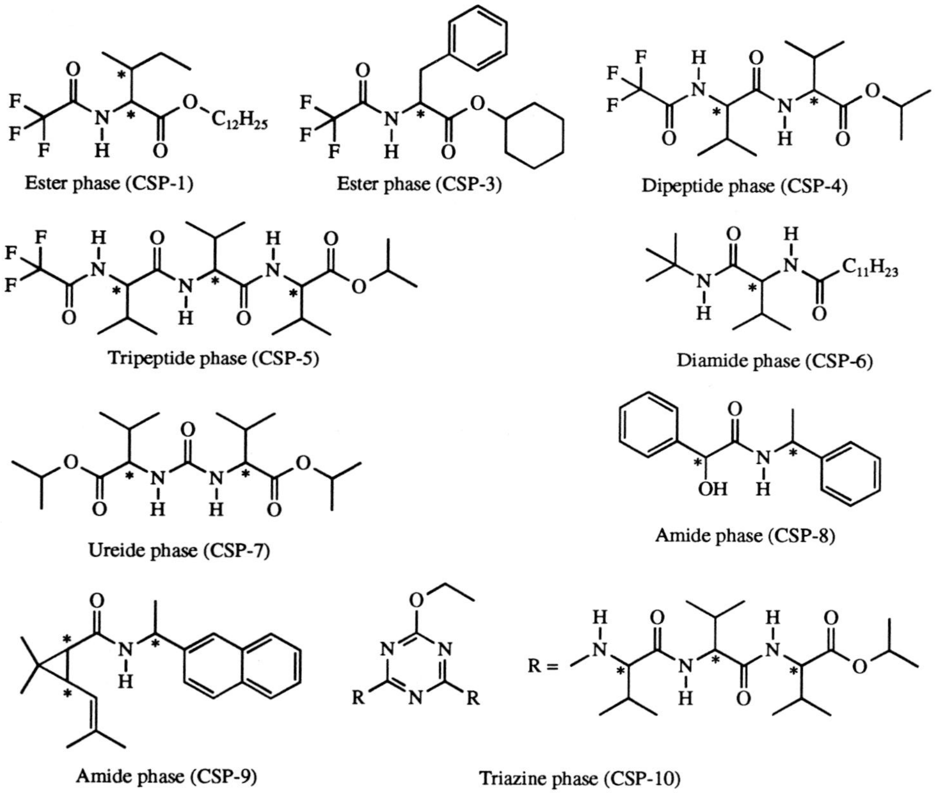

Ester phase (CSP-1) Ester phase (CSP-3) Dipeptide phase (CSP-4)

Tripeptide phase (CSP-5) Diamide phase (CSP-6)

Ureide phase (CSP-7) Amide phase (CSP-8)

Amide phase (CSP-9) Triazine phase (CSP-10)

Figure 3.5.5 Amide phases (selected examples).

The range of application of most of the lower molecular mass stationary phases (see Figure 3.5.5) is reduced at higher temperatures as a result of thermal degradation, racemization and bleeding of the column. Therefore, many low volatile compounds are not easily accessible to a racemate separation. In addition, these phases tend to crystallize at lower temperatures.

Table 3.5.4 Applications of commercially available amide-containing CSPs (written in parentheses stands for derivatives used).

OA-200 = N,N'-[2,4-(6-Ethoxy-1,3,5-triazine)diyl]-bis[*S*-valyl-*S*-valine isopropyl ester] (CSP-10) [48]	
Alcohols	exo-Norborneol, endo/exo-borneol (no) [109]
Amines	1-PEA (N-TFA) [110]; 1,1'-NEA (N-TFA) [48] or (N-PFP) [111]
Amino acids	Ala (N-TFA/tBE) [48]; met, lys (N-PFP/iPE) [111]; glu, phe (N-PFP/iPE) [112]
Miscellaneous	4-Hydroxy-3-methyl-2-(2-propynyl)-2-cyclopentenone = propargyllone (no) [113]
OA-300 = N,N'-[2,4-(6-Ethoxy-1,3,5-triazine)diyl]-bis[*S*-valyl-*S*-valyl-*S*-valine isopropyl ester] (CSP-23) [48]	
Acids	α-Alkylphenylacetic acids (iPA) [114]
Alcohols	1-Phenylethanol (iPC) [115]; 1,1'-naphthylethanol (no) [113]
Amines	1,1'-NEA (N-PFP) [48]; 2-(4-methylphenyl)-1-PEA (N-PFP)[110]; arylalkylamines (N-ac. or N-TFA or N-PFP) [116]
Amino acids	α-AA (N-TFA/iPE or N-TFA/iPA) [53]; pro (N-ac./iPA) [117]
Esters	Ethyl lactate, diethyl malate (iPC) [115]
Ketones	4-Hydroxy-3-methyl-2-(2-propenyl)-2-cyclopentenone = allethrolone (no) [113] or (iPC) [115]; 4-hydroxy-3-methyl-2-(2-propynyl)-2-cyclopentenone = propargyllone (iPC) [115]
Lactones	Pantolactone (iPC) [115]
Peptides	*E,Z*-Ala-ala, *E,Z*-pro-ala, *E,Z*-pro-val (N-TFA/iPE) [118]; 2,5-dimethyldiketopiperazine = cycloala-ala (no) [118]
Pesticides	2-(4-Chlorophenyl)isovaleric acid (iPA) [119]; O-ethyl-O-(5-methyl-2-nitrophenyl)-N-isopropylphosphoroamidothioate and related compounds (no) [120]; chrysanthemic acid (amide) [121]
Miscellaneous	α-Halocarboxylic acids (hexylamides) [122]
OA-310 = N,N'-[2,4-(6-Dioctylamino-1,3,5-triazine)diyl]-bis[(S)-valyl-*S*-valyl-*S*-valine isopropyl ester] (CSP-24) [123]	
Acids	2-Phenylpropanoic acid (= hydratropic acid) (tBA) [123]
Alcohols	1-Phenylethanol, 1,1'-napthylethanol (iPC) [52]
Amines	2-Octylamine (N-TFA) [52]; 1-PEA; 1,1'-NEA (N-PFP) [52]
Amino acids	α-AA (N-TFA/iPE) [52]; orn, lys (N-PFP/iPE) [52]
Esters	Hexyl lactate (no) [52]

OA-310 = N,N'-[2,4-(6-Dioctylamino-1,3,5-triazine)diyl]-bis[(S)-valyl-*S*-valyl-*S*-valine isopropyl ester] (CSP-24) [123]

Lactones	Pantolactone (no) [52]
Peptides	*E,Z*-Ala-ala (N-TFA/iPE) [52]
Pesticides	*E,Z*-Chrysanthemic acid, 2,2-dimethylcyclopropane carboxylic acid (no) [52]; O-ethyl-O-(3-trifluoromethylphenyl)-N-isopropylphosphoroamido-thioate (no) [52]
Miscellaneous	Flavanone (no) [52]; 2-bromo-3,3-dimethylbutanoic acid (tBA) [52]

OA-400 = N,N',N"-[2,4,6-(1,3,5-triazine)triyl]-tris[N-dodecanoyl-*S*-lysine tBA] (CSP-25) [51]

Alcohols	1-Phenylethanol (no) [113]; 1-phenyl-2-propyn-1-ol (no) [124]
Amino acids	α-AA (N-TFA/iPE) [51]; pro (N-TFA/iPA) [53]
Esters	Lactic acid esters, O-methyl lactic acid esters (no) [125]
Lactones	Pantolactone (no) [113]
Miscellaneous	2,2,2-trifluoro-1-phenylethanol (no) [124]

OA-500 = N-Dodecanoyl-*S*-proline *S*-1,1'-NEA (CSP-26) [29]

Acids	2-Phenylpropanoic acid (= hydratropic acid) (tBA) [29]; hydroxy-phenyl-acetic acid (= mandelic acid) (O-TFA/iPE) [29]
Amines	2-Octylamine, 1-PEA, 1,1'-NEA (N-TFA) [29]; 1-phenyl-2-(4-methyl-phenyl)ethylamine (N-PFP) [126]
Amino acids	Ala, val (N-TFA/iPE) [29]
Lactones	Pantolactone (no) [29]
Pesticides	*E*-3-(2,2-Dichloroethenyl)cyclopropanecarboxylic acid ethyl ester (no) [29]
Miscellaneous	2-(4-Chlorophenyl)-3-methylbutanonitrile (no) [29]; 2-bromo-3,3-di-methylbutanoic acid (tBA), 2-(4-chlorophenyl)-3-methylbutanoic acid (tBA) [29]

OA-510 = 1*R*,3*R*-*E*-Chrysanthemic acid *R*-1,1'-NEA (CSP-9) [43]

Acids	3,3-Dimethyl-2-ethylbutanoic acid (tBA) [43]
Alcohols	Menthol, 1-phenylethanol (no) [43]; 1,1'-naphthylethanol (O-ac.) or (O-TFA) [127]
Amines	2-Octylamine (N-TFA) [43]; 1-PEA (N-PFP) [43]
Amino acids	Ala, val, leu (N-TFA/iPE) [43]
Esters	1-Phenylethyl acetate (no) [127]

OA-510 = 1*R*,3*R*-*E*-Chrysanthemic acid *R*-1,1'-NEA (CSP-9) [43]

Lactones	Pantolactone (no) [43] or (O-ac.) or (O-TFA) [127]
Pesticides	*E*,*Z*-Chrysanthemic acid (tBA) [43]; *Z*-3-(2,2-dichloroethenyl)cyclopropanecarboxylic acid ethyl ester (no) [43]
Miscellaneous	2-(2-Fluorophenyl)-3-methylbutanonitrile, 2-(4-chlorophenyl)3-methylbutanonitrile (no) [29]; 2-bromo-3,3-dimethylbutanoic acid (tBA) [29]; 2,2,2-trifluoro-1-phenylethyl acetate (no) [127]

OA-520 = O-Dodecanoyl-*S*-mandelic acid *S*-1,1'-NEA (CSP-27) [29]

Acids	2-Phenylpropanoic acid (= hydratropic acid) (tBA) [29]; hydroxy-phenylacetic acid (= mandelic acid) (O-TFA/iPE) [29]
Amines	2-Octylamine, 1-PEA (N-TFA) [29]; 1,1'-NEA (N-PFP) [29]
Amino acids	Ala, val, leu (N-TFA/iPE) [29]
Lactones	Pantolactone (no) [29]
Miscellaneous	2-(2-Fluorophenyl)-3-methylbutanonitrile; 2-(4-chlorophenyl)-3-methylbutanonitrile (no) [29]; 2-bromo-3,3-dimethylbutanoic acid, 2-(4-chlorophenyl)-3-methylbutanoic acid (tBA) [29]

SP-300 = N-Dodecanoyl-*S*-valine tBA (CSP-6) [23]

Alkanolamines	Leucinol (N-TFA/pivaloyl ester) [128]
Amino acids	α-AA (N,O-TFA/iPE) [23] or (N,O-TFA/methyl ester) [25]; phe (N-ac./methyl ester) [129]
Lactones	Pantolactone (no) [130]

L-Chirasil-Val = N-2-Methylpropanoyl-*S*-valine tBA [statistical copolymer of dimethylsiloxane and (2-carboxypropyl)methylsiloxane] (CSP-11) [54] [D-Chirasil-Val (CSP-18) [72]]

Acids	2-Hydroxycarboxylic acids, malic acid, mandelic acid (3-PE) [131]; *E*,*Z*-2-hydroxy-3-methylpentanoic acid (3-PE) (also on D!) [72]; 3-hydroxybutanoic acid (iPE) or (PrE) or (3-PE) [131]; 2-hydroxycarboxylic acids, malic acid, mandelic acid, 3-phenyllactic acid (O-HFB/PrA) or (O-TMS/PrA) or (O-iPC/ME) or (O-iPC/iPA) [132]; 3-hydroxypentanoic acid, 3-hydroxy2-phenylpropanoic acid, 3-hydroxy-3-phenylpropanoic acid, 2,3-dihydroxybutanoic acid (3-PE) [97]; mandelic acid (iPA) or (iPC/iPE) [104]; *E*-tartaric acid(bis-acetonide) [97]
Alcohols	2,2-Dimethyl-3-hexanol [133], 2-methyl-3-heptanol [97], terpinen-4-ol, 1-phenylethanol and other 1-phenylalkanols (no) [97]; 2,2-dimethyl-3-hexanol, 1-indanol 1-(2,4,6-trimethylphenyl)ethanol, 1-(3-hydroxyphenyl)-ethanol, 1-(1-naphthyl)ethanol, 1-(9-anthryl)-2,2,2-trifluoroethanol (no) [133]; 2-methyl-6-methylen-7-octen-4-ol = ipsenol, 2-methyl-6-methylen-2,7-octadien-4-ol = ipsdienol (iCP) [134]
Aldehydes	Citronellal (oxazolidine der.) [136]; hydroxydanaidal (no) [82] 2-methylbicyclo[2.2.1]hept-2-ene-6-carbaldehyde (oxime) [82]

L-Chirasil-Val = N-2-Methylpropanoyl-*S*-valine tBA [statistical copolymer of dimethylsiloxane and (2-carboxypropyl)methylsiloxane] (CSP-11) [54] [D-Chirasil-Val (CSP-18) [72]]

Alkanolamines	Phenylalaninol, *E,Z*-threoninol, valinol (N,O-TFA) [89]
Amines	Aminoalkanes (N-PFP) [137]; 1-(2-aminoethyl)-2,2,3-trimethylcyclopropane (N-TFA) [138]; 1-PEA (N-PFP) [139]; 2,2'-diamino-1,1'-binaphthyl (TFA) [140] or (PFP)[97];1-amino-2-methoxymethylpyrrolidine (iPU) or (tBU) [141]
Amino acids	α-AA (N,O,S-PFP/iPE) [54] or (N,O,S-PFP/ME) [88] or (N-TFA/iPA) [142] or (N-pivaloyl/iPA) [142] or (N,O-PFP/PrE) [143] or (N,O,S-TFA/PrE) [144]; N-methyl-α-AA (N-TFA/iPA) [142] or (N-pivaloyl/iPA) [142]; α-alkyl-α-AA (N-TFA/2-BE) [145] or (N-TFA/3-methyl-2-butyl ester) [145]; *E,Z*-β-hydroxy-asp (N-PFP/O-TMS/ME) [146]; cys, *S*-methyl-cys (N-ac./ME) [147]; gln (pyrrolidone der./iPE) [148]; his (N-imidazol-ethoxycarbonyl der./N-TFA/IPE) [59] or (N-imidazolisobutyl-oxycarbonyl der./N-TFA/PrE) ; met-S-oxide, met-S-dioxide, 2-methyl--met-S-oxide, 2-methyl-met-S-dioxide (N-PFP/ME) [87] or (N-PFP/PrE) [149]; pro (N-iPU/iPE) [150]; *E,Z*-cystathionine, *E*-lanthionine (N-PFP/ME) [88]; fluoro-α-AA (N-TFA/ME) [151]; 3-halogen-ala (N-TFA/iPE) [152]
Carbohydrates	Ara, fuc, man, xyl, glc (HFB) [153]; glc, ara (boronate der.) [153]
Diols & polyols	1,2-Alkandiols, *E*-2,3-butanediol, *E*-2,4-pentanediol, 2-methyl-2,4-pentanediol, *E*-1,2-cyclohexanediol (N-TFA-gly-monoester) [154]; *E*-2,4-pen- tanediol, 2-methyl-2,4-pentanediol (N-TFA-gly-diester) [154]; 1,2-alkandiols, *E*-2,3-butanediol, *E*-2,4-pentanediol, *E*-1,2-cycloalkanediols (no) [155]; 1,2-alkandiols, *E*-2,3-butanediol (phosgene der.) [155]; *E,Z*-4-penten-1,2,3-triol (TFA) or (PFP) or (HFB) (also on D!) [156]; phenylglycol (TFA) [140] or (PFP) [97]; (4-hydroxy-3-methoxyphenyl)glycol (PFP) [60]; *E*-1,2-diphenylglycol (PFP) or (ac.) [140]; myo-inositol-1-phosphate (TMS/ME) [157]; chiro-inositol (HFB) [153]; chiro-inositol (boronate der.) [153]; 2,2'-dihydroxy-1,1'-binaphthyl (PFP) [140]
Esters	Methyl lactate (no) (also on D!) [72]; methyl esters of 2-hydroxy acids, malic acid, mandelic acid (no) [132]
Furans	2-Tetrahydrofurfurylic alcohol (no) [97]; *E*-sulcatoloxide (furanoid) = *E*-pityol (no) [158]
Ketones	*E*-3,4-Diphenyl-2,5-hexanedione (no) [140]
Lactones	Pantolactone (no) [97]; 4-alkanolide (no) [159]; 3-ethyl-3-hydroxy-5-pentanolide, 3-fluoromethyl-3-hydroxy-5-pentanolide; 4-(3-pentoxycarbonyl)-4-butanolide (no) [97]
Peptides	*E,Z*-Ala-ala, *E,Z*-ala-leu, *E,Z*-ala-phe (N-TFA/ME) [160]; gly-val (N-TFA/ME) or (N-TFA/EE) or (N-TFA/PrE) or (N-TFA/iPE) (D!) [161]; gly-leu, leu-gly (N-TFA/ME) (D!) [161]; E,Z-pro-val (N-TFA/ME) [162]; ala-ala-ala (N-TFA/ME) [90]

L-Chirasil-Val = N-2-Methylpropanoyl-S-valine tBA [statistical copolymer of dimethylsiloxane and (2-carboxypropyl)methylsiloxane] (CSP-11) [54] [D-Chirasil-Val (CSP-18) [72]

Pesticides	Bitertanol, diclobutrazol, triadimenol and related compounds (no) [163]; O-ethyl-N-isopropyl-S-phenyldithiophosphoric amide (no) [137]
Pharmaceuticals	1-Aryl-2-amino-1-alkanols (ephedrine, epinephrine, etilefrine, metaneprine, norephedrine, norepinephrine, norfenefrine, pseudoephedrine, synephrine, etc) (N,O-PFP) [55,60,164]; 2-amino-N-(2,6-dimethylphenyl)-propanoic acid amide = tocainide (N-HFB) [165] or (N-TFA) or (N-PFP) [166]; tocainide and related compounds (hydantoin der.) [167]; 5-ethyl-3-methyl-5phenylhydantoin = mephenytoin and related compounds (no); *E,Z*-2-amino-4-(methoxymethylphosphinoyl)-butanoic acid = phosphinotricin (N-PFP/iPE) [87]; 4-amino-3-(4-chlorophenyl)-butanoic acid = baclofen (N-HFB/iBE) [169]; 3-(2-chloroethyl)2-[(2-chloroethyl)amino-2H-1,3,2-oxazaphoshinane-2-oxide = ifosfamide (no) [170]; 6-Hydroxy-methtryptoline(N,O-PFP) [171]
Miscellaneous	1-Indanol [133]; lactic acid cyclohexylamide (O-PFP) [164]; γ-lactams (no) [137]; E-N,N'-2,3-O-isopropylidenetartaric acid bis(allylamide) 172]; 2-arylpyrrolidines (N-TFA) [173]; 2-(1-amino)alkylpyridines, 2-(1-amino)alkylarylpyridines (no) [174]; 3-aminopyrrolidine-3-carboylic acid =cucurbitine (N-TFA/alkyl ester) [175]; 4-halogenophenylglycol (PFP)[97]; 1-(9-anthryl)-2,2,2-trifluoroethanol [133]; 2- and 3-halogencarb-oxylic acids (N-tBA) or (N-amides) [176]; *E,Z*-1,2,2-trimethylpropyl-methylphosphonofluoridate = soman (no) [177]; *E,Z*-2-methyl-4-prop-yl-1,3-oxathian-3-oxide [178]; 2-aminoethyl methyl sulphoxide, phenylisopropyl sulphoxide, methylsulphinylmethyl methyl sulphide (N-PFP or (N-TFA) [149]; 3-aminopropyl methyl sulphoxide (N-PFP) [137]

RSL-007 = *S*-Valine tBA [bonded to OV-225] (CSP-12) [61]

Amino acids	α-AA (N,O-TFA or N,O-PFP or N,O-HFB/iPE or butyl ester) [61]; ala (N-HFB/isobutyl ester) [179]
Pharmaceuticals	Z-2-amino-1-phenyl-1-propanol = Norephedrine (N,O-PFP) [61]

XE-60-*S*-Valine-*S*-α-PEA = *S*-Valine *S*-1-phenylethylamide [bonded to XE-60] (CSP-13) [62]

Acids	2-Hydroxycarboxylic acids, malic acid, mandelic acid (iPC/iPE) [95]; 3-hydroxybutanoic acid (tBC/iPE) [95]; lactic acid, 2- and 3-hydroxybutanoic acid (iPC/iPA) [98]; 3-hydroxycarboxylic acids (tBC/tBA) [96]; 3-hydroxycarboxylic acids (iPC/iPA)[96]; 2-phenylbutanoic acid (iPA) [98]
Alcohols	2-Alkanols, *E,Z*-4-methyl-3-heptanol, sulcatol, 3-octanol, isoborneol, ipsdienol, terpinen-4-ol, *E*-pinocarveol, *E*-verbenol, 1-phenylethanol, 1-phenyl-1-propanol (iPC) [180;82]; 3-menthanole (iPC) [181]; *E,Z*-α-ionol (iCP) [182]; 1-octene-3-ol (reduction to 3-octanol), seudenol, *E,Z*-bisabolol (iPC) [183]; isopinocampheol (iPC) [105]; *E,Z*-fenchol (iPC) [184]; falcarinol, didehydrofalcarinol (reduction to 3-heptadecanol as iPC) [185]; *E,Z*-6-tridecen-2-ol (iPC) [186]; *E,Z*-3,7-dimethyl-2-pentadecanol = diprionol (iPC) [187]
Aldehydes	Citronellal (oxazolidine; cf. Chirasil-Val) [136]
Amines	2-Aminoalkanes, 1-PEA (iPU) [95]

XE-60-*S*-Valine-*S*-α-PEA	= *S*-Valine *S*-1-phenylethylamide [bonded to XE-60] (CSP-13) [62]
Amino acids	α-AA (N,O-TFA/iPE) [188] or (N-tert-butylhydantoins) [99] or (N-pivaloyl/alkyl ester) [189] or (N-BOC,O-TMS/ME) [82]; pipecolic acid (N-TFA/iPE) or (N-tBU/iPE) [183]; N-methyl-val (N-iPU/iPA) [98]; fluoro-α-AA (N-TFA/ME) [151]
Carbohydrates	Glc, man, gal, xyl, ara, fuc (TFA) or (TFA/methylglycoside) [62]; rib, lyx (TFA) or (TFA/methylglycoside) [45]
Diols & polyols	Glycerol (1,2-isopropylidene-3-iPC) [190]; arabitol (ac.) or (TFA) or (HFB) [191]; 6-deoxygalactitol = fucitol (TFA) [192]
Esters	Methyl mandelate (iPC) [82]; ethyl 2-methyl-3-oxobutanoate, ethyl 2-allyl-3-oxobutanoic acid (oximes) [101]; *E*-diisopropyl tartrate (cyclic carbonate via phosgene) [103]
Hydrocarbons	α-Pinene (hydroboration to isopinocampheol as iPC) [105]
Ketones	1,2-Ketols (e.g. 3-hydroxy-2-butanone = acetoin) (iPC) [193]; 4-methyl-3-heptanone; 2-methylcyclopentanone, 2-methylcyclohexanone, camphor, fenchone (oximes) [100,101]; 5,5-dimethyl-2-cyclohexenone (no) [194]; menthone (oxime) [82]; 1-methylbicyclo[4.4.0]dec-6-ene-2,8-dione (no) [195]
Lactones	Pantolactone (no) [196]
Pesticides	2-(2,4-Dichlorophenoxy)propanoic acid (iPA) [82]
Pharmaceuticals	Norephedrine (N,O-pivaloyl) [80]
Pyrans	6-Ethyl-2,3-dihydro-2-methyl-4H-pyran-4-one, 2,6-diethyl-2,3-dihydro-4H-pyran-4-one, 3-ethyl-1,8-dimethyl-2,9-dioxabicyclo[3.3.1]non-7-en-6-one (no) [197]
Miscellaneous	2-Fluorocyclohexanone, 2-TFM-cyclohexanone, 2-perfluorohexylcyclohexanone, 2-TFM-cyclopentanone (oximes) [198]; 1-amino-2-methoxymethylpyrrolidine (tBU) [198]; 2-amino-4-(4-methoxyphenyl)-3-methyl-4-butanolide (TFA) [199]; panthenol (O-TFA) [196]; (1-aminoethyl)phosphonic acid diethyl ester (N-TFA) [82]
XE-60-*S*-Valine-*R*-α-PEA	= *S*-Valine *R*-1-phenylethylamide [bonded to XE-60] (CSP-14) [45]
Acids	2-Hydroxycarboxylic acids, *E*,*Z*-2-hydroxy-3-methylpentanoic acid (1,3-dioxolane-2,4-diones via phosgene) [103]
Alkanolamines	2-Amino-1-alkanols (N,O-TFA) [45]
Amines	2-Aminoalkanes, 1-PEA (N-TFA) [188]
Amino acids	N-Methyl-α-AA (N-tBU/tBA) or (N-iPU/iPA) [98] or (N-methyloxazolidine-2,5-diones via phosgene) [103]; cys, ser, thr, allo-thr, penicillamine (via phosgene/iPE) [182]

XE-60-*S*-Valine-*R*-α-PEA = *S*-Valine *R*-1-phenylethylamide [bonded to XE-60] (CSP-14) [45]	
Diols & polyols	1,2-Alkanediols, 1,3-butanediol; *E,Z*-2-ethyl-1,3-hexanediol, phenylglycol, *E*-1,2-cycloalkanediols (cyclic carbonates via phosgene) [103]; 2-methyl-7,8-octadecanediol, 11,12-docosanediol, 2,13-dimethyl-7,8-tetradecanediol (TFA) [200], glycerol-1-alkylether (cyclic carbonates via phosgene) [190]; arabitol, mannitol (TFA) [192]
Ketones	1,2-Ketols (e.g. 3-hydroxy-2-butanone = acetoin) (iPC) [193]; 4-methyl-3-hexanone, 4-methyl-3-heptanone, 3-methyl-4-heptanone, 2-alkylcyclohexanone (oximes) [198]
Lactones	Pantolactone (iCP) [196]
Pharmaceuticals	*1-Aryl-2-amino-1-alkanols:* (aludrine, bametan, ephedrine, etilephrine, isoprenaline, metadrenaline, norephedrine, norfenefrine, octopamine, orciprenaline, pseudoephedrine, sympatol, salbutamol, synephrine, terbutaline,etc.) (N,O-TFA) [188] or (N,O-HFB) [202] or (oxazolidine-2-ones via phosgene) [202] or (oxazolidine-2-ones via phosgene) [102] ormethyl ether/oxazolidine-2-ones via phosgene) [202]. *1-Aryloxy-3-amino-2-propanols:*(alprenolol, bunitrolol, bupranolol, metipranolol, meto-prolol, oxprenolo, penbutolol, pindolol etc. (N,O-HFB) [202] or (oxa-zolidine-2-ones via phosgene [102, 202].
Miscellaneous	Photocyclodimers of 2-cycloalkenones, 3,5,5-trimethyl-2-cyclohexenone = isophorone, 2-butene-4-olide, 4-methyl-2-pentene-4olide (no) [194], pantothenic acid (O-TFA/ME) [196], 3,4-dimethyl-5-phenyl-1,3-imidazolidin-2-one (N-TFA) [82]

α-AA = amino acids, ac. = acetyl, BE = butyl ester, BOC = tert-butyloxycarbonyl, der. = derivative, EE = ethyl ester, HFB = heptafluorobutanoyl, iBE = isobutyl ester, iPA = isopropylamide, iPC = isopropylcarbamate (= isopropylurethane), iPE = isopropyl ester, iPU = isopropyl ureido, ME = methyl ester, NEA = 1,1'-naphthylethylamine, (*R/S*) = enantiomeric stationary phase, PE = pentyl ester, PEA = 1-phenylethylamine, PFP = pentafluoropropanoyl, PrA = propyl amide, PrE = propyl ester, tBA = tert-butylamide, tBC = tert-butylcarbamate (= tert-butylurethane), tBE = tert-butyl ester, tBU = tert-butyl ureido, TFA = trifluoroacetyl, TFM = trifluormethyl, TMS = trimethylsilyl.
ala = alanine, arg = arginine, asn = asparagine, asp = aspartic acid, cys = cysteine, gln = glutamine, glu = glutamic acid, gly = glycine, his = histidine, ile = isoleucine, leu = leucine, lys = lysine, met = methionine, orn = ornithine, phe = phenylalanine, pro = proline, ser = serine, thr = threonine, trp = tryptophan, tyr = tyrosine, val = valine.
ara = arabinose, frc = fructose, fuc = fucose, gal = galactose, glc = glucose, lyx = lyxose, man = mannose, rha = rhamnose, rib = ribose, sor = sorbose, xyl = xylose.

Polysiloxane-bonded amide phases

The first phase of this type (CSP-11) was synthesized by Frank et al. [54] in 1977 as follows: Copolymerization of dimethylsiloxane and (2-carboxypropyl) methylsiloxane, then reaction with valine tBA hydrochloride and N,N'-dicyclohexylcarbodiimide to an N-(2-methylpropanoyl)-*S*-valine tert-butylamide containing polymethylsiloxane, which is well-known under the name 'Chirasil-Val' (see Figure 3.5.6). In comparison with the above-described lower molecular mass phases,

'Chirasil-Val' shows an improved temperature stability with a working temperature from 70° to 250°C [55]. Previously, one to three hours were necessary to perform an analysis and broad and poorly resolved peaks were recorded [33]. Currently, baseline separations of 17 different enantiomeric pairs of α-amino acids (as N-PFP isopropyl esters) are achieved in less than 30 minutes [54]. It should be mentioned that 'Chirasil-Val' (CSP-11) contains an additional chiral centre in the side-chain, which is racemic (Figure 3.5.6).

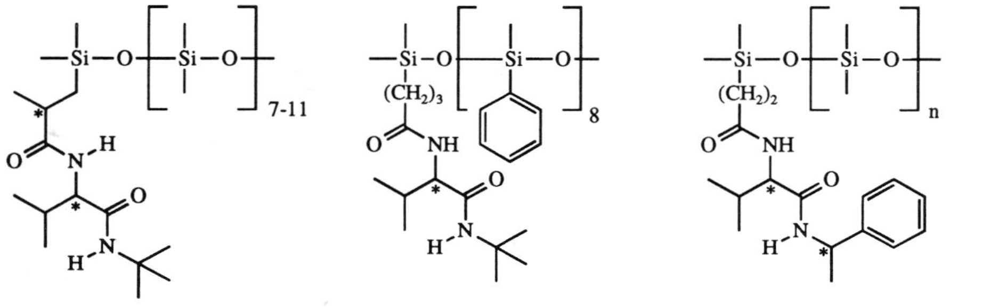

CSP-11 (Chirasil-Val) CSP-12 (RSL-007) CSP-13 (XE-60-*S*-Val-*S*-α-PEA)

Figure 3.5.6 Polysiloxane bonded amide phases (selected examples).

Due to the high thermal stability and the low tendency to column bleeding, these phases are very useful for GC-MS analysis [56], including single ion detection (single ion monitoring = SIM) [57]. The first application of a chiral phase ('Chirasil-Val') in multidimensional gas chromatography (MDGC) was described by Wang et al. [58].

Improving the surface by deactivating the glass and using fused silica capillaries, the separation of enantiomers with underivatized hydroxyl groups has finally been achieved [59]. Another important factor for better resolution is the number of dimethylpolysiloxane units between the chiral moieties. Phases with a smaller number of 'diluting' dimethylpolysiloxane groups are thermally less stable and, due to high melting temperatures, unsuitable as stationary phases [60].

A phase similar to 'Chirasil-Val' was described by Saeed et al. [61] (CSP-12). While 'Chirasil-Val' contains a polysiloxane similar to OV-101 (see Figure 3.5.6), Saeed et al. used OV-225, which exhibits, due to the phenyl groups in the polymer, a higher polarity in comparison to 'Chirasil-Val' (CSP-11). The cyano group of OV-225 was hydrolyzed to the carboxylic function and then converted to the acid chloride, which was connected later to *S*-valine tBA. In 1981, König et al. [62] introduced another polysiloxane-coupled phase, i.e. XE-60-*S*-valine-*S*-α-PEA (CSP-13). After hydrolyzation and conversion to the acid chloride of XE-60, they connected *S*-valine-*S*-1-phenylethylamide to it. With *S*-1-PEA, an additional chiral group was introduced. By immobilization (cross-bonding) of the stationary phase to the capillary wall the temperature stability was further improved. Besides reducing column bleeding, the stability against water-containing samples was clearly raised. Benecke and Schomburg [63] immobilized the phase XE-60-*S*-Val-*S*-α-PEA (CSP-13) and its diasteriomeric phase XE-60-*S*-Val-*R*-α-PEA (CSP-14); later 'Chirasil-Val' (CSP-11) was also immobilized [64; 65]. If produced under optimal conditions, 'Chirasil-Val' can endure temperatures above 300°C for a short time without any damage. Immobilized phases are also useful for SFC (see Section 3.6) [64; 66]. The most recent approach has been a more reproducible synthesis of 'Chirasil-Val' [67]. A number of monomeric chiral stationary phases [68] like 11-undecenoyl-*S*-valine tBA (CSP-15) have been immobilized (via the 10,11-epoxy derivative) [69]. Developing *R*,*R*-2,3-N-butyl-N'-isopropyltartramide (bonded on OV-1701) as a stationary phase (CSP-16) Nakamura et al. [70] showed that high separation efficiencies can also be achieved with phases free of amino acids.

It should be mentioned that nearly all amino acids present in amide phases show *S*-configuration. In the case of very high enantiomeric excess, overlapping of the second enantiomer can occur if the quantitatively dominating enantiomer elutes first, hampering an exact determination of the enantiomeric ratio. Consequently, a number of chiral phases with *R*-amino acids has been developed to reverse the order of elution - that is, the minor enantiomer is eluted first. Typical examples are the ureide phase carbonylbis[*R*-leucine isopropyl ester] (CSP-17) [71], as well as the polysiloxane-bonded phase 'D-Chirasil-Val' (CSP-18) [72]. As mentioned above, the phase CSP-14 is a diasteriomer of CSP-13. However, no reverse in the order of elution has been observed, since only the chiral centre in valine is responsible for chiral recognition, whereas the 1-phenylethylamide group contains another chiral centre [73].

The development of amide phases is still in progress. This is documented by some more recently published papers [74-79]. As shown by Oi et al. [78] with *S*-1,1'-binaphthyl-2,2'-bis(N-decylcarboxamide) a molecule with axial chirality has for the first time been used as a stationary phase (CSP-19).

The amide phases presented in this chapter comprise only a small part of the synthesized phases (there are more than 200). Due to the great variety of different phases, it is not easy to find a suitable one for one's purposes. Therefore Aichholz et al. [80] have developed a standard test mixture for amide phases (CHIRAL-Test I), to evaluate the separation efficiencies of the different phases.

The separation mechanisms

In the first interpretations of enantiomer separation Feibush and Gil-Av [81] suggested that association complexes were formed rapidly and reversibly with hydrogen bonds between carbonyl and amide functions of selector and selectand (Figure 3.5.7). Depending on the relative configuration of the selector and the selectand, the fusion to diastereomeric associates yields different stable complexes. By the introduction of dipeptide and diamide phases, two more hydrogen bonds are possible during the analysis of amino acids (Figure 3.5.7a).

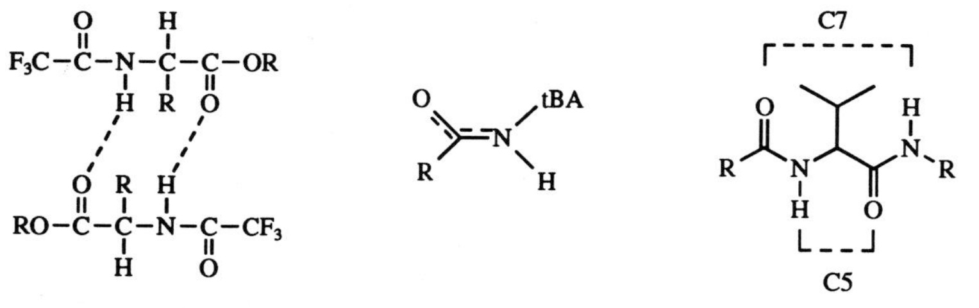

Figure 3.5.7 Model conceptions. a = Association of selector and selectand [81], b = Planar trans-conformation of a peptide bond [84], c = Intramolecular hydrogen bonds of diamides [26].

Besides hydrogen bonds, however, other enantiospecific association interactions can lead to the discrimination of enantiomers (e.g., dipole-dipole-interactions) [82]. Stölting and König [83] have observed a separation of a racemate of N-TFA-proline esters on N-TFA-*S*-prolyl *S*-proline cyclohexyl ester (CSP-20) as the stationary phase. In this selector/selectand system no amide hydrogen atoms are available to form hydrogen bonds.

More recent models are based on an understanding of the structure and conformation of peptides [84]. Assuming a planar trans-conformation of the amide function (Figure 3.5.7b), diamides in high dilutions are able to form intramolecular hydrogen bonds (Figure 3.5.7c). This leads to a planar five-membered ring (C_5-unit) or a folded seven-membered ring (C_7-unit). However, it has been suggested that under gas chromatographic conditions the melted diamides can oligomerize to form intermolecular hydrogen bonds to C_5-C_5-, C_5-C_7- and C_7-C_7-associates [26]. The pleated sheet structure of peptides is a typical example of this interaction.

According to Figure 3.5.8, associates with the ring size 5 + 5 = 10 and 7 + 7 = 14 will be formed if the arrangement of the chains is parallel, and associates with the ring size 5 + 7 = 12 and 7 + 5 = 12 will be formed if the arrangement of the chains is antiparallel. Many spectroscopic and structural investigations with N-acyl-α-amino acid alkylamides have confirmed this model. The preservation of the preferential crystal structure during melting is decisive for the structure of the diamide phases under gas chromatographic conditions. According to Beitler and Feibush [26], a selectand molecule is connected to a diamide selector with a 5 + 5 = 10-ring (Figure 3.5.9), similar to the pleated sheet structure of polypeptides. The observed enantioselectivity can be explained by the different arrangement of the isopropyl group in the diamide phase and the group R^3 of the selectand in the diastereomeric associates L,L' and L,D' (Figure 3.5.9). In accordance with experimental findings, the combination L,L' is more stable than L,D'. All amino acid derivatives with an NH_2-group, i.e. all except proline, can be resolved into their enantiomers, according to the same mechanism.

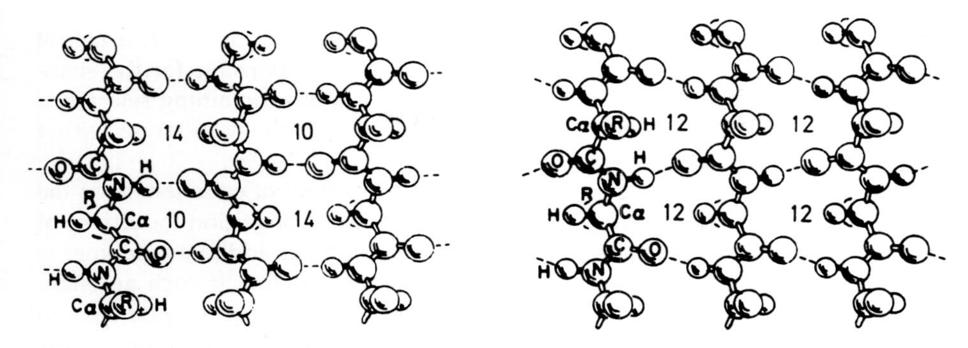

Figure 3.5.8 Pleated sheet structures of polypeptides [85] (the numbers stand for the ring size; left, parallel chains; right, antiparallel chains).

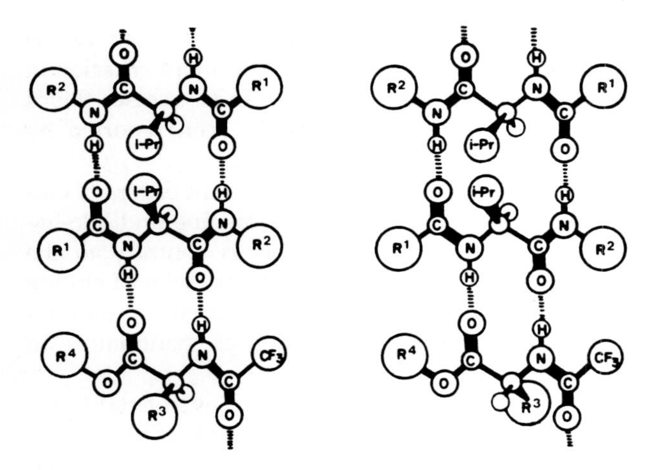

Figure 3.5.9 Association of N-TFA-amino acid derivatives and diamide phases (left, L,L'; right, L,D') [26].

Since models represent only the essentials of experimental findings, other association mechanisms may also be considered. Examples are associations by acyclic hydrogen bonds, van der Waals interactions as well as dipole-dipole attractions [86]. Besides 1:1-association complexes 1:2-interactions should be discussed as well, at which the selectand used (which exists in high dilution) is intercalated 'sand-wich'-like between the molecules of the undiluted selector (which exists in high excess) of the stationary phase. Such intercalation interactions have been suggested for amide phases [40].

In interpreting the enantioselectivity of optically active polysiloxanes (see Figure 3.5.6) the (statistical) distance of the chiral side-chain has to be considered, as well as the preferential conformation of the amino acid selector. According to calculations and statistical evaluations of the conformation of valine residues in polypeptides, the valine amides exist as pleated sheets or, preferentially, as an *R*-α-helical structure [87]. The latter structure especially is hold to be responsible for the enantioselectivity of 'Chirasil-Val' (CSP-11) of bifunctional, oxygen-containing selectands without amide functions. For nitrogen-containing selectands the association with a pleated sheet structure can be considered [87].

A substantial aspect of the gas chromatographic separation of enantiomers is the determination of the order of elution. This can be done by coinjection (addition) of one of the two enantiomers to a racemic mixture, in which the added enantiomer is optically pure (or enriched) and exhibits a known configuration. If such an enantiomer of a homologous compound is not available, it would seem to be easy to predict the order of elution, in analogy to the homologous enantiomeric pair determined before.

For a long time it was assumed that the order of elution within homologous series was always the same. Schurig wrote as late as 1984: "It needn't be especially emphasized that the correlation of configuration and retention behaviour of the selectand with the absolute configuration of the selector is an important criterion in the mechanical evaluation of chiral recognition" [86]. Although in most cases this is correct, there are a number of remarkable exceptions, as has been reported by several authors.

Küsters et al. have found that some allo-compounds show different orders of elution than the corresponding 'non'-allo-compounds (e.g., cystathionine and allo-cystathionine [88] or threoninol and allo-threoninol [89]). This is contrary to the previously established rule, i.e. that within a homologous series the order of elution does not change. Later, two different groups independently reported for the first time a temperature-dependent reversal of the order of elution of enantiomers on diamide phases.

Watabe et al. [31] separated N-TFA-valine tert-butylamide on several stationary phases, such as N-docosanoyl-*S*-leucine 1,1,3,3-tetramethylbutylamide (CSP-21) and Koppenhoefer et al. [90] were able to separate the tripeptide N-TFA-alanylalanylalanine methylester on 'Chirasil-Val'. More recently Lohmiller et al. [79] published similar results on a 'Chirasil-Val'-like phase (polysiloxane containing valine tetrapeptide; CSP-22). Schurig et al. [91] have observed the same phenomenon with spiro ketals on a metal complex phase (CSP-40; see Subsection 3.5.2.2). Last, but not least, this observation has also been reported by König et al. [92] using methyl lactate on a cyclodextrin phase.

It has to be mentioned that with the subsequently discussed cyclodextrin phases (Subsection 3.5.2.3) some examples of the reversal in the order of elution in homologous series are known (e.g., 1-alken-3-ols [93] and 5-alkanolides [94]). Therefore, conclusions drawn as to the order of elution in homologous series are limited.

Strategies of derivatization

For gas chromatographic enantiomer separation by association with hydrogen bonds in nearly all cases a derivatization of the racemic selectand is necessary (see

Table 3.5.4) [82]. By means of derivatization, on the one hand, low volatile and/or polar compounds become accessible to GC analysis; on the other, additional groups, which are able to associate with hydrogen bonds, are introduced into the molecule to improve chiral recognition. During derivatization neither racemization nor side or rearrangement reactions should occur. Moreover, a quantitative reaction with the selectand should be ensured. Acid groups are esterified to increase the volatility (e.g., as methyl or isopropyl esters) and the polar NH_2-, OH- and SH-groups are mostly perfluoroacylated (e.g., alanine as N-trifluoroacetyl methyl ester or serine as N,O-bis-pentafluoropropanoyl isopropyl ester). Even compounds with axial chirality such as 2,2'-diamino-1,1'-bis-naphthyl (as N,N'-bis-TFA-derivative) can be separated into their enantiomers [87]. Perfluoroacylation has often been used also for highly volatile compounds, e.g., short-chained alcohols, as these compounds tend to strong peak broadenings. Applying chiral polysiloxanes in fused silica capillaries, secondary alcohols, polar hydroxy carboxylic acids or diols have been analyzed without derivatization (see Table 3.5.4).

By derivatization with isopropyl isocyanate [95] or tert-butyl isocyanate [96] the chiral recognition of compounds from many different classes has been distinctly improved (e.g., sec. alcohols, diols, 2- and 3-hydroxy carboxylic acids, as well as alkyl branched carboxylic acids). However, it should also be mentioned that some compounds, e.g., mandelic acid, tend to racemize during derivatization with alkyl isocyanates [97].

(1) Isocyanate derivatization (according to [98])

(i) 2-Hydroxycarboxylic acids:

$$\text{lactic acid} + 2\ O{=}C{=}N{-}R \longrightarrow \text{bis(carbamoyl) derivative} + CO_2$$

The example shows the reaction of isocyanate with 2-hydroxypropanoic acid (lactic acid) (R = isopropyl).

(ii) Alkylcarboxylic acids:

$$\text{2-methylbutanoic acid} + O{=}C{=}N{-}R \longrightarrow \text{amide (C(=O)–NH–R)} + CO_2$$

The example shows the reaction of isocyanate with 2-methylbutanoic acid (R= isopropyl).

(iii) Secondary alcohols:

$$\text{2-pentanol} + O{=}C{=}N{-}R \longrightarrow \text{carbamate (O-C(=O)-N(H)-R)}$$

The example shows the reaction of isocyanate with 2-pentanol (R = isopropyl).

(iv) Secondary aminoalkanes:

$$\text{2-aminobutane } (NH_2) + O{=}C{=}N{-}R \longrightarrow \text{urea (N(H)-C(=O)-N(H)-R)}$$

The example shows the reaction of isocyanate with 2-aminobutane (R = isopropyl).

(v) α-Amino acids:

$$\text{amino acid } (NH_2, OH) + O{=}C{=}N{-}R \longrightarrow \text{imidazolidine-2,4-dione} + H_2O$$

The example shows the reaction of isocyanate with 4-methyl-2-aminopentanoic acid to 3-tert-butyl-5-(2-methylpropyl)-imidazolidine-2,4-dione (R=tert-butyl) [99].

(vi) N-Methyl-α-amino acids:

$$\text{N-methyl amino acid} + 2\ O{=}C{=}N{-}R \longrightarrow \text{product (N(H)-R, N-R, H)} + CO_2$$

The example shows the reaction of isocyanate with 3-methyl-2-methylamino-butanoic acid (R = isopropyl).

These reactions are also possible with other positional isomers, e.g., 3-hydroxy-carboxylic acids, 3-arylcarboxylic acids, tertiary alcohols, 3-aminoalkanes. Chiral aldehydes and ketones can be resolved on chiral amide phases after derivatization to oximes (see Table 3.5.3) [100]. A problem arises, as pairs of syn- and anti-oximes result; for instance, rac. fenchone gives two pairs of enantiomers. During storage or when higher reaction temperatures are used, the thermodynamically less stable syn-isomer disappears; for instance, rac. fenchone then yields only one pair of enantiomers [101].

+ HO-$NH_3^+Cl^-$ ⟶ N-OH + HCl + H_2O

The example shows the reaction of fenchone with hydroxylammonium chloride to fenchone oxime.

Aminoalcohols, which are found among important pharmaceutical compounds, N-methyl-α-amino acids, 2-hydroxy carboxylic acids and diols can also be derivatized with phosgene [102]. The results are cyclic derivatives (e.g., carbonates), which can easily be resolved on chiral amide phases.

(2) Phosgene derivatization [103]

(i) 1,2-Diols:

HO OH + Cl Cl O ⟶ O O O + 2 HCl

The example shows the phosgene reaction of 2,3-butanediol to 4,5-dimethyl-1,3-dioxolane-2-one.

(ii) 1,3-Diols:

OH OH + Cl Cl O ⟶ O O O + 2 HCl

The example shows the phosgene reaction of 1,3-pentandiol to 4-ethyl-1,3-dioxane-2-one.

(iii) 2-Hydroxycarboxylic acids:

The example shows the phosgene reaction of 2-hydroxybutanoic acid to 5-ethyl-1,3-dioxolane-2,4-dione.

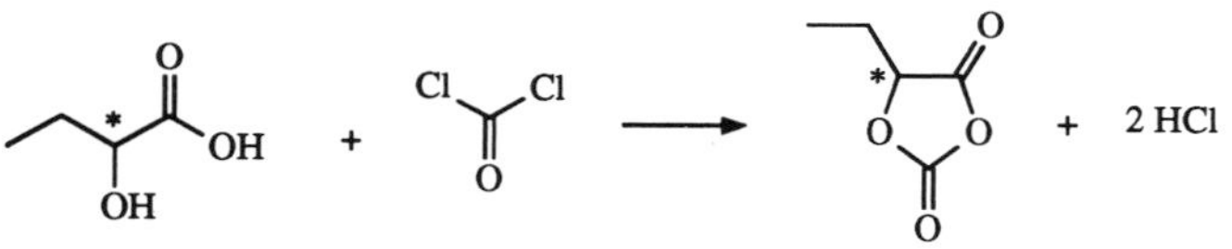

(iv) N-Alkyl-1,2-aminoalcohols:

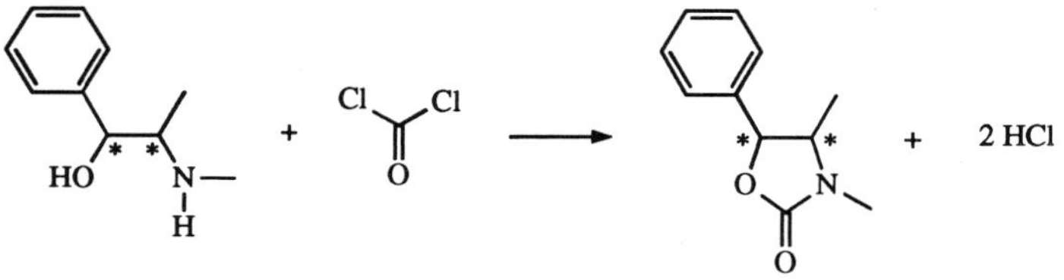

The example shows the phosgene reaction of 2-methylamino-1-phenyl-1-propanol (ephedrine) to 3,4-dimethyl-5-phenyl-1,3-oxazolidine-2-one.

(v) N-Methyl-α-amino acids:

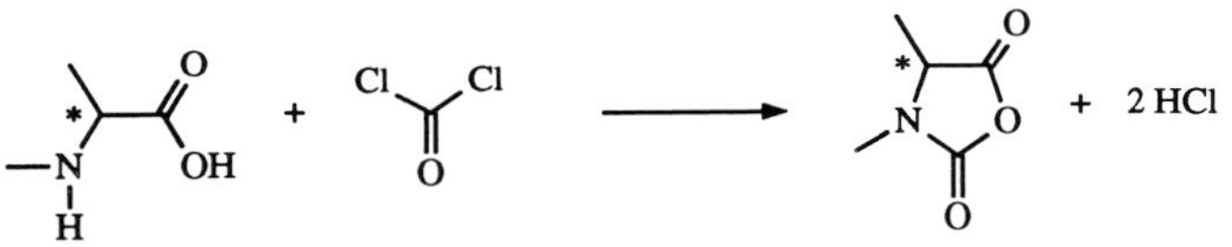

The example shows the phosgene reaction of 2-methylaminopropanoic acid (N-methylalanine) to 3,4-dimethyl-1,3-oxazolidine-2,5-dione.

(vi) α-Amino acid amides:

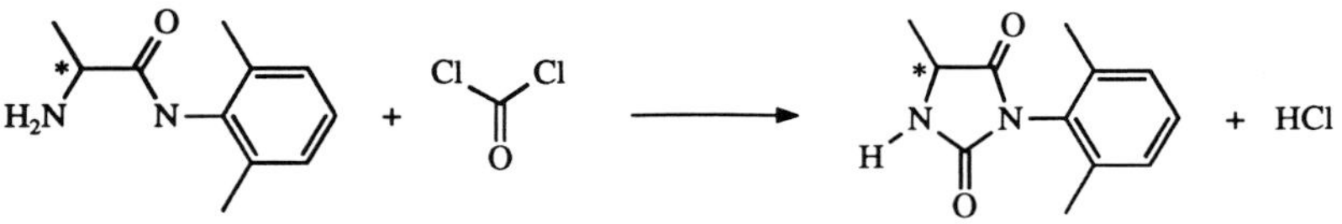

The example shows the phosgene reaction of 2-amino-N-(2,6-dimethylphenyl)-propanoyl amide (tocainide) to 3-(2,6-dimethylphenyl)-5-methylimidazoline-2,4-dione (tocainide-hydantoin).

The reaction principles and products of compounds with only one reactive functional group, e.g. sec. alcohols, sec. aminoalkanes or α-amino acid ester, are outlined in Section 3.5.1. Other suitable derivatives of hydroxycarboxylic acids are 3-pentyl esters. Koppenhoefer and Allmendinger [104] have shown that only 3-pentyl esters give satisfactory results in terms of racemization, side-reaction and chromatographic properties among all tested derivatives (e.g., methyl esters, isopropyl esters, isopropylamides).

Even terpenes, e.g., α-pinene, are accessible to chiral GC after hydroborating and subsequent derivatization with isopropyl isocyanate [105]. Nevertheless, for many compounds of various classes the above-mentioned techniques cannot be used or yield only poor results (e.g., ethers, esters, lactones, saturated hydrocarbons). In these cases, the use of other chiral phases is necessary (e.g., metal complex or inclusion chromatography; see Subsections 3.5.2.2 and 3.5.2.4).

In Table 3.5.4 many examples are listed of racemates successfully separated on commercially available amide phases, with priority given to flavour compounds, pheromones, pharmaceutical compounds, pesticides, as well as to amino acids and carbohydrates. This table gives an overview about the properties and application possibilities of these phases. For additional or more comprehensive data about the abilities of amide phases, a database such as CHIRBASE-GC™ [106-108] will deliver sufficient information.

References

[1] Karagounis G., Lippold G. *Naturwissenschaften* (1959), *46*, 145

[2] Goldberg G., Ross W.A. *Chem. Ind. (London)* (1962), 657

[3] Coleman C.B., Cooper D.C., O'Donnell J.F. *J. Org. Chem.* 1966, *31*, 975

[4] Irurre Pérez J., Julia Aréchaga S., Sanz Burata M. *Anales Fís. Quím.* 1966, *62B*, 595

[5] Lynch C.T. *Diss. Abstracts* 1961, *21*, 3609

[6] Goeckner A. *Diss. Abstracts* 1959, *19*, 3127

[7] Karagounis G., Lemperle E. *Angew. Chem.* 1962a, *74*, 185

[8] Karagounis G., Lemperle E. *Fresenius Z. Anal. Chem.* 1962b, *189*, 131

[9] Sievers R.E., Moshier R.W., Morris M.L. *Inorg. Chem.* 1962, 1, 966

[10] Murai A., Tachikawa Y. *J. Chromatogr.* 1964, *14*, 100

[11] Gil-Av E., Feibush B., Charles-Sigler R. *Tetrahedron Lett.* 1966, *7*, 1009

[12] Gil-Av E., Feibush B., Charles-Sigler R. in: *6th Int. Symp. on Gas Chromatography and Associated Techniques, Rome 1966,* Littlewood A.B. (Ed.), Inst. of Petroleum: London, 1967, p. 227

[13] Parr W., Yang C., Bayer E., Gil-Av E. in: *Advances in Chromatography 1970, Proc. 6th Int. Symp.,* A. Zlatkis (Ed.), Chromatogr. Symp. Dep. Chem. Univ. Houston: Houston, 1970, p. 287

[14] Corbin J.A., Rhoad J.E., Rogers L.B. *Anal.Chem.* 1971, *43*, 327

[15] Gil-Av E., Feibush B. *Tetrahedron Lett.* 1967, *8*, 3345

[16] König W .A., Parr W., Lichtenstein H.A., Bayer E., Oró J. *J.Chromatogr. Sci.* 1970, *8*, 183

[17] Parr W., Howard P. *J. Chromatogr.* 1972, *66*, 141

[18] König W.A., Nicholson G.J. *Anal. Chem.* 1975, *47*, 951

[19] Andrawes F., Brazell R., Parr W., Zlatkis A. *J. Chromatogr.* 1975, *112*, 197

[20] Brazell R., Parr W., Andrawes F., Zlatkis A. *Chromatographia* 1976, *9*,57

[21] Smith G.G., Wonnacott D.M. *Anal. Biochem.* 1980, *109*, 414
[22] Gil-Av E., Feibush B. US 3,494,105 (Cl.B01D15/08), 10 Feb 1970
[23] Feibush B. *J.Chem.Soc Chem. Commun.* 1971, 544
[24] Grohmann K., Parr W. *Chromatographia* 1972, *5*, 18
[25] Charles R., Beitler U., Feibush B., Gil-Av E. *J. Chromatogr.* 1975, *112*, 121
[26] Beitler U., Feibush B. *J. Chromatogr.* 1976, *123*, 149
[27] Charles R., Gil-Av E. *J. Chromatogr.* 1980, *195*, 317
[28] Chang S.-C., Charles R., Gil-Av E. *J. Chromatogr.* 1980, *202*, 247
[29] Ti N., Kitahara H., Inda Y., Doi T. *J. Chromatogr.* 1982, *237*, 297
[30] Faklam C., Kortus K, Oehme G. *Z. Chem.* 1985, *25*, 438
[31] Watabe K., Charles R., Gil-Av E. *Angew. Chem.* 1989, *101*, 195
[32] Lou X., Zhan Z., Zhou L. Sepu 1990, 8, 240-243 (C.A. 1991, 114: 114392g)
[33] Feibush B., Gil-Av E. *J. Gas Chromatogr.* 1967, *5*, 257
[34] Feibush B., Gil-Av E., Tamari T. *J. Chem. Soc. Perkin Trans. II* 1972, 1197
[35] Lochmüller C.H., Harris J.M, Souter R.W. *J. Chromatogr.* 1972, *71*, 405
[36] Lochmüller C.H., Souter R.W. *J. Chromatogr.* 1973, *87*, 243
[37] Lochmüller C.H., Souter R.W. *J. Chromatogr.* 1974, *88*, 41
[38] Souter R.W. *J. Chromatogr.* 1975, *114*, 307
[39] Stan'kov I.N., Tarasov S.N., Agureyev V.G., Beresnev A.N., Lysenko V.V., Lanin S.N. *Anal. Chim. Acta* 1991, *251*, 223
[40] Weinstein S., Feibush B. Gil-Av E. *J. Chromatogr.* 1976, *126*, 97
[41] König W.A., Sievers S., Schulze U. *Angew. Chem.* 1980, *92*, 935
[42] König W.A., Sievers S. *J. Chromatogr.* 1980, *200*, 189
[43] Ti N., Kitahara H., Inda Y., Doi T. *J. Chromatogr.* 1981, *213*, 13
[44] Ti N., Doi T., Kitahara H., Inda Y. *Bunseki Kagaku* 1981, *30*, 552 (C.A. 1981, 95: 196853j)
[45] König W.A., Benecke I., Sievers S. *J. Chromatogr.* 1981, *217*, 71
[46] Ti N., Kitahara H., Doi T. *J. Chromatogr.* 1983, *254*, 282
[47] Horinouchi T., Hobo T., Suzuki S., Watabe K., Gil-Av E. in: *7th Int. Symp. on Capillary Chromatography, Gifu, Japan 1986*, Ishii D, Jinno K., Sandra P. (Eds.), Univ. of Nagoya Press: Nagoya, 1986, p. 499
[48] Ti N., Takeda H., Shimada H. DE 27 20 995 (Cl.C07D251/26), 24 Nov 1977
[49] Ti N., Moriguchi K., Matsuda M. US 4,104,040 (Cl.B01D15/08), 1 Aug 1978
[50] Ti N., Moriguchi K., Matsuda M., Shimada H., Hiroaki O. *Bunseki Kagaku* 1978, *27*, 637 (C.A. 1979, 90:152563g)
[51] Ti N., Hiroaki O., Shimada H., Horiba M., Kitahara H. *Bunseki Kagaku* 1980, *29*, 270 (C.A. 1980, 93:186758b)
[52] Ti N., Kitahara H., Matsushita Y. in: *12th Int. Symp. on Capillary Chromatography, Kobe, Japan 1990*, Jinno K., Sandra P. (Eds.), Ind. Publ. Consult.: Tokyo ,1990, p. 828
[53] Ti N., Horiba M., Kitahara H. *J. Chromatogr.* 1980, *202*, 299
[54] Frank H., Nicholson G.J., Bayer E. *J. Chromatogr. Sci.* 1977, *15*, 174
[55] Bayer E., Frank H. DE 27 40 019 (Cl.C08G77/40), 22 Mar 1979
[56] Woiwode W., Frank H., Nicholson G.J., Bayer E. *Chem. Ber.* 1978, *111*, 3711
[57] Farmer P.B., Bailey E., Lamb J.H., Connors T.A. *Biomed. Mass Spectrom.* 1980, *7*, 41
[58] Wang C., Frank H., Wang G., Zhou L., Bayer E., Lu P. *J. Chromatogr.* 1983, *262*, 352

[59] Nicholson G.J., Frank H., Bayer E. *J. High Resol. Chromatogr. Chromatogr. Commun.* 1979, *2*, 411
[60] Frank H., Nicholson G.J., Bayer E. *J. Chromatogr.* 1978, *146*, 197
[61] Saeed T., Sandra P., Verzele M. *J. Chromatogr.* 1979, *186*, 611
[62] König W.A., Benecke I., Bretting H. *Angew. Chem.* 1981, *93*, 688
[63] Benecke I., Schomburg G. *J. High Resol. Chromatogr. Chromatogr. Commun.* 1985, *8*, 191
[64] Frank H. in: *9th Int. Symp. on Capillary Chromatography, Monterey, 1988,* Sandra P. (Ed.), Hüthig: Heidelberg, 1988, p. 124
[65] Lai G., Nicholson G., Bayer E. *Chromatographia* 1988, *26*, 229
[66] Brügger R., Krähenbühl P., Marti A., Straub R., Arm H. *J. Chromatogr.* 1991, *557*, 163
[67] Frank H., Abe I., Fabian G. *J. High Resol. Chromatogr. Chromatogr. Commun.* 1992, *15*, 444
[68] Fabian G., Schierhorn M., Pörschmann J., Kraus G., Altmann H., Zaschke H. DD 241 410 A1 (Cl.C07C103/50), 10 Dec 1986
[69] Fabian G. DD 266 741 A1 (Cl.B01D15/08), 12 Apr 1989
[70] Nakamura K., Hara S., Dobashi Y. *Anal. Chem.* 1989, *61*, 2121
[71] Lochmüller C.H., Souter R.W. *J. Chromatogr.* 1973, *87*, 243
[72] Bayer E., Allmendinger H., Enderle G., Koppenhoefer B. *Fresenius Z. Anal. Chem.* 1985, *321*, 321
[73] König W.A., Benecke I., Sievers S. *J. Chromatogr.* 1981, *217*, 71
[74] Koppenhoefer B., Abdalla S., Hummel M. *Chromatographia* 1991, *31*, 31
[75] Marti A., Krähenbühl P., Brügger R., Arm H. *Chimia* 1991, *45*, 13
[76] Petersson P., Markides K.E., Johnsson D.F., Eguchi M., Reese S. L., Curtis J., Rossiter B.E., Bradshaw J.S., Lee M.L. in: *13th Int. Symp. on Capillary Chromatography, Riva del Garda 1991,* Sandra P. (Ed.), Hüthig: Heidelberg, 1991, p. 242
[77] Lou X., Liu X., Zhou L. *13th Int. Symp. on Capillary Chromatography, Riva del Garda 1991,* Sandra P. (Ed.), Hüthig: Heidelberg , 1991, p. 169
[78] Ti S., Ochiai Y., Miyano S. *Chem. Lett.* 1991, 1575
[79] Lohmiller K., Bayer E., Koppenhoefer B. *J. Chromatogr.* 1993, *634*, 65
[80] Aichholz R., Bölz U., Fischer P. *J. High Resol. Chromatogr. Chromatogr. Commun.* 1990, *13*, 234
[81] Feibush B., Gil-Av E. *Tetrahedron* 1970, *26*, 1361
[82] König W.A. *The Practice of Enantiomer Separation by Capillary Gas Chromatograph,* Hüthig: Heidelberg , 1987
[83] Stölting K., König W.A. *Chromatographia* 1976, *9*, 331
[84] Gil-Av E. *J.Mol.Evol.* 1975, *6*, 131
[85] Feibush B., Balan A., Altman B., Gil-Av E. *J. Chem. Soc. Perkin Trans. II* 1979, 1230
[86] Schurig V. *Angew. Chem.* 1984, *96*, 733
[87] Bayer E. *Z. Naturforsch.* 1983, *38b*, 1281
[88] Küsters E., Allgaier H., Jung G., Bayer E. *Chromatographia* 1984, *18*, 287
[89] Küsters E., Rahmen W. *Z. Naturforsch.* 1988, *43b*, 619
[90] Koppenhoefer B., Lin B., Muschalek V., Trettin U., Willisch H., Bayer E. in: *Peptides 1988, Proc. 20th Eur. Peptide Symp., Tübingen 1988,* Jung G., Bayer E. (Eds.), de Gruyter: Berlin, New York, 1989, p. 109
[91] Schurig V., Ossig J., Link R. *Angew. Chem.* 1989, *101*,197

[92] König W.A., Icheln D., Runge T., Pfaffenberger B., Ludwig P., Hühnerfuss H. *J. High Resol. Chromatogr. Chromatogr. Commun.* 1991, *14*, 530

[93] Mosandl A., Rettinger K., Fischer K., Schubert V., Schmarr H.-G., Maas B. *J. High Resol. Chromatogr. Chromatogr. Commun.* 1990, *13*, 382

[94] Bernreuther A., Bank J., Krammer G., Schreier P. *Phytochem. Anal.* 1991, 2, 43

[95] König W.A., Benecke I., Sievers S. *J. Chromatogr.* 1982, *238*, 427

[96] König W.A., Benecke I., Lucht N., Schmidt E., Schulze J., Sievers S. *J. Chromatogr.* 1983, *279*, 555

[97] Koppenhoefer B., Allmendinger H., Nicholson G. *Angew.Chem.* 1985, *97*, 46

[98] Benecke I., König W.A. *Angew. Chem.* 1982, *94*, 709

[99] König W.A. in: *Drug Stereochemistry,* Wainer I.W., Drayer D.E. (Eds.), Dekker: New York, 1988, p. 113

[100] König W.A. *J. High Resol. Chromatogr. Chromatogr. Commun.* 1982, 5, 588

[101] König W.A., Benecke I., Ernst K. *J. Chromatogr.* 1982, *253*, 267

[102] König W.A., Ernst K., Vessman J. *J. Chromatogr.* 1984, *294*, 423

[103] König W.A., Steinbach E., Ernst K. *Angew. Chem.* 1984, *96*, 516

[104] Koppenhoefer B., Allmendinger H. *Fresenius Z. Anal. Chem.* 1987, *326*, 434

[105] Lindstrom M., Norin T., Birgersson G., Schlyter F. *J. Chem. Ecol.* 1989, *15*, 541

[106] Roussel C., Piras P. *Pure Appl. Chem.* 1993, *65*, 235

[107] Koppenhoefer B., Nothdurft A., Stiebler M., Trettin U., Pierotanders J., Piras P., Popescu D., Roussel C. *GIT Fachz. Lab.* 1993, *37*, 225

[108] Koppenhoefer B., Nothdurft A., Pierrot-Sanders J., Piras P., Popescu C., Roussel C., Stiebler M., Trettin U. *Chirality* 1993, *5*, 213

[109] Hashimoto K., Takahashi M., Inoue T., Sumida Y., Terada S., Watanabe C. in: *7th Int. Symp. on Capillary Chromatography, Gifu, Japan 1986,* Ishii D, Jinno K., Sandra P. (Eds.), Univ. of Nagoya Press: Nagoya, 1986, p. 523

[110] Horiba M., Kitahara H., Yamamoto S., Ti N. *Agr. Biol. Chem.* 1980, *44*, 2987

[111] Ti N., Takeda H., Shimada H., Hiroaki O. *Bunseki Kagaku* 1979, *28*, 69 (C.A. 1979,90: 214747f)

[112] Ti N. in: *Amino Acid Analysis by Gas Chromatography. Vol. II,* Zumwalt R.W., Kuo K.C.T., Gehrke C.W. (Eds.), CRC Press: Boca Raton, 1987, p. 47

[113] Ti N., Doi T., Kitahara H., Inda Y. *J. Chromatogr.* 1981, *208*, 404

[114] Ti N., Horiba M., Kitahara H. *Bunseki Kagaku* 1979, *28*, 607 (C.A. 1980, 92: 333393h)

[115] Horiba M., Kida S., Yamamoto S., Ti N. *Agr. Biol. Chem.* 1982, *46*, 281

[116] Horiba M., Yamamoto S., Ti N. *Agr. Biol. Chem.* 1982, *46*, 1219

[117] Horiba M., Kitahara H., Yamamoto S., Ti N. *Agr. Biol. Chem.* 1980, *44*, 2989

[118] Ti N., Horiba M., Kitahara H., Shimada H. *J. Chromatogr.* 1980, *46*, 302

[119] Horiba M., Kitahara H., Takahashi K.-I., Yamamoto S., Murano A., Ti N. *Agr. Biol. Chem.* 1979, *43*, 2311

[120] Ti N., Shimada H., Hiroaki O., Horiba M., Kitahara H. *Bunseki Kagaku* 1979, *28*, 64 (C.A. 1979, 90: 179657c)

[121] Ti N., Horiba M., Kitahara H. *Agr. Biol. Chem.* 1981, *45*, 1509

[122] Ti N., Kitahara H., Inda Y., Horiba M., Doi T. *Bunseki Kagaku* 1981, *30*, 254 (C.A. 1981, 95:54356a)

[123] Ti N., Kitahara H., Matsushita Y. *J. High Resol. Chromatogr. Chromatogr. Commun.* 1990, *13*, 720
[124] Ti N., Doi T., Kitahara H., Inda Y. *Bunseki Kagaku* 1981, *30*, 79 (C.A. 1981, 94:149716m)
[125] Ti N., Kitahara H., Horiba M., Doi T. *J. Chromatogr.* 1981, *206*, 143
[126] Ti N., Doi T., Kitahara H., Inda Y. *J. Chromatogr.* 1982, *239*, 493
[127] Ti N., Takai R., Kitahara H. *J. Chromatogr.* 1983, *256*, 154
[128] Rubinstein H., Feibush B., Gil-Av E. *J. Chem. Soc. Perkin Trans. II* 1973, 2094
[129] Dang T.P., Poulin J.-C., Kagan H.B. *J. Organomet. Chem.* 1975, *91*, 105
[130] Wilken D.R., Dyar R.E. *Anal. Biochem.* 1981, *112*, 9
[131] Koppenhoefer B., Allmendinger H., Nicholson G.J., Bayer E. *J. Chromatogr.* 1983, *260*, 63
[132] Frank H., Gerhardt J., Nicholson G.J., Bayer E. *J. Chromatogr.* 1983, *270*, 159
[133] Koppenhoefer B., Allmendinger H. *Chromatographia* 1986, *21*, 503
[134] Francke W., Pan M.-L., Bartels J., König W.A., Vité J.P., Krawielitzki S., Kohnle U. *J. Appl. Ent.* 1986, *101*, 453
[135] Takeichi T., Shimura S., Toriyama H., Takayama Y., Morikawa M. *Chem. Lett.* 1992, 1069
[136] Taylor W.G., Schreck C.E. *J. Pharm. Sci.* 1985, *74*, 534
[137] Koppenhoefer B., Bayer E. *J. Chromatogr. Sci.* 1985, *32*, 1
[138] Brunner H., Becker R. *Angew. Chem.* 1985, *97*,7 13
[139] Koppenhoefer B., Bayer E. *Chromatographia* 1984, *19*, 123
[140] Koppenhoefer B. Doctoral Thesis, University of Tübingen, Germany 1980
[141] Günther K., Martens, J., Messerschmidt M. *J. Chromatogr.* 1984, *288*, 203
[142] Hosten N., Anteunis M.J.O. *Bull. Soc. Chim. Belg.* 1988, *97*, 45
[143] Brückner H., Schäfer S., Bahnmüller D., Hausch M. *Fresenius Z. Anal. Chem.* 1987, *327*, 30
[144] Link R., Faller J. in: *8th Int. Symp. on Capillary Chromatography, Riva del Garda 1987*, Sandra P. (Ed.), Hüthig: Heidelberg, 1987, p. 1252
[145] Brückner H., Langer M. *J. Chromatogr.* 1991, *542*, 161
[146] Demange P., Abdallah M.A., Frank H. *J. Chromatogr.* 1988, *438*, 291
[147] Frank H., Thiel D. *J. Chromatogr.* 1984, *309*, 261
[148] Frank H., Eimiller A. *J. Chromatogr.* 1981, *224*, 177
[149] Bayer E., Küsters E., Nicholson G.J., Frank H. *J. Chromatogr.* 1985, *320*, 393
[150] Jemal M., Cohen A.I. *J. Chromatogr.* 1987, *392*, 442
[151] Vlasáková V., Tolman V., Zivny *J. Chromatogr.* 1993, *639*, 273
[152] Wagner J., Wolf E., Heintzelmann B., Gaget C. *J. Chromatogr.* 1987, *392*, 211
[153] Leavitt A.L., Sherman W.R. *Carbohydr. Res.* 1982, *103*, 203
[154] Koppenhoefer B., Walser M., Bayer E. *J. Chromatogr.* 1986, *358*, 159
[155] Koppenhoefer B., Lin B. *J. Chromatogr.* 1989, *481*, 17
[156] Koppenhoefer B., Walser M., Schröter D., Häfele B., Jäger V. *Tetrahedron* 1987, *43*, 2059
[157] Sherman W.R., Leavitt A.L., Honchar M.P., Hallcher L.M., Phillips B.E. *J. Neurochem.* 1981, *36*, 1947
[158] Francke W., Pan M.-L., König W.A., Mori K., Puapoomchareon P., Heuer H., Vité J.P. *Naturwissenschaften* 1987, *74*, 343
[159] Bricout J. in: *Flavour Science and Technology*, Martens M., Dalen G.A. (Eds.), Wiley: Chichester, 1987, p. 87
[160] Koppenhoefer B., Allmendinger H., Chang L.P., Cheng L.B. *J. Chromatogr.* 1988, *441*, 89

[161] Koppenhoefer B., Allmendinger H., Bayer E. *J. High Resol. Chromatogr. Chromatogr. Commun.* 1987, *10*, 324
[162] Liu D.-M. *J. Chromatogr.* 1992, *589*, 249
[163] Clark T., Deas A.H.B. *J. Chromatogr.* 1985, *329*, 181
(164) Frank H, Nicholson G.J., Bayer E. *Angew.Chem.* 1978, *90*, 396
[165] McErlane K.M., Pillai G.K. *J. Chromatogr.* 1983, *274*, 129
[166] Antonsson A.-M., Gyllenhaal O., Kylberg-Hanssen K., Johansson L., Vessman J. *J. Chromatogr.* 1984, *308*, 181
[167] Gyllenhaal O., Vessman J. *J. Chromatogr.* 1987, *411*, 285
[168] Wedlund P.J., Sweetman B.J., McAllister C.B., Branch R.A., Wilkinson G.R. *J. Chromatogr.* 1984, *307*, 121
[169] Sioufi A., Kaiser G., Leroux F., Dubois J.P. *J. Chromatogr.* 1988, *450*, 221
[170] Frank H. in: *Chirality and Biological Activity. Proc. Int. Symp., Tübingen 1988,* Holmstedt B., Frank H., Testa B. (Eds.), Liss: New York, 1990, p. 33
[171] Beck O., Repke D.B., Faull K.F. *Biomed. Environ. Mass Spectrom.* 1986, *13*, 469
[172] Koppenhoefer B., Allmendinger H., Peters B. *Liebig's Ann. Chem.* 1987, 991
[173] Becker R., Brunner H., Mahboobi S., Wiegrebe W. *Angew. Chem.* 1985, *97*, 969
[174] Wu L., Liu G., Mi A., Jiang Y. *Sepu* 1992, *10*, 285
[175] Schenkel E., Duez P., Hanocq M. *J. Chromatogr.* 1992, *625*, 289
[176] Koch E., Nicholson G.J., Bayer E. *J. High Resol. Chromatogr. Chromatogr. Commun.* 1984, *7*, 398
[177] Jung G., Brückner H. *Hoppe Seyler's Z. Physiol. Chem.* 1981, *362*, 275
[178] Singer G., Heusinger G., Mosandl A., Burschka C. *Liebig's Ann. Chem.* 1987, 451
[179] Odham G., Tunlid A., Larsson L., Mårdh P.-A. *Chromatographia* 1982, *16*, 83
[180] König W.A., Francke W., Benecke I. *J. Chromatogr.* 1982, *239*, 227
[181] Benecke I., König W.A. *Angew. Chem. Suppl.* 1982, 1605
[182] Schomburg G., Husmann H., Hübinger E., König W.A. *J. High Resol. Chromatogr. Chromatogr. Commun.* 1984, *7*, 404
[183] König W.A. in: *Analysis of Volatiles,* Schreier P. (Ed.), de Gruyter: Berlin, 1984, p. 77
[184] Satterwhite D.M., Croteau R.B. *J. Chromatogr.* 1988, *452*, 61
[185] Bruhn G., Faasch H., Hahn H., Hausen B.M., Bröhan J., König W.A. Z. *Naturforsch.* 1987, *42b*, 1328
[186] Bergstrom G., Wassgren A.-B., Hogberg H.-E., Hedenstrom E., Hefetz A., Simon D., Ohlsson T., Lofqvist J. *J. Chem. Ecol.* 1992, *18*, 1177
[187] Hogberg H.-E., Hedenstrøm E., Wassgren A.-B., Hjalmarsson M., Bergstrom G., Lofqvist J., Norin T. *Tetrahedron* 1990, *46*, 3007
[188] König W.A., Sievers S., Benecke I. in: *Proc. 3rd Int. Symp. Glass Capillary Chromatography, Hindelang 1979,* Kaiser R.E. (Ed.), Inst. Chromatogr.: Bad Dürkheim, 1979, p. 703
[189] Aichholz R., Fischer P. *J. High Resol Chromatogr. Chromatogr. Commun.* 1989, *12*, 213
[190] Schmidt N., Gercken G., König W.A. *J. Chromatogr.* 1987, *410*, 458
[191] Wong B., Castellanos M. *J. Chromatogr.* 1989, *495*, 21
[192] König W.A., Benecke I. *J. Chromatogr.* 1983, *269*, 19
[193] Rieck M., Hagen M., Lutz S., König W.A. *J. Chromatogr.* 1988, *439*, 301
[194] Anklam E., König W.A., Margaretha P. *Tetrahedron Lett.* 1983, *24*, 5851
[195] Tsai W.-L., Hermann K., Hug E., Rohde B., Dreiding A.S. *Helv. Chim. Acta* 1985, *68*, 2238

[196] König W.A., Sturm U. *J. Chromatogr.* 1985, *328*, 357
[197] Schulz S., Francke W., König W.A., Schurig V., Mori K., Kittmann R., Schneider D. *J. Chem. Ecol.* 1990, *16*, 3511
[198] König W.A., Schmidt E., Krebber R. *Chromatographia* 1984, *18*, 698
[199] Zimmermann G., Hass W., Faasch H., Schmalle H., König W.A. *Liebig's Ann. Chem.* 1985, 2165
[200] Graham S.M., Prestwich G.D. *Experentia* 1992, *48*, 19
[201] König W.A., Ernst K. *J. Chromatogr.* 1983, *280*, 135
[202] König W.A., Gyllenhaal O., Vessman J. *J. Chromatogr.* 1986, *356*, 354

3.5.2.2 Metal complex phases

First attempts to use chiral metal coordination compounds as enantioselective stationary phases for gas chromatographic enantiomer separation were undertaken by Schurig and Gil-Av in 1971 [1]. They applied dicarbonylrhodium(I)-[1*R*-3-TFA-camphorate] dissolved in squalane (CSP-28), but were unable to achieve a successful separation of chiral compounds. It was only in 1977 that Schurig [2] was able to separate the first compound, 3-methylcyclopentene, into its enantiomers on this phase. He was successful using a 200 m stainless steel column; the separation time was approximately 3 hours. To confirm the observed results, the reversal of the order of elution on the mirror-image phase (CSP-29) was performed.

Golding et al. [3] used an europium-containing NMR-shift reagent (see Section 3.2) as a separating phase (CSP-30). The authors achieved a partial separation of 1,2-epoxypropane. Schurig et al. [4] developed a nickel camphorate phase (CSP-31), exhibiting the best separation properties of all the metal complex phases described to date. Later, this phase was dissolved in several polysiloxanes, such as OV-101 (CSP-32) or SE-54 (CSP-33). The latter is commercially available as 'Chirasil-Metal' from CC & CC (D-72138 Kirchenstellinsfurt).

Schurig et al. [7] were the first to separate a threefold coordinated compound into its invertomers (CSP-31), i.e. the sterically hindered N-chloro-2,2-dimethylaziridine. Subsequently, β-diketonates were also used as stationary phases (e.g., carvonates [8], menthonates [9], pinanonates [10], pulegonates [11], thujonates [9]). In addition, other metals, such as cobalt [12], copper [13], manganese [14] and zinc [15], have also been investigated. Among them, besides the already-mentioned nickel camphorate phase, especially the manganese camphorate phases (CSP-34 [14], CSP-35 [16] and CSP-36 [12]; dissolved in squalane, OV-101 and SE-54) are useful for enantiomeric separations. In some cases they are much better than the Ni phases (e.g., for pheromones). In Figure 3.5.10, several structures of β-diketonate phases are presented.

Schurig and Bürkle [17] have observed peak coalescence in the course of the separation of 1,6-dioxaspiro[4.4]nonane on CSP-31. Partial racemization occurred during elution on the stationary phase (inter alia due to unstable chiral centres). Schurig called this dynamic process 'enantiomerization' (Figure 3.5.11; for definitions see Section 3.3). This phenomenon can also occur on cyclodextrin phases (see Subsection 3.5.2.3).

In principle, complexation gas chromatography is suitable for a large variety of chiral substances, such as flavour compounds and pheromones as well as halogen compounds and oxygen-containing heterocycles (e.g., oxiranes, oxetanes, furans, pyrans, dioxolanes, oxaspiranes etc.) (Table 3.5.5). A significant disadvantage, however, is the low temperature stability: the metal complexes are stable only below 150° C. On the other hand, there is no need to derivatize, except for diols, which can be separated as cyclic boronates or ketals (acetonides) [18].

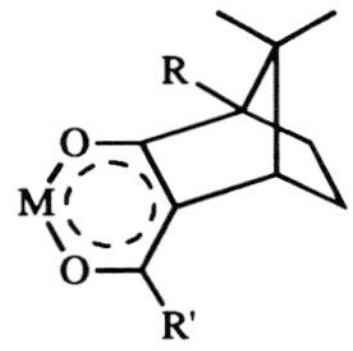

CSP-30: M = Eu/3; R = CH_3; R' = CF_3
CSP-31: M = Ni/2; R = CH_3; R' = C_3F_7
CSP-34: M = Mn/2; R = CH_3; R' = C_3F_7
CSP-38: M = Ni/2; R = C_2H_5; R' = C_3F_7
CSP-40: M = Ni/2; R = C_2H_3; R' = C_3F_7

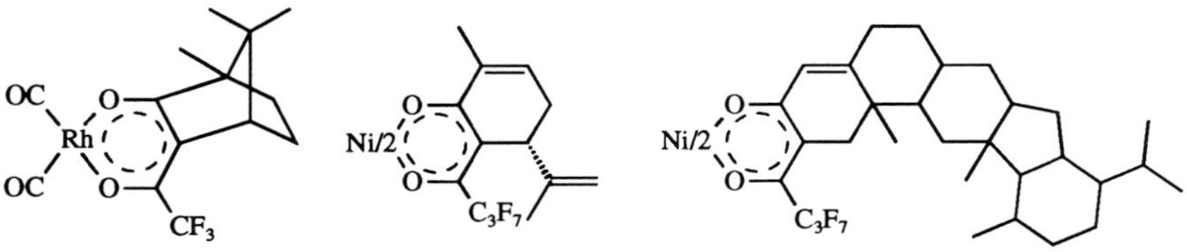

CSP-28 CSP-37 CSP-39

Figure 3.5.10 Structures of metal complex phases based on β-diketonates.

The possible use of such phases in packed columns has been shown by Schurig [19]. For example, CSP-31 or Nickel(II)-bis-[4*S*-3-HFB-carvonate] (dissolved in squalane; CSP-37 [20]) and others have been used for semi-preparative separations of 1-chloro-2,2-dimethylaziridine or oxetone and other spiro ketals. A multidimensional GC-system with capillary columns has been described by Mosandl and Schubert [21].

By variation of the nickel camphorate phases the scope of separable compounds has been extended, e.g., 2-methylbutyl acetate has been separated on Nickel(II)-bis-[1*R*-3-HFP-10-methylcamphorate] (CSP-38). The exchange of camphor by 4-cholesten-2-one (CSP-39) has proved to be unfavourable; it was only possible to separate tetrahydro-2-methylpyran [9]. A further development consists in the covalent bonding of Nickel(II)-bis-1*R*-3-HFB-10-methylencamphorate (CSP-40) [48] on polysiloxane OV-101 (CSP-41; Figure 3.5.12) and immobilization of this phase to the inner surface of fused silica capillaries [45]. Beyond an essential increase in temperature stability (up to 190° C) this phase is even suitable for SFC (see Section 3.6).

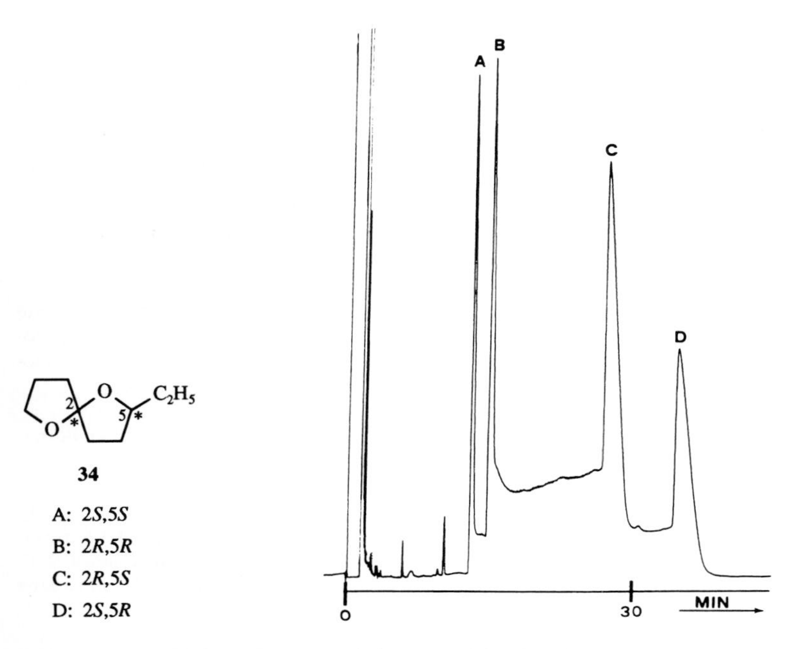

Figure 3.5.11 Peak coalescence of the second kind and coincident epimerization of the diastereomers of the pheromone *E*,*Z*-2-ethyl-1,6-dioxaspiro[4.4]nonane on CSP-32 [12].

Table 3.5.5 Applications of important CSPs for complexation gas chromatography.

Nickel(II)-bis-[1*R*-3-HFB-camphorate] (diss. in squalane) (CSP-31) [4]	
Alcohols	3,3-Dimethyl-2-butanol [17]
Epoxides	1,2-Epoxyalkanes, *E*,*Z*-2,3-epoxyalkanes [4;22]
Ketones	*E*-2,5-Dimethylcyclopentanone [17]
O-Heterocycles	*E*-2,3-Dimethyloxetane [23]; tetrahydro-2-methylfuran, *E*-2,5-tetrahydro-dimethylfuran, tetrahydro-2-methylpyran [17]; 1,6-dioxaspiro[4.4]nonane [23]
Miscellaneous	1-Chloro-2,3-epoxypropane [17]; 1-chloro-2,2-dimethylaziridine [7]; *E*-2,3-dimethylthiirane, *E*-2,4-dimethylthietane [17]
Nickel(II)-bis-[1*R*-3-HFB-camphorate] (diss. in OV-101) (CSP-32) [5]	
Alcohols	2-Methyl-3-alkanols, *E*,*Z*-3-methyl-2-alkanols, *E*,*Z*-4-methyl-3-hexanol,6-methyl-3-heptanol [6]; *Z*-4-methyl-3-heptanol, *E*,*Z*-5-methyl-3-heptanol [10]; 3-methyl-2-cyclohexen-1-ol [10]; iso-/menthol, neomenthol, α-fenchol [5]; isoborneol/borneol, 1-phenylethanol [10]
Alkanediols	1,2-Alkanediols (as butylboronate) [18]

Nickel(II)-bis-[1*R*-3-HFB-camphorate] (diss. in OV-101) (CSP-32) [5]	
Epoxides	1,2-Epoxyalkanes, *E*-2,3-epoxybutane [5;24]; 1-methylcyclopentene oxide, 1-methylcyclohexene oxide, 3-phenyl-1,2-epoxyalkanes [12]; styrene oxide [25]
Esters	1-Octen-3-yl alkanoates [26]; 1-alken-3-yl acetates [27]; 1-alken-3-yl propanoates, linalyl acetate [28]
Ethers	Methyl alkenylethers, methyl alkynylethers, methylcyclohexyl alkynylethers, methylaryl alkynylethers [29]
Ketones	3-Methyl-2-pentanone, 4-methyl-3-hexanone [6]; 4-methyl-3-heptanone [10]; 5-methyl-3-heptanone [6]; *E,Z*-4,6-dimethyl-3-octanone = dihydromanicon [30]; alkylcyclopentanones, alkylcyclohexanones [10], isomenthone/menthone [31]
Lactones	4- and 5-Alkanolides [very long retention times for the higher homologues!], *Z*-3-methyl-4-pentanolide [32]
O-Heterocycles	Tetrahydro-2-methylfuran [33]; *E*-2-ethoxytetrahydro-5-pentylfuran [34]; 2-alkyltetrahydropyrans [33]; *E*-rose oxide [5]; *E,Z*-2-ethyl-1,6-dioxaspiro[4.4]nonane = chalcogran [35]; 2-alkyl-4-methyl-1,3-dioxolanes, 2-alkyl-4,5-dimethyl-1,3-dioxolanes [36]; 2-(2-butyl)-4-methyl-1,3-dioxolane [27]
Miscellaneous	3,3,3-Trichloro-1,2-epoxypropane [24]; *E,Z*-methyl 2-chloro-3-methylpentanoate, methyl 2-chloro-4-methylpentanoate [37]; 1,2,2-trimethylpropylmethylphosphonofluoridate = somane [38]; 2-methyl-1,3-oxathiane, *E,Z*-2,4-dimethyl-1,3-oxathiane, 4-propyl-1,3-oxathiane [26]; *E,Z*-2-methyl-4-propyl-1,3-oxathiane [39]

Chirasil-Metal = Nickel(II)-bis-[(1R)-3-HFB-camphorate] (diss. in SE-54) (CSP-33) [6]	
Alcohols	3,3-Dimethyl-2-butanol, 2-methyl-3-heptanol [12]; 1-octen-3-ol [32]; menthol, isoborneol [12]; 4-terpinenol [6]
Epoxides	1,2-Epoxyalkanes, *E,Z*-2,3-epoxybutane [40]
Ketones	Isomenthone/menthone [40]
Lactones	4-Hexanolide [41]
O-Heterocycles	*E,Z*-2-ethyl-1,6-dioxaspiro[4.4]nonane = chalcogran [41]; endo-/exo-2-hydroxycineole, 2-oxocineole [42]; nerol oxide, 10-homonerol oxide [43]
Miscellaneous	3-Halogeno-2-hydroxybutanes, 4-halogeno-3-hydroxyhexanes, 2-halogenocyclopentanols, 2-halogenocyclohexanols [44]; 1-bromo- and 1-chloro-1,2-epoxypropane [12]

Chirasil-Nickel = Nickel(II)-bis-[1*R*-3-HFB-10-methylencamphorate] (bonded on OV-101 and immobilized) (CSP-41) [45]	
Alcohols	2-Alkanols, 3,3-dimethyl-2-butanol, 4-methyl-2-pentanol, *E*-2-methylcyclopentanol, *E,Z*-2-methylcyclohexanol, bicyclo[2.2.1]heptan-2-ol

Chirasil-Nickel = Nickel(II)-bis-[1*R*-3-HFB-10-methylencamphorate] (bonded on OV-101 and immobilized) (CSP-41) [45]

Alcohols	*E,Z*-4-methyl-3-heptanol, 6-methyl2-heptanol, 6-methyl-5-hepten-2-ol = sulcatol, iso-/borneol, 1-phenylethanol, 1-phenyl-1-propanol [46]
Epoxides	1,2-Epoxyalkanes, *E,Z*-2,3-epoxybutane, 2,2-dimethyl-3-phenyloxirane [46]
Esters	2-Octyl acetate, 1-phenyl-1-propyl acetate [46]
Ketones	2- and 3-Methylcyclohexanone, bicyclo[2.2.1]heptan-2-one, *E*-5-methyl-2-hepten-4-one = filbertone, bicyclo[3.2.1]octan-2-one, 3-butylcyclohexan-one, camphor, fenchone [46]
Lactones	4-Alkanolides, 5-hexanolide [46]
O-Heterocycles	2-Methyloxetane, tetrahydro-2-methylfuran [45]; homofuran [47]
Miscellaneous	1-Cyanobicyclo[2.2.1]heptan-2-ol, 2-cyano-2-phenylpentane [46]

Of particular interest are the short analysis times using microbore columns (1.5 m x 0.05 mm i.d.). For instance, tetrahydro-2-methylfuran has been separated into its enantiomers only after 12 seconds.

The mechanism of separation of olefines has been explained by the different π-complexation strength of both enantiomers [2]. All compounds separated on metal complex phases (alcohols, cyclic ethers, ketones, acetals, esters, lactones, thioethers, N-chloroaziridines, etc.) exhibit a common property; they are all Lewis bases and have at least one free pair of electrons, enabling complexation with the selector. Separation is achieved due to the different stability of the diastereomeric 1:1-association complexes between selector and selectand. The enantioselectivity of the selector depends on the applied metal and the applied β-terpene diketone used [9].

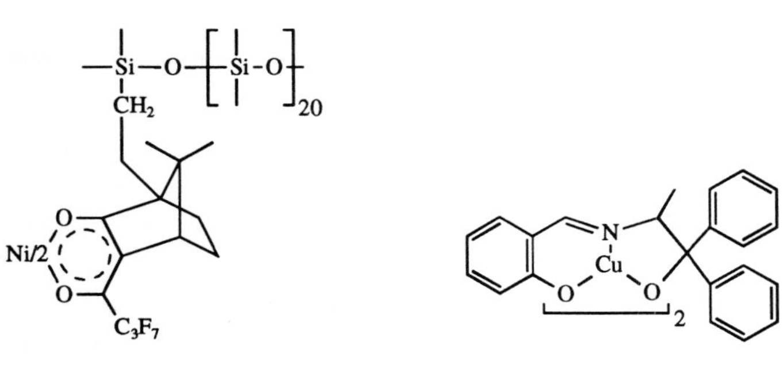

CSP-41 CSP-42

Figure 3.5.12 Polysiloxane anchored nickel-phase (CSP-41) [48] and copper(II) complex of a Schiff base (CSP-42) [54].

The temperature-dependent reversal in the order of elution of *E*-chalcogran [49] and isopropyloxirane [25] using CSP-40 should be briefly mentioned. Comprehensive reviews dealing with complexation gas chromatography have been provided by Schurig [50] and Schurig and Wistuba [51].

A quite different type of metal complex phase has been described by Oi et al. [52-55]. These authors employed optically active copper(II) chelates of Schiff bases as stationary phases (Figure 3.5.12), for instance, copper (II)-bis-[*S*-N-salicyliden-2-amino-1,1-diphenyl-1-propanol] (CSP-42). In this way, e.g., lactic acid esters have been separated. The separation mechanism is not yet understood.

References

[1] Schurig V., Gil-Av E. *J. Chem. Soc. Chem. Commun.* 1971, 650
[2] Schurig V. *Angew. Chem.* 1977, *89*, 113
[3] Golding B.T., Sellars P.J., Wong A.K. *J. Chem. Soc. Chem. Commun.* 1977, 570
[4] Schurig V., Koppenhöfer B., Bürkle W. *Angew. Chem.* 1978, *90*, 993
[5] Schurig V. DE 32 47 714 (Cl.B01L3/32), 27 Oct 1983
[6] Schurig V., Leyrer U., Weber R. in: *6th Int.Symp. on Capillary Chromatography, Riva del Garda 1985*, Sandra P. (Ed.), Hüthig: Heidelberg, 1985, p. 18
[7] Schurig V., Bürkle W., Zlatkis A., Poole C.F. *Naturwissenschaften* 1979, *66*, 423
[8] Schurig V. *Naturwissenschaften* 1987, *74*, 190
[9] Schurig V., Bürkle W., Hintzer K., Weber R. *J. Chromatogr.* 1989, *457*, 23
[10] Schurig V., Weber R. *Angew. Chem. Suppl.* 1983, 1130
[11] Koppenhoefer B., Hintzer K., Weber R., Schurig V. *Angew. Chem.* 1980, *92*, 473
[12] Schurig V. *J. Chromatogr.* 1988, *441*, 135
[13] Weber R., Schurig V. *Naturwissenschaften* 1981, *68*, 330
[14] Weber R., Hintzer K., Schurig V. *Naturwiss*enschaften 1980, *67*, 453
[15] Schurig V., Schleimer M., Jung M., Grosenick H., Fluck M. in: *15th Int. Symp. on Capillary Chromatography, Riva del Garda 1993*, Sandra P. (Ed.), Hüthig: Heidelberg, 1993, poster presentation
[16] Schurig V., Weber R., Nicholson G.J., Oehlschlager A.C., Pierce H. Jr., Pierce A.M., Borden J.H., Ryker L.C. *Naturwissenschaften* 1983, *70*, 92
[17] Schurig V., Bürkle W. *J. Am. Chem. Soc.* 1982, *104*, 7573
[18] Schurig V., Wistuba D. *Tetrahedron Lett.* 1984, *25*, 5633
[19] Schurig V. DE 34 10 801 (Cl.C07B57/00), 10 Oct 1985
[20] Schurig V. *Naturwissenschaften* 1987, *74*, 190
[21] Mosandl A., Schubert V. *J. Ess. Oil Res.* 1990, *2*, 121
[22] Kagan H.B., Mimoun H., Mark C., Schurig V. *Angew. Chem.* 1979, *91*, 511
[23] Schurig V. in: *Asymmetric Synthesis. Volume 1. Analytical Methods*, Morrison J.D. (Ed.), Academic Press: New York, 1983, p. 59
[24] Schurig V., Wistuba D. *Angew. Chem.* 1984, *96*, 808
[25] Schurig V., Wistuba D., Laderer H., Betschinger F. in: *Organic Synthesis via Organometallics*, Dotz H.K., Hoffmann R.W. (Eds.), Vieweg: Wiesbaden, 1991, p. 241

[26] Deger W., Gessner M., Heusinger G., Singer G., Mosandl A. *J. Chromatogr.* 1986, *366*, 385

[27] Schurig V., Laderer H., Wistuba D., Mosandl A., Schubert V., Hagenauer-Hener U. *J. Ess. Oil Res.* 1989, *1*, 209

[28] Mosandl A., Schubert V. *J. Ess. Oil Res.* 1990, *2*, 121

[29] Halterman R.L., Roush W.R., Hoong L.K. *J. Org. Chem.* 1987, *52*, 1152

[30] Bestmann H.J., Attygalle A.B., Glasbrenner J., Riemer R., Vostrowsky O. *Angew. Chem.* 1987, *99*, 784

[31] Schurig V., Weber R. *Angew. Chem.* 1983, *95*, 797

[32] Schurig V. in: *Bioflavour '87*, Schreier P. (Ed.), de Gruyter: Berlin, New York, 1988, p. 35

[33] Keinan D., Seth K.K., Lamed R. *J. Am. Chem. Soc.* 1986, *108*, 3474

[34] Mosandl A., Palm U. *Z. Lebensm. Unters. Forsch.* 1991, *193*, 109

[35] Schurig V., Weber R. *J. Chromatogr.* 1984, *289*, 321

[36] Mosandl A., Hagenauer-Hener U. *J. High Resol. Chromatogr. Chromatogr. Commun.* 1988, *11*, 744

[37] Schurig V., Ossig A., Link R. in: *8th Int.Symp. on Capillary Chromatography, Riva del Garda 1987*, Sandra P. (Ed.), Hüthig: Heidelberg, 1987, p. 196

[38] Degenhardt C.E.A.M., Van Den Berg G.R., De Jong L.P.A., Benshop H.P., Van Genderen J., Van De Meent D. *J. Am. Chem. Soc.* 1986, *108*, 8290

[39] Mosandl A., Heusinger G., Wistuba D., Schurig V. *Z. Lebensm. Unters. Forsch.* 1984, *179*, 385

[41] Schurig V., Link R. in: *Chiral Separations*, Stevenson D., Wilson I.D. (Eds.), Plenum Press: New York, London, 1988, p. 91

[41] Hogberg H.-E., Hedenstrom E., Isaksson R., Wassgren A.-B. *Acta Chem. Scand.* 1987, *B41*, 694

[42] Carman R.M., MacRae I.C., Perkins M.V. *Aust. J. Chem.* 1986, *39*, 1739

[43] Bergstrom G., Wassgren A.-B., Hogberg H.-E., Hedenstrom E., Hefetz A., Simon D., Ohlsson T., Lofquist J. *J. Chem. Ecol.* 1992, *18*, 1177

[44] Joshi N.N., Srebnik M. *J. Chromatogr.* 1989, *462*, 458

[45] Schurig V., Schmalzing D., Schleimer M. *Angew. Chem.* 1991, *103*, 994

[46] Schleimer M., Schurig V. *J. Chromatogr.* 1993, *638*, 85

[47] Schurig V., Jung M., Schleimer M., Klärner F.-G. *Chem. Ber.* 1992, *125*, 1301

[48] Schurig V., Ossig A., Link R. *J. High Resol. Chromatogr. Chromatogr. Commun.* 1988, *11*, 89

[49] Schurig V., Ossig J., Link R. *Angew. Chem.* 1989, *101*, 197

[50] Schurig V. *Kontakte (Darmstadt)* 1986, *(1)*, 3

[51] Schurig V., Wistuba D. in: *InCom '90*, Günther W., Matthes J.P., Perkampus H.-H. (Eds.), GIT Verlag: Darmstadt, 1990, p. 346

[52] Oi N., Horiba M., Kitahara H. *Koen Yoshishu-Seitai Seibun no Bunseki Kagaku Shinpojumu*, 4th 1979, 69 (C.A. 1980, 92: 176703v)

[53] Oi N., Horiba M., Kitahara H. *Bunseki Kagaku* 1980, *29*, 156

[54] Oi N., Horiba M., Kitahara H., Doi T., Tani T., Sakakibara T. *J. Chromatogr.* 1980, *202*, 305

[55] Oi N., Shiba K., Tani T., Kitahara H., Doi T. *J. Chromatogr.* 1981, *211*, 274

3.5.2.3 Cyclodextrin phases

Structure and properties of cyclodextrins

Cyclodextrins (CDs) are cyclic glucans (cyclomaltooligoses) with at least six to twelve α-D-glucopyranose units in α-1,4-glycosidic connection. As early as 1891 they were isolated by Villiers as degradation products of starch, but only in 1904 did Schardinger characterize them as cyclic oligosaccharides [1]. Freudenberg and Cramer [2] recognized their ability to form inclusion complexes. CDs are produced biotechnologically by bacterial degradation of starch under the influence of CD glycosyl transferases of *Klebsiella pneumoniae, Bacillus macerans* and other bacillus species [1,3]. The most important CDs (Figure 3.5.13) contain 6, 7 or 8 glucose units and are called α-, β- and γ-cyclodextrins.

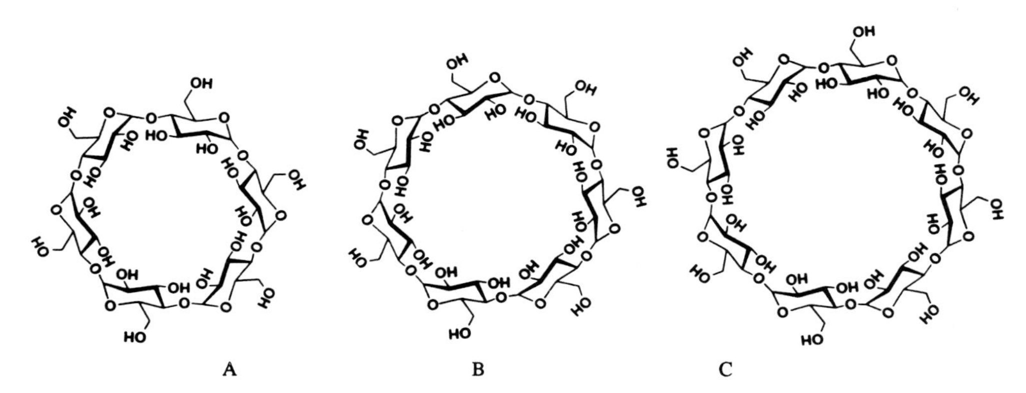

Figure 3.5.13 Cyclodextrin structures [4]. A = α-cyclodextrin (cyclomaltohexaose), B = β-cyclodextrin (cyclomaltoheptaose), C = γ-cyclodextrin (cyclomaltooctaose).

The complexation power of CDs, which can be considered as empty capsules of different sizes, has been systematically investigated [1, 4-10]. The inclusion is favoured by hydrophobic interactions. Besides 1:1 inclusion complexes, 2:1 host-guest associates can be formed as well, in which two CD molecules are connected via hydrogen bonds of the secondary hydroxyl groups. Investigations of heptakis-(2,6-O-dimethyl)-β-CD showed that protection of the free hydroxyl groups prevents the forming of CD dimers. The presence of the methoxy groups can extend the available space for a hydrophobic guest molecule (the cavity is bigger than that of an underivatized CD) [7]. This example clearly shows that the conformation and the readiness to complexation of the CDs can be varied by chemical modification.

The inherent molecular asymmetry of the CDs is due to the occurrence of D-glucose units in the cyclooligomers. The molecular intercalation of chiral guests is in many cases enantioselective. The first separation of racemates with CDs was described as early as 1959 [11]. MacNicol and Rycroft [12] employed β-CD as the chiral shift-reagent (CSR) for ^{19}F NMR techniques in order to induce chemical shifts of compounds with CF_3-groups, which are intercalated in the β-CD molecule. Direct evidence of enantioselective complexation of *R*- and *S*-2-(3-phenoxyphenyl)-propionic acid (fenoprofen) with β-CD has been demonstrated by crystal structure determination [13].

Application of cyclodextrins in chromatography

Cyclodextrins have often been used in chromatographic methods for the separation of enantiomers. In Table 3.5.6 the first chromatographic enantiomer separations applying CDs are outlined. For further information refer to the special literature (Table 3.5.1).

Table 3.5.6 Application of cyclodextrins in chromatography (M = as mobile phase; S = as stationary phase).

Chromat. technique	Phase	Cyclodextrin	Author	Year	Ref.
GPC	S	α-CD; β-CD	Harada et al.	1978	14
HPLC	M	β-CD	Debowski et al.	1982	15
GLC	S	α-CD	Koscielski et al.	1983	16
HPLC	S	β-CD	Armstrong/DeMond	1984	17
MEKC	M	β-CD der.	Terabe et al.	1985	18
HPTLC	S	β-CD	Alak/Armstrong	1986	19
SubFC	S	β-CD	Macaudière et al.	1987	20
ITP	M	β-CD	Snopek et al.	1988	21
HPTLC	M	β-CD	Armstrong/He/Han	1988	22
HPCZE	S	β-CD	Gutman et al.	1988	23
SFC	S	β-CD	Macaudière et al.	1989	24
Displacement chromat.	S	β-CD	Vigh et al.	1990	25
EC	S	β-CD der.	Mayer, Schurig	1992	26

Cyclodextrins in gas chromatography

In the early sixties peracetylated and perpropionylated α- and β-CD were used above their melting points (>220°C) as stationary phases in packed GLC columns for the selective separation of fatty acid methyl esters [27]. Using the retention data, inclusion effects could not be conclusively deduced. However, permethylated α-and β-CDs above their melting points (α-CD: >200°C; β-CD: >155°C [28]) or in polysiloxane solution (DC-710) on silanized Chromosorb W in packed columns cause a retention increase of isoalkanes in comparison to n-alkanes, which was explained by inclusion [29]. Besides the 2,3,6-tri-O-methyl-CDs, the 2,6-di-O-methyl derivatives were also investigated. α- and β-CDs, as well as 6-deoxy-α-CD and β-CD polyurethane resins, were used in gas solid chromatography (GSC) for the separation of positionally isomeric aromates, e.g., xylenes and cresols [30,31]; the researchers also assumed that there were inclusion effects. Tanaka et al. [32] described the covalent linkage of peracetylated and perpropionylated β-CDs to the polysiloxane OV-225 for separation of achiral compounds.

The first enantiomer separation on CD phases was achieved with a packed column. Koscielski et al. [16] investigated celite coated with an aqueous formamide solution of α- or β-CD; only with the α-CD phase (CSP-43) the separation of α- and β-pinene was achieved. Previously, α- and β-CD phases were synthesized via aqueous dimethylformamide solutions and used to coat Chromosorb W [33], but for the separation of achiral compounds. In the same manner, Smolková-Keulemansová et al. [34] synthesized a γ-CD phase.

Despite good separation factors, packed columns exhibit low efficiency. Therefore, one could not have expected increased applications before the introduction of high resolution capillary columns. Nevertheless, the first attempt by Kim, which was already carried out in 1981, failed. He tried the enantiomer separation of saturated aliphatic hydrocarbons, such as 2,2,3-trimethylheptane, on permethylated β-CD, diluted in polysiloxane OV-101 (20 %) (taken from ref. [33]).

The first enantiomer separation by Juvancz et al. [36] succeeded on deactivated glass capillaries with molten permethylated β-CD (TRIMEB; CSP-44) as the stationary phase. The same group has developed additional permethylated CD phases, in particular TRIMA (α-CD; CSP-45) [37] and TRIMEG (γ-CD; CSP-46) [28]. The high melting points of all permethylated CDs are disadvantageous, i.e. α-CD: >200°C; β-CD: >155°C; γ-CD: >200°C. Indeed, enantiomer separations are possible below the respective melting points (GSC!), but the chromatographic conditions are not ideal. This leads in most cases to peak broadening and peak deformation, which makes these phases unusable for the separation of highly volatile compounds. Venema and Tolsma [38] were able to overcome this problem by conditioning a permethylated β-CD phase above 200°C. The authors found out that the stationary phases exist in a kind of 'super-cooled' melt, even below 80°C. Thus, it was possible to separate volatile compounds like methyl lactate or 2-pentanol into their enantiomers at 90°C. A similar phase with a degree of methylation >70 % (CSP-47) has been synthesized by Knoll et al. [39]; the melting point has been decreased to approximately 90°C. This phase is also commercially available as 'Peraldex' (cf. Table 3.5.3)

In 1987, i.e. a little later than the first permethylated CD capillary column was reported, Schurig and Nowotny [35] introduced several polysiloxane-dissolved CDs exhibiting better enantioselectivity. In 1988, König et al. [40] synthesized several peralkylated and acylated/alkylated CDs also exhibiting very good enantioselectivity. These CD derivatives can be used undiluted and dissolved in polysiloxane [41]. Similar undiluted CDs were reported by Armstrong et al. [42] in 1990. Meanwhile, several research groups have developed new CD derivatives for gas chromatographic use. In total, far more than 100 different CD phases have been developed up to 1994; more than 30 of these are commercially available.

Besides the above-mentioned modified CDs, permethylated dextran (CSP-48) [43] as well as perpentylated amylose (CSP-49) [44] have been synthesized. While no enantioseparation took place on CSP-48, on CSP-49 a few chiral compounds were separated, including the important flavour compound *E*-rose oxide (the *Z*-isomer could not be separated!). Schurig et al. [45] have investigated the use of seven alkyl amylose and cellulose carbamates as chiral stationary phases in gas chromatography. One of them, amylose tris(butylcarbamate) dissolved in OV-61 (CSP-50), can be used for enantiomer separation of chiral compounds such as menthol, γ-lactones, and oxiranes, as well as some esters. This phase combines the presence of the -CO-NH- function as in Chirasil-Val (cf. Subsection 3.5.2.1) with a polysaccharide as in perpentylated amylose. The results obtained by Schurig et al. [45] show that molecular inclusion is not a prerequisite for enantiomer separation using carbohydrate stationary phases.

Undiluted cyclodextrin derivatives

In contrast to the permethylated and partially methylated derivatives perpentylated and partially pentylated CD derivatives are liquid even at room temperature. These thermally stable stationary phases, which are largely lipophilic, were developed by König et al. [40]. They can be applied within a temperature range of 30 to 220°C (the degradation temperature is 400°C) and they are excellently suitable for the coating of deactivated glass or fused silica capillaries. Some of these phases are commercially available (Table 3.5.3). Besides flavour compounds and pheromones a large number of compounds of other classes can be separated into their enantiomers, such as O-TFA derivatives of sugars and polyols, N-TFA derivatives of amines and amino acid methyl esters, halogenic compounds (e.g., anaesthetica and pesticides), nitrogen- and phosphorous-containing pesticides, pharmacologically active compounds, as well as intermediate and final products of a number of asymmetric syntheses [46]. The separation of O-TFA monosaccharides on hexakis-(2,3,6-tri-O-pentyl)-α-CD (CSP-51) has shown that not only inclusion effects are responsible for enantioseparations; the cavities of this phase are too small for these derivatized carbohydrates. Probably only a part of the large molecules projects into the cavity [46]. It is possible that the chiral recognition during the separation of spiroacetals on hexakis (2,3,6-tri-O-dodecyl)-α-CD (CSP-52) works without any inclusion.

By variation of substitution patterns further separation improvement of chiral compounds has been achieved. While, e.g., limonene could not be separated completely on heptakis-(2,3,6-tri-O-pentyl)-β-CD (CSP-53), the substitution of a pentyl group at C3 by a methyl group (CSP-54) resulted in a distinctly improved enantioselectivity [47]. Using the same phase the first successful gas chromatographic separation of an axially asymmetric compound (1-ethyliden-4-tert-butylcyclohexane) was carried out by König [48]. Using octakis-(6-O-methyl-2,3-di-O-pentyl)-γ-CD (CSP-55) the separation of limonene was further improved; in comparison to the β-CD homologue, the reverse order of elution was found (CSP-56) [49]. Noteworthy is the separation of the simplest saturated chiral hydrocarbon, 3-methylhexane, on this phase [49]. More recently published studies deal with substitution at the C2 position [44,50,51]. Substituting acetyl groups for alkyl groups at the C3 position, polar compounds such as γ-lactones can be separated excellently. The separation power and the order of elution are nearly similar on 3-O-acetylated α-, β- and γ-CD derivatives, i.e. the influence of the ring size of the CDs and, therefore, the cavity size is not important for the separation of γ-lactones [52]. This is quite different with δ-lactones; only octakis-(3-O-butyryl-2,6-di-O-pentyl)-γ-CD fits as a separation phase (CSP-57). On heptakis-(3-O-acetyl-2,6-di-O-pentyl)-β-CD (CSP-58) the separation of almost all of the δ-lactones is only moderate [53]. CDs with C6 acyl groups exhibit practically no enantioselectivity [48], except for heptakis-(2,3-di-O-pentyl-6-O-acetyl)-β-CD (CSP-59) concerning the flavour compound 3,7,11-trimethyl-1,6,10-dodecatrien-3-ol (nerolidol) (taken from ref. [54]).

On octakis-(3-O-butyryl-2,6-di-O-pentyl)-γ-CD (CSP-57) temperature-dependent peak coalescence of the second kind of the important flavour compound 4-hydroxy-2,5-dimethyl-3[2H]-furanone (furaneol) has been observed [52] (cf. also Sections 3.3 and 3.5.2.2). In contrast, using heptakis-(2,3,6-tri-O-pentyl)-β-CD [52] the enantiomer separation of furaneol was achieved without racemization (cf. analytical data Table 3.5.8 [52]). Subsequently, two further examples showing this phenomenon on CD columns were reported [55,56].

Octakis-(3-O-butyryl-2,6-di-O-pentyl)-γ-CD (CSP-57) is a very versatile phase for many different compounds. It can be used for the enantioseparation of cyclic acetals, acids, alcohols, N-TFA derivatives of amines and α-amino acid methyl ester, diols and polyols, esters, ketones, lactones, pesticides (e.g., lindane), and pharmaceuticals [57] (for more details cf. Table 3.5.8). Using the same column, König et al. [58,59] have described the temperature-dependent inversion of elution order of the herbicides mecoprop and dichlorprop (as methyl esters), as well as of methyl lactate. Kobor et al. [60] also observed this phenomenon with 2,2,2-trifluorophenylethanol on octakis-(2,3-di-O-methyl-6-O-tert-butyldimethylsilyl)-γ-CD (CSP-60).

Another interesting phenomenon was observed by Schmarr et al. [61] on hexakis-(3-O-acetyl-2,6-di-O-pentyl)-α-CD (CSP-61). At temperatures of 130 to 150°C the authors noticed unusual peak broadening ('defocusing') of γ-lactones and of achiral compounds such as alkanes and methyl esters, which has not been shown by other CD phases. For instance, this phenomenon could not be observed for the similar phase pentakis-(3-O-acetyl-2,6-di-O-pentyl)-mono(2,3,6-tri-O-pentyl)-α-CD ('over-pentylated' phase; CSP-62). The authors discussed a possible transition from the smectic to the isotropic (liquid) phase (behaviour of liquid crystals).

Several other CD derivatives, which are also liquid at room temperature, have been introduced by Armstrong et al. [42]. Most of the following phases are commercially available (cf. Table 3.5.3). None of these derivatives is pure. The reaction with the derivatization reagents was incompletely carried out, since pure 2,6-di-O-pentyl-CDs are solid at room temperature. By using the ^{13}C NMR technique Armstrong et al [62] observed that not all 2,6-O-positions of these CD derivatives are occupied with pentyl groups; pentylated 3-O-positions also exist. In these mixtures the melting point is decisively decreased. Dipentyl phases are available as α-, β- and γ-CD derivatives. They are especially suited for O-TFA carbohydrates and N-TFA amines and nicotine derivatives, as well as for O-TFA alcohols and N,O-TFA amino alcohols [42] (for more details cf. Table 3.5.8).

According to Armstrong, the α-, β- and γ-derivatives of the O-*S*-2'-hydroxypropyl-per-O-methyl-CD type show the most universal separation properties of all CD phases [42]. They are hydrophilic, relatively polar and soluble in water. Most of the successful separations reported have been carried out on heptakis-(O-*S*-2'-hydroxypropyl-per-O-methyl)-β-CD (CSP-63). The synthesis of these phases starts with an incomplete reaction of the free hydroxyl groups of the CDs with *S*-1,2-epoxypropane (propylene oxide); only every third hydroxyl group reacts. Subsequently, the derivatives are permethylated. The degree of substitution has been checked by ^{252}Cf-plasma desorption mass spectrometry [63]. With the 2'-hydroxypropyl groups, these CD derivatives carry additional chiral centres, but this is without influence on enantioselectivity, which was proved by Armstrong et al. [63]. These researchers synthesized a phase analogous to the phase CSP-63, i.e. heptakis-(O-*R*-2'-hydroxypropyl-per-O-methyl)-β-CD (CSP-64), where the 2'-hydroxypropyl parts are *R*-configurated instead of *S*, as with CSP-63. The tests performed showed no differences in the order of elution comparing both CD phases. Similar results have been achieved by Schurig and Nowotny [7]. With these hydroxypropyl phases a great variety of compounds from many different classes can be separated into their enantiomers, such as TFA-derivatives of alcohols, amines, amino alcohols, carbohydrates, diols, furans, halo compounds, pyrans, etc. (for more details cf. Table 3.5.7).

The 2,6-di-O-pentyl-3-O-TFA-CDs (α-, β- and γ-CD) show a moderate polarity, like the 3-O-acyl-2,6-di-O-pentyl-CDs. For instance, lactones such as γ-jasmine lactone [64] can be separated, whereas especially octakis-(2,6-di-O-pentyl-3-O-TFA-)-γ-CD (CSP-65) is suitable for the enantiomer separation of δ-lactones [65,66]. These TFA phases show enantioselectivity also for TFA derivatives of alcohols (e.g., α-bisabolol [66]), aldehydes (e.g., citronellal [67]), carbohydrates, diols, esters (e.g., ethyl 2-methylbutanoate [68]), halo compounds, pyrans (e.g., rose oxide [66]) and many more (for more details see Table 3.5.7).

All chiral phases made by Armstrong et al. are stable over a wide temperature range (30-300°C). Nonetheless, at working temperatures > 150°C an irreversible loss of efficiency of 20 to 30 % has been observed [62]. In addition, it was found that the order of elution of one CD derivative cannot be transferred to another CD derivative, not even when only the ring size of the CD, but not the substitution pattern is changed [42]. In addition, Armstrong and Jin [69] investigated the most suitable derivatization reagent for OH, NH_2 and NH groups of low volatile compounds with the goal of increasing their volatility. Besides the most frequently used method (i.e. trifluoroacetylation), in addition, acetylation, monochloroacetylation and trichloroacetylation were also tested. The results showed that the most suitable reagent can only be empirically determined, since suitability is dependent on the

CD column used. It is important to know that some compounds tend to racemization during the derivatizing reaction. For instance, trifluoroacetylation of 1-indanol leads to a degree of racemization of about 20 % [46]. Preferably, derivatization should be avoided if ee values have to be determined.

In 1992, Takeichi et al. [70] introduced three carbamate-containing cyclodextrins which are liquid at room temperature. Heptakis-(2,6-di-O-pentyl-3-O-phenylcarbamoyl)-β-CD (CSP-66) showed the best result, in particular for TFA-phenylethylamine. However, in most cases the α values were rather low. Similar phases have been developed by Stoev and Gancheva [71], e.g., heptakis-(2,6-di-O-methyl-3-O-naphthylcarbamoyl)-β-CD (CSP-67). These phases show high enantioselectivity for acids, alcohols, and amines.

Polysiloxane dissolved cyclodextrin derivatives

Further progress has been achieved by dissolving peralkylated and partially alkylated CDs in polysiloxanes [7,33]. By this means, the strong enantioselectivity of the CD derivatives was brought together with the high separation power of the polysiloxane derivatives. The most suitable polysiloxane was found to be OV-1701. Polysiloxane mixed-phases have been produced that are liquid even at room temperature. As a result it has become possible to analyze even highly volatile compounds (working temperature: 20 to 250°C). Due to the low CD concentration, e.g., 10 %, the analysis times are kept short, while the separation efficiency is kept [7].

Permethylated β-CD, dissolved in OV-1701 (10 %; CSP-68), has proven to be a versatile mixed-phase [35]. This is shown by the rapidly increasing number of published enantiomer separations applying this phase (Table 3.5.8-D). Besides compounds with chiral centres, also some substances with chiral axes have been separated into their enantiomers, e.g., ethyl 3,4-hexadienoate [7]. Due to the high degree of deactivation of the fused silica capillary columns, even underivatized alcohols or acids can be chromatographed (cf. Table 3.5.8). This phase (and others) are commercially available (cf. Table 3.5.3).

In no case does extending the dimensions of the CD cavity affect the enantioselectivity positively. For instance, permethylated γ-CD, dissolved in OV-1701 (CSP-69), does not exhibit enantioselectivity with saturated cyclic hydrocarbons. On the other hand, these compounds can be separated on α-, β- and γ-CDs [7].

The separation efficiency is also dependent on the purity of the modified CDs used. 'Impurities' and 'decomposition products' originating from the CD synthesis often contribute to the separation. Alkylation, for instance, is not at all a mild reaction and purification is of the utmost importance [73]. With extremely pure CDs better separation results can be achieved [73,74]. By double recrystallization from methyl tert-butyl ether highly purified heptakis-(2,6-di-O-methyl-3-O-TFA)-β-CD (11.2 %; dissolved in OV-1701) has been obtained. An extreme resolution (R_S) of 14.3 for the flavour compound γ-decalactone was obtained (after 40 min). Baseline separation of δ-lactones was also achieved, which was not possible with less pure stationary phases of the same type [52,53].

In contrast, some separations were improved by using 'dirty' CD derivatives [75]. In addition, an undiluted phase containing a mixture of permethylated α-, β- and γ-CDs (α:β:γ = 6:3:1; CSP-70) showed better results than phases coated with the pure single CDs [76]. One important reason for this behaviour is the decreased melting point, which allows lower working temperatures. Enantioselectivity is strongly dependent on liquid CD derivatives; solid CDs show only poor selectivity.

Bicchi et al. [77] demonstrated the importance of the conditioning temperature. While a mixture of γ- and δ-lactones could not be separated on permethylated α-CD (CSP-71) after conditioning at 190°C, a partial separation of the γ-lactones as well as baseline separation of some δ-lactones was achieved on the same phase after conditioning at 220°C. The authors have tried to explain this phenomenon by relating the melting point of permethylated α-CD (200°C) to the high conditioning temperature. If conditioning is performed above the melting point of the cyclodextrin, probably the crystalline structure will be modified. A similar phenomenon has already been described at the beginning of this chapter [38].

The importance of the solvent has been studied by several groups. In many cases OV-1701 showed the best properties as a solvent for CDs [35]; for instance, for polar compounds the separation power and the peak shapes were improved by the use of the less polar polydimethylsiloxanes and polyphenylmethylsiloxanes [75,78,79]. Even the background deactivation might influence the separation, e.g., persilylation or carbowax deactivation. The following classes of polysiloxanes have been used as solvents for CDs:

- polydimethylsiloxanes: OV-1 [60]; OV-101 [35]; PS-255 [79]; PS-268 [79]; PS-347.5 [80]; PS-537 [81]; SE-30 [79]
- polyphenylmethylsiloxanes: CP-Sil 13 [82]; DC-550 [35]; OV-7 [83]; OV-17 [35]; PS-086 [84]; SE-52 [85]; SPB-35 [86]
- polyphenylvinylmethylsiloxanes: SE-54 [60]
- polycyanopropylphenyl(vinyl)methylsiloxanes: OV-1701 and similar products [35,72]
- polycyanopropylmethylphenylmethylsiloxanes: OV-225 [87].

The concentration of the CD derivative in the polysiloxane solution is also very important for the separation power of the particular column. For instance, Bicchi et al. [77] were only able to separate 1,8-epoxy-9-(3-methyl-3-buten-1-yl)-p-menthane on a permethylated β-CD, which was dissolved in OV-1701 by 30 %. But the attempt to separate the same compound on a permethylated β-CD phase (CSP-68) failed, when its concentration in OV-1701 was only 10 %. Higher CD concentrations seem to be better for compounds with high volatility. But according to Jung and Schurig [89] no improvement in the separation was achieved with CD concentrations above 30 % using CSP-68. According to Buda et al. [88] the optimal CD concentration for heptakis-(2,3-di-O-acetyl-6-O-tert-butyldimethylsilyl)-β-CD (CSP-72) is 50 % in OV-1701. Nevertheless, the optimal concentration must be empirically determined for each CD type [90] and, in rare cases, it is also dependent upon the compound to be analyzed [91]. Most commercially available polysiloxane-dissolved columns contain approximately 10 % of the CD in OV-1701 (cf. Table 3.5.3).

The influence of the column length has been investigated by Lindstrom [92]. Short columns (2 to 8 m), coated with CSP-68, showed better separation properties than 50 m columns; in addition, the retention times were much shorter. α-Damascone, for example, was separated completely (100 %) on a short column (2 m x 0.25 mm i.d.) after 6 min, but on a long column (50 m x 0.25 mm i.d.) it took at least 52 min and the separation was only 95 %. Bicchi et al. [93] successfully separated α-ionone with a 0.19 m column. They observed for increasing length slightly decreasing α values and an increasing resolution. Another advantage of shorter columns is that lower working temperatures are possible. However, reducing the length of the capillary column is only a solution for low-volatile compounds in solute mixtures which are not too complex, since separation efficiency is lost due to the reduced number of theoretical plates [94].

By changing the alkyl parts different selectivities can be achieved. With perethylated β-CD, dissolved in OV-1701 (CSP-73), Askari et al. [95], as well as Rettinger et al. [96], obtained better separation results for the flavour compounds linalool, methyl 2-methylbutanoate and ethyl 2-methylbutanoate, as compared with those obtained with the similar permethylated phase (CSP-68). Substitution of methyl groups with deuterated methyl groups did not effect the enantioselectivity [91].

More polar are the 2,6-di-O-methyl-3-O-perfluoroacylated CD phases [7] (Table 3.5.8). They show various enantioselectivities, due to the fact that the perfluoroacyl groups are able to form additional hydrogen bonds and dipole-dipole interactions. γ-Lactones can be separated substantially better on the β-CD derivative CSP-74 than on CSP-68, while δ-lactones can be preferentially separated on the homologous γ-CD CSP-75, due to the larger cavity. This phase does not exhibit enantioselectivity for hydrocarbons, e.g., limonene [7]. A drawback of the respective β-CD phase CSP-74 is its 'aging' which is caused by a degradation of the CD (hydrolysis of the TFA group) during storage [93,97,98]. Even the storage of permethylated CDs (dissolved in OV-1701) can lead to a reduced separation power and a higher minimal working temperature. By means of reconditioning the loss of quality can be prevented and the performance of these phases can be maintained [93]. Other CD derivatives such as perpentylated β-CD (CSP-53) can be stored for a long time without any, or only a little, decrease in separation power [52].

A more recent development is the 2,3-diacetylated and 2,3-dimethylated CDs with a 6-tert-butyldimethylsilyl group (acetylated α-, β-, γ-CDs: CSP-76 [99]; CSP-72 [100]; CSP-77 [101]; methylated α-, β-, γ-CDs: CSP-78 [60]; CSP-79 [85]; CSP-60 [102]). On the corresponding acetylated β-CD-derivative (CSP-72) underivatized flavour compounds such as lavandulol, citronellol and linalyl acetate can be separated into their enantiomers. These results could not be achieved with other CD phases, even after derivatization. Comparatively good results are obtained using the methylated CD-derivatives. It should be mentioned that somewhat earlier a similar phase, heptakis-(tri-O-tert-butyldimethylsilyl)-β-CD (20%; dissolved in PS-086; CSP-80), was successfully used as the chiral stationary phase [84].

A large number of CD phases have been diluted in polysiloxanes [41,50,94,103, 104]. The most suitable phase for the resolution of camphor is CSP-57 dissolved in OV-1701 (60%). Using the undiluted CD-derivative, the resulting peaks are much broader. This result shows that in many cases for CD-derivatives that are liquid at room temperature, the combination of CD-inclusion (chiral re-cognition) and separation efficiency (polysiloxane) also gives the better results [94]. More recently, Ruge and coworkers [105] presented new CD-derivatives with de-oxygroups at the C6 position, which are dissolved in OV-1701 (50%). The enantioselectivity increased from γ- to α-CD (CSP-81); e.g., it was possible to separate even δ-lactones.

Polysiloxane-anchored cyclodextrin derivatives

Further progress in increasing phase stability was achieved by Fischer et al. [106] and Schurig et al. [107]. They bonded permethylated β-CD onto OV-101 (cf. Figure 3.5.14 and Table 3.5.7), with 2 to 5 of the 7 existing C_6-OH groups being connected to a long alkenyl rest. The synthesis starts with the reaction of native β-CD with alkenyl bromide (7-octenyl bromide has mainly been used, but allyl bromide or 4-pentenyl bromide as well). Then, after methylation with methyl iodide, the alkenylated and permethylated β-CD reacts with a polydimethylsiloxane containing 5% Si-H-groups (activator: H_2PtCl_6).

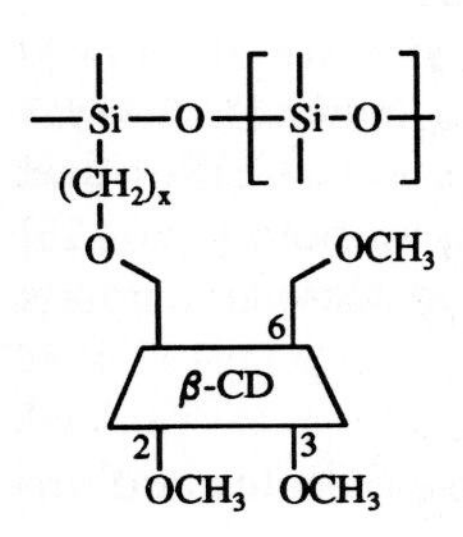

Advances of chemically bonded cyclodextrin phases:

- higher temperatur stability (0° to 280° C)
- reduced column bleeding
- reduced sensibility against thermal shocks
- no tendancy to droplet formation
- short conditioning times
- longer life-time
- direct injections (on-column; splitless) are possible

Figure 3.5.14 Permethylated β-cyclodextrin (bonded on OV-101) [109]. (CSP-84: x = 3; CSP-85: x = 5; CSP-82: x = 8)

Schurig et al. [108] reported the immobilization of polysiloxane-anchored CD on the capillary wall ('Chirasil-Dex'; CSP-82). The loss of separation power is only about 5 to 10%, compared to the unbonded phase (CSP-68). The rate of immobilization is approximately 75% [109]. Using 'Chirasil-Dex' the successful separation of γ-lactones (γ-pentalactone to γ-dodecalactone) was reported, which was not possible on the same - but unbonded - phase (CSP-68) [110]. Using this phase, Juvancz et al. [111] observed an unusual overloading phenomenon, i. e. peak fronting of the second eluted peak. 'Chirasil-Dex' can be used for HRGC and SFC, as well as for CZE [110].

Recently, Jung and Schurig [89] improved the synthesis and isolated mono-octenylated CD by means of column chromatography. They investigated the CD weight percentage that was polysiloxane-anchored and the percentage of hydromethylsiloxane units contained in the polydimethylsiloxane used for hydrosilylation. The same group also reported the polysiloxane-bonding and immobilization of heptakis-(2,6-di-O-methyl-6-O-TFA)-β-CD ('Chirasil-Dex-TFA'; CSP-83) using a C_8-spacer [112]. In contrast to the permethylated 'Chirasil-Dex', the thermostability of this new phase is lower; only below 190°C is the separation power stable for a longer time.

Armstrong et al. [81] also synthesized permethylated β-CD-derivatives which are anchored via propyl or pentyl groups to a polydimethylsiloxane (PS-537) and then immobilized to the surface of a fused silica column, whereby alkenyl groups were substituted for approximately 5 of 21 possible hydroxyl groups (randomly, not only C_6-OH) with all other hydroxyl groups having been methylated. With such a β-CD (ratio of CD:PS-537 = 1:4; CSP-86) baseline separation of the enantiomers of α-ionone and other flavour compounds has been achieved. This phase can also be used for SFC [81]. A completely different way of synthesizing siloxane-anchored CDs was described by Bradshaw et al. [113] and Yi et al. [114]. They synthesized novel CD-oligosiloxane copolymers which can be used as chiral stationary phases for SFC, although some of these phases seem to be applicable for GC as well. In Figure 3.5.15 a typical example of such a CD-derivative (CSP-87) is given.

The separation mechanisms

No uniform explanation exists for separation mechanisms, i.e. there are often only hypotheses that have been partly confirmed by X-ray structure analysis. Besides NMR measurements, 3D computer models are also helpful for the investigation and explanation of the separation mechanisms. According to several authors [7,46,115] the following selector-selectand interactions are to be considered (covalent bonds are excluded!):

- conformational-dependent inclusion effects ('fit' at the guest molecules and 'induced fit' at the CD molecules)
- loose, probably external, multiple association (top, side and/or bottom)
- hydrophobic interactions
- van-der-Waals interactions
- electrostatic interactions
- charge-transfer interactions
- hydrogen bonds (not possible with peralkylated CDs)
- (induced) dipole-dipole interactions (not possible with peralkylated CDs).

These interactions are strongly dependent on temperature, since at higher temperatures the flexibility of the CD derivatives increases (enhancement of vibraional and/or rotational energies) and, consequently, the inclusion decreases [116].

Since there is no case where all the compounds of a homologue series can be separated (e.g., underivatized 2-alkanols on CSP-53 [52]; for other examples cf. [117]), some members of a series can form either inclusion or association complexes. Many compounds show intermediary behaviour, with the type of complexation probably being temperature-dependent [116].

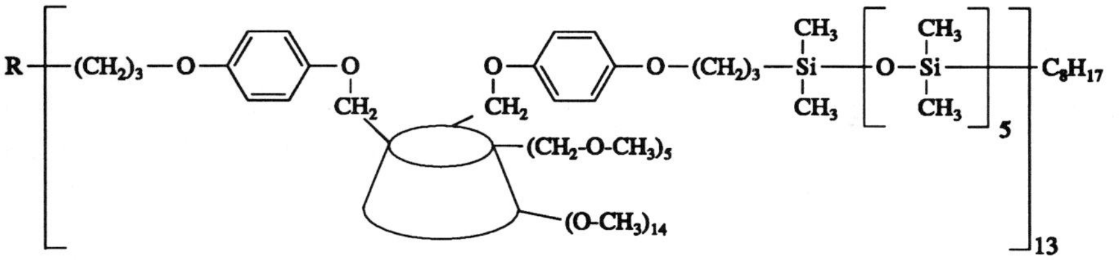

Figure 3.5.15 Example of a permethylated β-CD-oligosiloxane copolymer [113] (MW = 25000; CSP-87); (R = $C_8H_{17}\text{-}[Si(CH_3)_2\text{-}O]_5\text{-}Si(CH_3)_2\text{-}$).

In principle it can be said: the lower the temperature of elution, the better. Low elution temperatures are more important than conditions fitting van Deemter curves [118]. Among other reasons this is why low volatile compounds are derivatized to lower the boiling points. Highly volatile compounds elute too fast, so that the interactions with the stationary phase do not last long enough.

The linear carrier gas velocity should be optimized for lower temperatures, i.e. even excessive gas velocities might be better if the elution temperatures can be maintained lower [118]. However, Bicchi et al. [77] discovered that the separation factor for menthol on permethylated β-CD (30%, dissolved in OV-1701; CSP-68) decreased with increasing velocity.

Reducing film thickness decreases the minimal working temperature of permethylated β-CD (30% dissolved in OV-1701; CSP-68), which permits the separation of volatile compounds (e.g., 2-butanol) [93]. Increasing film thickness results in an increase of the retention times; therefore, for less volatile compounds the use of thinner films is recommended, leading to shorter analysis times [88;119]. Columns with thicker films can be used for semi-preparative separations, due to their higher capacity. Werkhoff et al. used thick-film capillaries for semi-preparative separations of several important flavour compounds (e.g., *E-/Z*-theaspirane [66]; γ- and δ-jasmine lactone [67]). Even packed columns coated with CDs have been prepared. A packed column with 5% of a 1:1 mixture of heptakis-(2,6-di-O-methyl-3-O-pentyl)-β-CD and OV-1701 (CSP-88) on Chromosorb W/HP was used for the preparative separation of several compounds (e.g., methyl jasmonate and hexobarbital) [120].

The optimal carrier gas is usually hydrogen [121], but for packed columns helium is recommended [120]. When the efficiency is critical carbon dioxide or nitrogen might be better than hydrogen [88]. At their optimal velocity, no difference in the separation efficiency between nitrogen and helium has been observed [52]. Gyllenhaal et al. [122] used ammonia as a carrier gas to improve peak symmetry and the detection limit of underivatized aliphatic amines on peralkylated α-CDs.

Summary of properties and behaviour of CD-phases

For selecting the appropriate CD-phase and optimization of the chromatographic system, the following points should be taken into account:

CD derivative

It has to be empirically chosen; the following CD-derivatives have proven to be useful for enantioselection:

Heptakis (2,3,6-tri-O-pentyl)-β-cyclodextrin
Octakis (3-O-butyryl-2,6-di-O-pentyl)-γ-cyclodextrin
Heptakis [O-*S*-2'-hydroxypropyl-per-O-methyl]-β-cyclodextrin
Octakis (2,6-di-O-pentyl-3-O-TFA)-γ-cyclodextrin
Heptakis (2,3,6-tri-O-methyl)-β-cyclodextrin (diss. in OV-1701)
Heptakis (2,6-di-O-methyl-3-O-pentyl)-β-cyclodextrin (diss. in OV-1701)
Heptakis (2,3-di-O-acetyl-O-6-tert-butyldimethylsilyl)-β-cyclodextrin (dissolved in OV-1701).

Polysiloxane type

For most applications OV-1701 shows the best properties, but for more polar compounds less polar polysiloxanes (e.g., SE-52) might be better (strongly dependent on the analyte).

CD concentration

Between 10 and 50% CD is optimal (dependent on the analyte and the solubility of the CD-derivative in the chosen polysiloxane). For highly volatile compounds higher CD concentrations are better. In some cases a dilution of CD derivatives improves the separation efficiency.

Derivatization method

If the background of the columns is sufficiently deactivated then often a derivatization is not necessary. When it is necessary, it is dependent on the compound; in many cases trifluoroacetyl or acetyl derivatives are the best. Possibly, racemization should be taken into account.

Column conditioning

Most CD derivatives can be conditioned by lower temperatures, but few CD derivatives have to be conditioned with a temperature which is close to the temperature limit of the column to achieve a better enantioselectivity.

Column storage

Some CD derivatives can be stored under 'normal' laboratory conditions for a long time with very small loss of separation effiency, but others should be stored under a stream of carrier gas to prevent rapid decrease in separation efficiency.

Column dimensions

In most cases standard lengths (e.g., 25 m) are optimal, but in some cases very short columns (e.g., 2 m) can lead to better results than long columns.

Film thickness

Thicker films increase the capacity, which is useful for semipreparative separations, but do not influence the separation efficiency. With normal film thicknesses the concentration of the analytes should not be very high to avoid peak overloading due to the low capacities of CD phases.

Carrier gas and linear carrier gas velocity

H_2, He, N_2, CO_2, NH_3 can be used; in most cases, hydrogen is the best carrier gas. The velocity can deviate from recommended van Deemter curves for achiral columns; for lower elution temperatures (with moderate retention times) faster gas velocities are necessary.

Working temperature

In almost all cases the temperature should be as low as possible (if not, derivatization may lead to more volatile compounds). Working at a temperature usually above 150 °C or even higher, can lead to a loss of separation (this is also dependent on the CD derivative).

With the CD-columns listed above, approximately 25-35% of the tested chiral compounds can be separated, but no column shows universal enantioselectivity. It is virtually impossible to predict a successful enantiomer separation on a given CD column. For checking which classes of compounds can be separated by a chosen CD column, a test-mix ('Schurig Test Mixture' [123-125]) has been developed, which contains several selected chiral compounds for the evaluation of CD phases: α-pinene; *E*- and *Z*-pinane; *E*-2,3-butanediol; 4-pentanolide; 1-phenylethylamine; 1-phenylethanol; 2-ethylhexanoic acid and one achiral compound (*Z*-2,3-butanediol). Another test-mix for evaluation of column performance was applied by Bicchi et al. [77]. This mixture contains ten chiral compounds: limonene, 3-octanol, *Z*-3-hexenyl 2-methylbutanoate, menthol, 7-hydroxycitronellal, 1,8-epoxy-9-(3-methyl-2-butenyl)-p-menthane, *E*-2-(2-butylcyclopropyl)acetic acid, γ-decalactone, δ-decalactone and 6-methyl-2-(4-methylphenyl)-2-heptanol.

Applications using cyclodextrin phases

Table 3.5.7 presents a compilation of successful enantiomer separations using commercially available stationary phases that contain CD derivatives. Additionally, some of the most promising, recently developed stationary phases are also included (e.g., tert-butyldimethylsilyl CD derivatives and polysiloxane-anchored CDs). In Table 3.5.8 the stationary achiral phases (to dissolve cyclodextrins) are listed.

Table 3.5.7 Applications of commercially available cyclodextrin-containing CSPs (written in parentheses stands for derivative used: *TFA-der.) (if a compound class is given, it might be that not all compounds of a class can be successfully separated; cf., in particular, higher homologues!)

A) UNDILUTED PERMETHYLATED CYCLODEXTRINS [36]	
CSP-47	Per-O-methyl-β-cyclodextrin (degree of methylation >70 %) [39] (Peraldex)
Esters	Z-4-Hydroxy-2-cyclopenten-1-yl alkanoates [39]
Ketones	Bicyclo[3.2.0]hept-2-en-6-one [39]
Lactones	2-Oxabicyclo[3.3.0]oct-6-en-3-one [39]
B) UNDILUTED CYCLODEXTRIN DERIVATIVES [40]	
CSP-51	Hexakis(2,3,6-tri-O-pentyl)-α-cyclodextrin [40] (FS-Lipodex A; FS-CYCLODEX alpha-II)
Acetals	Spiroacetals (e.g., 1,7-dioxaspiro[5.5]undecane and related compounds) [126]
Acids	Lactic acid, glyceric acid, 2-/3-hydroxybutanoic acid, *E*-tartaric acid, galactonic acid (TFA/ME) [127]; mannonic acid (TFA/ME) [128]; mandelic acid (TFA/ME) [129]
Alcohols	2-Methyl-1-butanol, 3-alkanols [130]; 2-methyl-1-penten-3-ol, 3-methyl-3-alkanols [46]; *endo-/*exo-5-norbornen-2-ol [129]; *2-heptanol [131]; 2-ethyl-1-hexanol [132]; 4-octanol [130]; 1-octen-3-ol [132]; 4-nonanol [130]; 1-phenylethanol [46]; *1-phenyl-1-alkanols [127;130]; *1-(3-/*1-(4-methylphenyl) ethanol [133]; 2-phenyl-1-propanol [132]; 1-indanol (no or TFA) [46;131]
Alkanolamines	*2-Amino-1-butanol [127]; *sphingosine [46]
Amines	1-PEA (no or TFA) [40;122]
Amino acids	Ala, phe (TFA) [127]; thr (N,O-TFA/iPE) [40]; 3,4-dihydroxyphe = DOPA (TFA/ME) [134]
Carbohydrates	*Glyceraldehyde, *erythrose, *α-/*β-aldopentoses [127;128]; *α-/*β-aldohexopyranoses, *α-methyl hexosides, *β-fucofuranose [40]

B) UNDILUTED CYCLODEXTRIN DERIVATIVES [40]

Diols & polyols	*1,2-Alkanediols [127]; 1,3-butanediol [130]; *1,2,4-butanetriol [127]; *E*-1,2-cyclohexanediol [131]; **E*-1,2-cycloalkanediols [127]; *arabinitol, *hexitols, *1,4-anhydropentitols, *1,5-anhydrohexitols [40]; **E*-/*Z*-1,2,4-heptanetriol [129]; *phenylglycol [127]
Epoxides	1,2-Epoxypropane [132]; 1,2-epoxy-3-butene [135]; glycidol (no or TFA) [134;136]; glycidyl methylether, glycidyl butanoate [132]; 2,3-epoxy-3-methyl-1-butanol [134]; **E*-/**Z*-2-epoxy-3-methyl-4-pentyn-1-ol [137]; 3,4-epoxy-4-methyl-2-pentanone [130]; **E*-4,5-epoxy-1-hexanol [136]; *E*-/*Z*-2,3-epoxyoctane, *E*-/*Z*-1,2-epoxy-3-octanol; **E*-2,3-epoxy4-octyn-1-ol [136]; styrene oxide [131]
Esters	Methyl lactate, methyl cyclopentanone-3-carboxylate, methyl mandelate [131]
Furans & pyrans	**E*-Sulcatol furanoxide = *E*-pityol, tetrahydro-3-methyl-2H-pyran [46]; dihydrorose oxide [132]
Halo-compounds	epichloro-/epibromohydrin [132]; *2-chloro-/*2-bromo-1-phenylethanol [126]; 2-bromobutane, methyl 2-bromopropanoate [135]; 2-tert-butyl-5-bromomethylene-1,3-dioxane [131]
Hydrocarbons	3-Methyl-1-cyclohexene [131]; 1,2-cycloalkanedienes [46]; dimerization products of 1,2-cyclooctadienes [134]
Ketones	3-Hydroxy-2-butanone = acetoin, 4-hydroxy-3-methylene-2-pentanone [46]; 2-alkylcyclohexanones [138]; 4-methyl-3-heptanone [133]; carvone [127]
Lactones	2-Methyl-4-butanolide, *E*-/*Z*-whisky lactone [131]
Pesticides	α-1,2,3,4,5,6-Hexachlorocyclohexane = lindane [46]
Pharmaceuticals	Bromochlorofluoromethane [46]; 1-bromo-1-chloro-2,2,2-trifluoroethane = halothane, 1-chloro-2,2,2-trifluoroethyldifluoromethylether = isoflurane [139]; hexobarbital, methylphenobarbital[140]; 3-ethyl-5-phenylhydantoin = ethotoin, 4-ethyl-1-methyl-4-phenylhydantoin = mephenytoin [134]; 3-ethyl-3-phenyl-2,6-piperidinedione = gluthetimide [141]
Miscellaneous	*4-(Hydroxymethyl)-2,2-dimethyl-1,3-dioxolane = solketal [127]; Z-lactide [129]; *1-ethyl-2-aminomethylpyrrolidine [122]; ethyl methylsulphoxide, butyl methylsulphoxide [46]
CSP-53	Heptakis(2,3,6-tri-O-pentyl)-β-cyclodextrin [137] (FS-Lipodex C; FS-CYCLODEX beta-II) (additional data, cf. Table 3.5.8)
Acetals	1,1-Dimethoxy-2-methylbutane [142]
Acids	2-Hydroxyalkanoic acids, mandelic acid (TFA/ME) [143]
Alcohols	2-Butanol [116]; 3-butyn-2-ol [143]; 6-methyl-5-hepten-2-ol = sulcatol [46]

B) UNDILUTED CYCLODEXTRIN DERIVATIVES [40]

Alcohols	*Z-1-(2-hydroxyethyl)-2-isopropenyl-1-methylcyclobutane = grandisol [143]; 3-menthanols, *E*-/*Z*-carveol, *E*-pinocarveol, *isoborneol, iso-pinocampheol, *E*-verbenol, ipsenol, ipsdienol, hotrienol [46]; linalool [47]; *E*-nerolidol [46]
Amines	*1-PEA [70]
Carbohydrates	*α-Methyl rhamnopyranoside, α-methyl quinovopyranoside [143]
Diols	E-2,3-Butanediol (cyclic carbonates via phosgene) [70]
Epoxides	1,2-Epoxy-3-butene, styrene oxide [132]
Esters	Methyl 2-methylbutanoate [144]
Ethers	2-Methoxybutane [116]; 1-alkoxyglycerols [129]
Furans & Pyrans	*E*-/*Z*-Sulcatol furanoxide = *E*-/*Z*-pityol [46]; 3,9-epoxy-1,4(8)-p-menthadiene = linden ether [145]; *E*-sulcatol pyranoxide = *E*-vittatol [46]
Halo-compounds	2-Chloroalkanes [143]; 1,2-halopropanes [146]; 1-halo-1-phenylethane [142]; methyl 2-chloropropanoate [144]; 2-bromoalkanes [116]; 1-(2-bromoethylidene)-4-tert-butylcyclohexane [46]; methyl 2-bromoalkanoates [135;142]; 2-iodobutane [116]
Hydrocarbons	3-Methyl-1-hexene, 3-methyl-1-cyclohexene [143]; 3-methyl-Z-1,4-hexadiene, 3,5-dimethyl-1,4-hexadiene [137]; *E*-1,2-/E-1,3-dimethylcyclohexane, 1,2-cycloalkanedienes, 1-ethylidene-4-tert-butylcyclohexane [46]; Z-pinane [147]; limonene, terpenoid cyclopropane hydrocarbons [143]; terpenoid cyclic hydrocarbons (algal pheromones) [148]; 3-phenyl-1-butene [149]
Pesticides	α-1,2,3,4,5,6-Hexachlorocyclohexane = lindane [46]; dichlorprop, fenoprop [58]
Miscellaneous	4-Alkyl-/4-phenyl-1,3-dioxolan-2-ones [41;132]; *alkylcyanohydrins [138]; *mandelonitrile [46]
CSP-89	Octakis (2,3,6-tri-O-pentyl)-γ-cyclodextrin [150] (FS-CYCLODEX gamma II)
Esters	Methyl/ethyl mandelate [132]
Hydrocarbons	2,4-Dimethyl-1-heptene, *E*-/*Z*-pinane [147]; α-phellandrene [47]; 1,2-cyclooctadiene [46]; *E*-3,3,8,8-tetramethylcyclooctene [147]; 5-ethyl-3-methyl-enenonane [151]
Miscellaneous	Tricarbonyl(h^4-2-methyl-1,3-butadiene)iron(0), tricarbonyl (h^4-2-methyl-1,3-butanediyl)iron(0) [150]
CSP-61	Hexakis (3-O-acetyl-2,6-di-O-pentyl)-α-cyclodextrin [126] (FS-Lipodex B)
Aldehydes	*3-Hydroxybutanal, *2-methyl-3-hydroxypentanal [152]

B) UNDILUTED CYCLODEXTRIN DERIVATIVES [40]

Alkanolamines	*1-Amino-2-propanol, *alaninol, *2-amino-1-butanol, *leucinol [138]
Carbohydrates	*Glyceraldehyde, *threose, *α-/*β-aldopentopyranoses, *α-/*β-aldohexopyranoses, *allofuranose, *gulofuranose (as methyl glycosides), *fructopyranose [46;138]
Diols	1,2-Alkanediols (as cyclic carbonates) [152]; 1-O-alkylglycerols (as cyclic carbonates) [46]
Lactones	4-Alkanolides, 2-alkyl-4-butanolides [126;137;152]; pantolactone (no or TFA) [52;129]; *E-/Z*-3-methyl-4-pentanolide [46]; *E-/Z*-3-methyl-4-octanolide = whisky lactone [152]; 4-pentadecanolide [153]; 2-decen-5-olide [154]; *Z*-6-dodecen-4-olide [155]; *Z*-3-oxabicyclo[3.1.0]hexan-2-one [156]; 3-oxabicyclo[4.3.0]non-7-en-2-one [126]; 3-phenyl-/3-benzyl-4-butanolide [133]; 4-phenyl-4-butanolide [52]
Pharmaceuticals	Mesuximide, phensuximide [152]; *ephedrine, *norephedrine [157]
Miscellaneous	*Mandelonitrile [133]
CSP-58	Heptakis (3-O-acetyl-2,6-di-O-pentyl)-β-cyclodextrin [127] (FS-Lipodex D)
Acetals	Spiroacetals (e.g. endo-brevicomin, frontalin and related compounds) [133]
Acids	3-Hydroxybutanoic acid (TFA/ME) [129]
Alcohols	2-Alkanols [46]; vulgarol [158]; *E-/Z*-nerolidol [54]
Aldehydes	3-Methylalkanals [46]; citronellal [67]
Alkanolamines	*Alaninol, *valinol [159]
Amines	*2-Aminoalkanes [127]; *3-amino-4-methyl-1-pentene [46]; *endo-/*exo-5-aminomethylnorbornene [159]; *1-PEA [127]
Amino acids	3-Amino-2-methylpropanoic acid, 2-/3-aminobutanoic acid (TFA/ME) [159]
Diols & polyols	*1,2-Propanediol, *1,3-butanediol, **E*-1,2-/*E-1,3-cycloalkanediols [159]; 1,2-propanediol, *E*-2,3-butanediol (as cyclic carbonates) [70]; *mannitol [46]; *E*-sobrerol [52]
Esters	1-Octen-3-yl acetate [160]; ethyl 2,3-epoxybutanoate [52]
Epoxides	**E*-/**Z*-Epoxycinnamyl alcohol [146]
Furans	Tetrahydro-2-furanmethanol [52]
Halo-compounds	*1-(4-Chlorophenyl)ethylamine [161]; methyl 2-chloropropanoate [134]

B) UNDILUTED CYCLODEXTRIN DERIVATIVES [40]

Ketones	Camphor, carvone [52]
Lactones	4-Alkanolides [162]; pantolactone [141]; 2-hexen-4-olide [163]; *E*-/*Z*-solerol [164]; **E*-/**Z*-solerol, solerone [165]; 2-nonen-4-olide [166]; *Z*-7-decen-4-olide = γ-jasmine lactone [67]; 3-oxabicyclo[4.3.0]nonan-2-one [129]; dihydroactinidiolide [167]; 5-alkanolides [53]; mevalolactone [46]; *Z*-7-decen-5-olide = δ-jasmine lactone [67]; 2-decen-5-olide = massoi lactone [52]
Pesticides	α-1,2,3,4,5,6-Hexachlorocyclohexane = lindane [168]; O,S-dimethyl thiophosphoric acid amide = methamidophos [58]
Pharmaceuticals	*Amphetamine, *mexiletine, *pholedrine, *tranylcypromine [159]; *ephedrine, *norephedrine, mesuximide, phensuximide[141]; α-lipoic acid (ME) [169]
Miscellaneous	*2-(Methoxymethyl)pyrrolidine [159]; 1-phenylethylisocyanate [52]
CSP-57	Octakis (3-O-butyryl-2,6-di-O-pentyl)-γ-cyclodextrin [57] (FS-Lipodex E) (#60 % dissolved in OV-1701)
Acetals	Cyclic acetals (e.g. endo-/exo-brevicomin, lineatin, frontalin and related compounds) [57]
Acids	2-/3-Hydroxyalkanoic acids, malic acid, tartaric acid, mandelic acid (TFA/ME) [57]; 2-methylbutanoic acid, 2-methylsuccinic aicd (ME) [46]; #abscisic acid, #phaseic acid (ME) [51]
Alcohols	2-Alkanols (no or TFA) [131;170]; 1-cyclopropylethanol [52]; 2-methyl-1-penten-3-ol, Z-4-hepten-2-ol [171]; 6-methyl-5-hepten-2-ol = sulcatol (no or TFA) [52;57]; *3-octanol, *1-octen-3-ol [57]; Z-3,3,5-trimethyl-1-cyclohexanol (no or TFA) [131]; *2-(1-methylenecyclopentyl)-1-butanol [46]; *Z-1-(2-hydroxyethyl)-2-isopropenyl-1-methylcyclobutane = *grandisol, *4-terpinenol, *iso-/*borneol, *myrtenol, *citronellol [57]; menthol [52]; *E*-/*Z*-carveol [134]; **E*-pinocarveol, *2-(4-tert-butylcyclohexylidene)ethanol [46]; 1-phenylethanol (no or TFA) [57;161]; *1-(4-methylphenyl)ethanol, *1-phenyl-1-propanol, *1-phenyl-2-propanol [57]
Aldehydes	Myrtenal [57]
Amines	*2-Aminoalkanes [57]; *1,2-diaminopropane [46]
Amino acids	α-Amino acids (TFA/ME) [57]; N-methyl-α-amino acids, α-alkyl-α-amino acids, cyclic sec. α-amino acids, β-amino acids (TFA/ME) [134]; 2-amino-3-(3,4-dihydroxyphenyl)-2-methylpropanoic acid = aldomet, 4-amino-3-hydroxy-5-phenylpentanoic acid (TFA/ME) [141] *E*-/*Z*-3-hydroxyaspartic acid (TFA/ME) [46]
Diols & polyols	*1,2-Alkanediols, *1,3-butanediol, **E*-2,3-butanediol, **E*-2,4-pentanediol [57]; *E*-2,3-butanediol [52]; *1,4-pentanediol [46]; *1,2,3-/*1,2,6-hexanetriol [57]; **E*-1,2-cyclohexanediol [129]; *1,2,3-heptanetriol [57]

B) UNDILUTED CYCLODEXTRIN DERIVATIVES [40]

Epoxides	1,2-Epoxybutane, 1,2-epoxyhexane, 3,3-dimethylglycidol [161]; styrene oxide [170]
Esters	*E*-/*Z*-3-Hydroxy-2-butyl acetate, *E*-2,3-butanediol diacetate, 3-oxo-2-butyl acetate = acetoin acetate, 2-alkyl acetates [52]; *Z*-4-hepten-2-yl acetate[[171]; 1-octen-3-yl acetate [52]; linalyl acetate [50]; nopyl acetate [52]; *Z*-4-hepten-2-yl-propanoate [171]; methyl/ethyl lactate, ethyl 2-hydroxybutanoate [52]; methyl 3-hydroxybutanoate [172]; ethyl-3-acetoxybutanoate, ethyl 2,3-epoxybutanoate [52]; methyl *E*-/*Z*-3-(3-hydroxybutanoyloxy)butanoate [172]; dimethyl *E*-1,2-cyclopropanedicarboxylate [57]; ethyl 3-hydroxyalkanoates [46]; methyl 2-methylbutanoate [96]; ethyl 2-methylbutanoate [131]; dimethyl *E*-1,2-cycloalkanedicarboxylates [46]; methyl/ethyl mandelate [52]
Ethers	3-Hydroxy-2-methylene-1-butyl methylether (TFA or ac.) [173]; 7-(methoxymethyl)-5,7-dimethyl-1,3,5-cycloheptatriene, 7-(methoxymethyl)-6,7-dimethyl-1,3,5-cycloheptatriene [134]; methyl 2-(2-methylphenyl)-1-propylether [174]
Furans & pyrans	Dihydro-2-methyl-3(2H)-furanone, *E*-/*Z*-tetrahydro-5-methyl2-furanmethanol, 4-hydroxy-2,5-dimethyl-3(2H)-furanone = furaneol, *Z*-sulcatol furanoxide = *Z*-pityol [52]; **E*-/**Z*-sulcatol furanoxide [57]; *E*-/*Z*-sulcatol pyranoxide = *E*-/*Z*-vittatol, *Z*-linalool furanoxide [52]; *2-methyl-2-(4-methyl-2-furanyl)-3-buten-2-ol [46]; *E*-rose oxide [52]
Halo-compounds	Ethyl 3,3,3-trifluorolactate [175]; 2-methoxy-2-(trifluoromethyl)-phenylacetic acid (ME) [176]; E-/Z-3-chloro-2-butylacetate [52]; methyl 2-chloropropanoates [46]; isobutyl 2-chloropropanoate [161]; 4,4,4-trichloro-3-butanolide [57]; 2-bromoalkanes [46]; 1-bromo-1-phenylethane [170]; methyl 2-bromoalkanoates [46]
Hydrocarbons	α-Thujene [177]; Z-pinane [170]; 3-carene [178]
Ketones	3-Hydroxy-2-butanone = acetoin [52]; *E*-/*Z*-3-alkyl-2-methylcyclopentanones [161]; 2-alkylcyclohexanones [57]; 5-methyl-3-heptanone [46]; isomenthone/menthone, piperitone, carvone, verbenone, fenchone, camphor [57]; isopinocamphone [179]; verbenone [46]; α-ionone [52]; *Z*-dihydro-α/γ-irone [46]; *E*-/*Z*-α-irone [66]; β-irone, *Z*-γ-irone [46]
Lactones	2-/3-Alkyl-4-butanolides [46]; 4-/5-alkanolides [57]; 2-alkyl-4-alkanolides [46]; 2-hexen-4-olide, 4-methyl-5-hexen-4-olide [52]; pantolactone [161]; *E*-/*Z*-whisky lactone [131]; 4-phenyl-4-butanolide [161]; 4-ethoxycarbonyl-4-butanolide, bicyclic 4-alkanolides, 2-/4-alkyl-5-pentanolides [46]; 5-alkyl-2-penten-5-olides [180]; 6-octanolide [57]
Pesticides	α-1,2,3,4,5,6-Hexachlorocyclohexane = lindane [57]; mecoprop, dichlorprop (ME) [58]; 7-oxabicyclo[2.2.1]heptane-2,3-dicarboxylic acid, *E*-/*Z*-permethric acid (ME) [46]
Pharmaceuticals	*2,4-Dihydroxy-N-(3-hydroxypropyl)-3,3-dimethylbutaneamide = panthenol [141]; enflurane, ethotoin [157]; baclofen, tacainide (TFA/ME) [157]; 3-ethyl-3-phenyl-2,6-piperidinedione = gluthetimide [46]

B) UNDILUTED CYCLODEXTRIN DERIVATIVES [40]	
Pharmaceuticals	*Deoxyephedrine [131]
Miscellaneous	1-Cyano-3-hydroxy-2-methylenealkanes (TFA or Ac.) [173]; dihydro-2-methyl-3(2H)-thiophenone [52]; #methyl phenylsulphoxide, #phenyl 2-propylsulphoxide [51]
C) UNDILUTED CYCLODEXTRIN DERIVATIVES [42]	
CSP-90	Hexakis (2,6-di-O-pentyl)-α-cyclodextrin [42] (#CSP contains 0.1 % PEG 400) (Chiraldex A-DA)
Acids	O-Acetylmandelic acid (TMS) [62]
Alcohols	#exo-2-(Hydroxymethyl)norbornane, #4-terpinenol, #1-phenylethanol, 1-phenyl-2-propanol [62]; 1-tetralinol (ac. or TFA) [42;69]
Alkanolamines	**E*-/*Z-3-Aminomethyl-3,5,5-trimethyl-1-cyclohexanol [62]
Amines	#*2-Aminopentane[62]; #endo-2-aminonorbornane[62]; 1-cyclohexyl-ethylamine (ac. or #TFA) [42;62]; #*1-PEA 62]; *1-aminoindan [69]; 1-aminotetralin (MCA) [62]
Carbohydrates	*Ara, *gal, *glc, *lyx, *man, *rib, *tal, *methyl-β-arabinopyranoside [181]
Epoxides	#1,2-Epoxyhexane, #1,2-epoxy-7-octen, #alkyl glycidylethers, #*E*-/*Z*-limonene oxide, #styrene oxide [62]
Furans & pyrans	#2-Ethoxytetrahydrofuran, #2-ethoxy-3,4-dihydro-2H-pyran, #tetra-hydro-2-(2-propynyloxy)-2H-pyran [62]
Halo-compounds	#exo-2-Chloro-/exo-2-bromonorbornane, #tetrahydro-2-chloromethyl-2H-pyran, #tetrahydro-2-bromomethyl-2H-pyran [62]
Hydrocarbons	#endo-/exo-2-Benzylnorbornane [62]
Ketones	#endo-/exo-2-Acetylnorbornane [62]
Miscellaneous	#*1-Amino-1-methoxypropane, **E*-/*Z-3'-methylnornicotine, *N-[1-(3-pip-eridyl)ethyl]-ethylamine [62]; *1-ethyl-2-aminomethylpyrrolidine [122]; #**E*-1,2-dithiane-4,5-diol [62]
CSP-91	Heptakis(2,6-di-O-pentyl)-β-cyclodextrin [42] (Chiraldex B-DA)
Acids	O-Acetylmandelic acid (TMS) [62]
Alcohols	1-Tetralinol [62]
Alkanolamines	2-Amino-1-propanol (MCA), 2-amino-1-pentanol (ac.), **E*-/*Z-3-amino-methyl-3,5,5-trimethyl-1-cyclohexanol [62]
Amines	endo-2-Aminonorbornane, *1-PEA, *1-aminoindan [62]; 1-aminotetralin (ac. or TFA) [62;69]; *1,1'-NEA [42]

C) UNDILUTED CYCLODEXTRIN DERIVATIVES [42]

Amino acids	Tyr, a-methyl-tyr (N,O-TFA/ME) [62]
Carbohydrates	*Erythrose, *ara, *lyx, *rib, *sor, *methyl-b-arabinopyranoside [42]; *gal, *glc, *man, *xyl [181]
Esters	Methyl mandelate (no or TFA) [42;62]; *ethyl mandelate [62]
Ethers	*2-Methoxy-2-phenylethanol [62]
Halo-compounds	N'-(2,2-Difluoro-/N'-(2,2,2-trifluoroethyl)nornicotine, 2,2,2-trichloro-1-phenylethyl acetate [62]
Lactones	Pantolactone [62]
Pharmaceuticals	*2-Amino-1-phenylethanol = octopamine, *2-amino-3-phenyl-1-propanol, *3-amino-2-methyl-1-phenyl-1-propanol = pseudoephedrine [62]; hexobarbital, methylphenobarbital [62]
Miscellaneous	2-Amino-1-methoxypropane (TCA) [69]; *2-phenylpyrrolidine, *2-phenethylpyrrolidine, *1-(3-pyridyl)ethylamine, *2-(4-/*2-(2-pyridyl)-pyrrolidine, *4-methylnornicotine, *anabasine, *2-(3-pyridyl)-1-azacycloheptane, N'-benzylnornicotine, *E-1,2-dithiane-4,5-diol, N-acetylhomocysteinthiolactone [62]
CSP-92	Octakis (2,6-di-O-pentyl)-γ-cyclodexdrin [42] (Chiraldex G-DA)
Amines	*1-Aminoindan, *1-aminotetralin[62]; *1,1'-NEA [42]
Diols	*E*-1,2-Cycloalkanediols (no or ac. or MCA) [62]
Miscellaneous	*2-(3-Pyridyl)-1-azacycloheptane, *anatabine [62]
CSP-93	Hexakis[O-S-2'-hydroxypropyl-per-O-methyl]-α-cyclodextrin[42] (Chiraldex A-PH)
Alcohols	*Isopinocamphenol [63]; *1-tetralinol [42]
Alkanolamines	1-Amino-2-propanol (ac.), 2-amino-1-alkanols (ac.) [63]
Amines	*Sec. aminoalkanes, *endo-/*exo-2-aminonorbornane, *1-PEA, *1-aminotetralin [63]; *1,1'-NEA [42]
Carbohydrates	*All, *rib [63]
Epoxides	Glycidol and related compounds (no or TFA) [62]; *E*-/*Z*-limonene oxide [42]
Esters	*Methyl mandelate [42]; methyl/ethyl mandelate [63]
Furans & pyrans	2-Ethoxytetrahydrofuran, *E,E-/E,Z-/Z,E-/Z,Z*-tetrahydro-2,5-dimethoxyfuran-2-carbaldehyde, 3,4-dihydro-2-methoxy-2H-pyran, 2-ethoxy-3,4--dihydro-2H-pyran [63]

C) UNDILUTED CYCLODEXTRIN DERIVATIVES [42]

Halo-compounds	exo-2-Bromonorbornane [63]
Hydrocarbons	Limonene [63]; 2,6,10-trimethyl-7-(3-methyl-1-butyl) dodecane, alkyldihydroindans, alkyltetralins and related compounds [182]
Ketones	Carvone [63]
Lactones	4-Phenyl-4-butanolide [63]
Miscellaneous	4-Phenyl-1,3-dioxane, 2-amino-1-methoxypropane (ac.), **E*-1,2-dithiane-4,5-diol [63]
CSP-63	Heptakis[O-(S)-2'-hydroxypropyl-per-O-methyl]β-cyclodextrin[42] (Chiraldex B-PH)
Acids	Lactic acid, malic acid (TMS) [65]
Alcohols	*exo-Norborneol [183]; *5-norbornen-2-ol, *isoborneol, *isopinocamphenol [63]; *1-indanol [69]; *1-tetralinol [42]; *E-/Z*-nerolidol [54]
Alkanolamines	1-Amino-2-propanol (ac.) [63]; *2-amino-1-propanol [42]; 2-amino-1-butanol (ac.) [63]; 2-amino-3-methyl-1-butanol (no or ac.) [63;69]; 2-amino-1-hexanol (N-ac.) [69]
Amines	*Sec. aminoalkanes [63]; 1-cyclohexylethylamine (ac. or TFA) [42;63]; 1-PEA (ac. or TFA), 1-aminoindan (ac. or TFA) [69]; 1-aminotetralin (ac. or TFA), 1,1'-NEA (ac.) [69]
Carbohydrates	Glyceraldehyde (ac.)[63]; *erythrose, *ara, *lyx, *rib, *sor, *methyl arabinopyranoside [42]; *gal, *man [63]; *xyl [181]
Diols	*E*-1,2-Cycloalkanediols [63]; *E*-1,2-cyclohexanediol (MCA) [69]; **E*-/*Z-2-ethyl-1,3-hexanediol = ethohexadiol [184]
Epoxides	1,2-Epoxyalkanes, alkyl glycidylether, glycidyl methacrylate [63]; *E-/Z*-limonene oxide [42]; styrene oxide [63]
Esters	Methyl 2-hydroxyalkanoates, alkyl lactats [65]; butyl lactate (no or TFA) [63]; **E*-dimethyl/**E*-diisopropyl tartrate, methyl/ethyl mandelate [65]
Furans & pyrans	2-Ethoxytetrahydrofuran [63]; *E-/Z*-2-ethoxytetrahydro-5-pentylfuran [185]; *E,E-/E,Z-/Z,E-/Z,Z*-tetrahydro-2,5-dimethoxyfuran-3-carbaldehyde, 3,4-dihydro-2-methoxy-2H-pyran, 2-ethoxy-3,4-dihydro-2H-pyran; tetrahydro-2-(2-propynyloxy)2H-pyran [63]
Halo-compounds	N'-(2,2-Difluoroethyl)nornicotine, 3-chloro-2-norbornanone [63]; methyl 2-chloropropanoate [42]; 2,2,2-trichloro-1-phenylethyl acetate [63]; dimethyl chloro-/dimethyl bromosuccinate [65]; exo-2-bromonorbornane [63]; methyl 2-bromoalkanoates [65]
Hydrocarbons	Limonene [63]; 1,1,3-trimethyl-2-(3-methyloctyl)cyclohexane [182]; endo-*E*-6-methyl-5-phenyl-2-norbornene [63]; alkyltetralins and related compounds [182]

C) UNDILUTED CYCLODEXTRIN DERIVATIVES [42]

Ketones	3-Hydroxy-2-butanone = acetoin [65]; 5-acetyl-2-norbornene [63]
Lactones	3-Butanolide, 2-methyl-4-butanolide, 2-acetyl-2-methyl-4-butanolide, pantolactone [63]
Pesticides	Mitotane [184]
Miscellaneous	4-(Hydroxymethyl)-2,2-dimethyl-1,3-dioxolane = solketal [63]; 2-amino-1-methoxypropane (ac. or MCA), 2-amino-4-butanolide (ac.) [69]; *nor-nicotine, *4'-methylnornicotine, N-acetylhomocysteinthiolactone, **E*-1,2--dithiane-4,5-diol [63]
CSP-94	Octakis[O-(S)-2'-hydroxypropyl-per-O-methyl]γ-cyclodextrin[63] (Chiraldex G-PH)
Alkanolamines	2-Amino-1-propanol (ac.) [63]
Amines	*Sec. aminoalkanes, *1-PEA [63]
Epoxides	*E*-/*Z*-Limonene oxide [63]
Halo-compounds	2-(Bromomethyl)-tetrahydrofuran [63]
Pyrans	2-Ethoxy-3,4-dihydro-2H-pyran [63]
Miscellaneous	*4-(Hydroxymethyl)-2,2-dimethyl-1,3-dioxolane = solketal, **E*-1,2-dithiane-4,5-diol [63]
CSP-95	Hexakis(2,6-di-O-pentyl-3-O-TFA)-α-cyclodextrin [181] (Chiraldex A-TA)
Alkanolamines	*1-Amino-2-propanol, *2-amino-1-alkanols [186]
Amines	*Sec. aminoalkanes, *1-aminoindan, *1-aminotetralin [186]
Carbohydrates	*All, *ara, *gul, *lyx, *xyl[181]; *glc[187]; *methyl-β-arabinopyranoside [181]
Diols	*1,2-Alkanediols, *1,3-butanediol [186]
Epoxides	1,2-Epoxyalkanes, allyl glycidylether [186]
Furans & pyrans	2-Ethoxytetrahydrofuran, 3,4-hydroxy-2-methoxy-2H-pyran, 2-ethoxy-3,4-dihydro-2H-pyran [186]
Halo-compounds	2-Haloalkanes, 2-chloro-3-butene, 2-bromo-1-chloropropane, 1,2-/1,3-dibromobutane [186]
Ketones	Carvone [186]
Lactones	2-methyl-4-butanolide, *pantolactone [186]

C) UNDILUTED CYCLODEXTRIN DERIVATIVES [42]

CSP-96	Heptakis(2,6-di-O-pentyl-3-O-TFA)-β-cyclodextrin [42] (Chiraldex B-TA)
Alcohols	*2-Alkanols, 3-butyn-2-ol, 3-hexanol (no or TFA) [188]; E-2-methyl-1-cyclohexanol [189]; *E-/*Z-2- and *E-/*Z-3-methyl-1-cyclohexanol, *exo-2-norbornanol [188]; 2-octanol, 1-cyclohexylethanol [189]; *isopinocampheol, 1-phenylethanol (no or TFA) [188]
Aldehydes	Citronellal [67]; 2-phenylpropanal [188]
Alkanolamines	*2-Amino-1-propanol [42]; *2-amimo-1-alkanols, *3-amino-1-hexanol [188]; *3-amino-1,2-propanediol [186]
Amines	Sec. aminoalkanes (ac. or MCA or TFA), *endo-/*exo-2-aminonorbornane, 1-cyclohexylethylamine (ac. or TFA) [188]; *4-phenyl-2-aminobutane [190]; *1-aminotetralin [188]; 1-PEA (no or TFA) [83;186]
Amino acids	Ala, leu (N-TFA/ME), phe (N-TFA/EE) [188]
Carbohydrates	*All, *ara, *gal, *glc, *gul, *lyx, *man, *rib, *sor, *tal, *xyl, *methyl-β-arabinopyranoside [181]
Diols & polyols	*1,2-Alkanediols, *1,3-butanediol [188]; *1,4-pentanediol [65]; *1,2,4-butanetriol [186]; *arabinitol [181]; *1,2,6-hexanetriol [186]
Epoxides	1,2-Epoxyalkanes, 1,2-epoxyalkenes, E-2,3-epoxybutane, alkylglycidylether, glycidyl methacrylate, E-/Z-limoneneoxide, styrene oxide [188]
Esters	*E-Dimethyl-/*E-diisopropyl tartrate [186]; Z-methyljasmonate [67]; *methyl-/*ethyl mandelate [186]
Furans & pyrans	*Tetrahydro-2-hydroxyfuran, E-tetrahydro-2,5-dimethoxyfuran, E-tetrahydro-2,5-dimethylfuran, 2-ethoxytetrahydrofuran, tetrahydro-2-(2-propynyloxy)furan, 3,4-dihydro-2-methoxy-2H-pyran, 2-ethoxy-3,4-dihydro-2H-pyran; tetrahydro-2-(2-propynyloxy)-2H-pyran [63]
Halo-compounds	2-Chloro-/2-bromoalkanes [188]; 1,2-dichloro/1,2-dibromobutane [186]; E-2,3-dichlorobutane, exo-2-chloro-/exo-2-bromonorbornene, 3-chlorobicyclo[3.2.1]oct-2-ene [188]; *1-chloro-2,3-propanediol [186]; 2-chloro-3-butanone [188]; 2-chlorocyclopentanone, 2-chlorocyclohexa-none [186];epichlorohydrin [188]; 2-chloro-/2-bromoproprionitrile [186]; methyl 2-chloropropanoate [42]; ciprofibrate (ME), 4,4,4-trichloro-3-butanolide [186]; tetrahydro-2-chloromethyl-2H-pyran, 2-bromo-1-chloropropane[188]; 1,3-dibromobutane [186]; 1,2-dibromo-3-chloropropane [191]; 2-bromo-1-phenylpropane [188]; alkyl-2-bromopropanoates, E-/Z-2-alkyl 2-bromopropanoates, alkyl-2-bromobutanoates [186]; epibromohydrin, tetrahydro-2-bromomethyl-2H-pyran, 2-bromo-1-cyanopropane[188]; 2-iodobutane [186]
Hydrocarbons	4-Methyl-1-cyclohexene [188]; 3-carene [67]
Ketones	2-Methylcyclohexanone [188]; endo-/exo-5-acetyl-2-norbornene [186]; carvone [65]; E-/Z-α-irone [66]

C) UNDILUTED CYCLODEXTRIN DERIVATIVES [42]

Lactones	2-methyl-4-butanolide [188]; *pantolactone, 2-acetyl-2-methyl-4-butanolide, 5-hydroxy-3,3-dimethyl-4-butanolide, 4-nonanolide [186]; mintlactone, dehydromintlactone [67]; 4-undecanolide [83]
Miscellaneous	*4-(Hydroxymethyl)-2,2-dimethyl-1,3-dioxolane = solketal [186]; 2-amino-1-methoxypropane (ac. or MCA or TFA) [188]; lactonitrile (no or TFA) [65;188]; *2-methylpiperidine [188]; 1-methyl-2-phenylpyrrolidine, *2-benzylpyrrolidine, nicotine[186]; *E*-/*Z*-2-methyl-4-propyl-1,3-oxathiane [66]
CSP-65	Octakis(2,6-di-O-pentyl-3-O-TFA)-γ-cyclodextrin [65] (Chiraldex G-TA)
Acids	2-Ethylhexanoic acid [68]
Alcohols	2-Alkanols [192]; *2-alkanols, *3-butyn-2-ol [186]; sec. alkyl-1-alkanols, sec. alkanols [193]; 3-alkanols [192]; 3-methyl-1-pentyn-3-ol, 1-hexen-3-ol, 1-hexyn-3-ol [191]; *3-heptanol [186]; 6-methyl-5-hepten-2-ol = sulcatol [193]; *isoborneol [186]; dihydro-β-ionol [66]; 1-phenylethanol [68]; α-bisabolol [66]
Alkanolamines	*1-Amino-2-propanol, *2-amino-1-alkanols [186]
Amines	*Sec. aminoalkanes, *1-aminotetralin [186]
Diols	*1,2-Alkanediols, *1,3-butanediol [186]; **E*-2,3-butanediol [65]; *1,4-pentanediol, **E*-1,2-cycloalkanediols [186]
Epoxides	1,2-Epoxyhexane, *E*-1,2:3,4-diepoxybutane, *glycidol, alkyl glycidyl ethers, glycidyl acrylate, glycidyl methacrylate, *E*-/*Z*-limonene oxide, styrene oxide, *Z*-2-methyl-1-phenyl-1,2-epoxybutane, *E*-stilbene oxide, *E*-/*Z*-ethyl 3-phenylglycidate [186]
Esters	*Methyl/*butyl lactate [186]; ethyl 2-methylbutanoate, 1-phenylethyl acetate [68]; *ethyl mandelate [186]
Furans & pyrans	2-Ethoxytetrahydrofuran, *E*-tetrahydro-2,5-dimethoxyfuran, *E*-tetrahydro-2,5-dimethylfuran [186]; 4-methoxy-2,5-dimethyl-3(2H)-furanone = methoxyfuraneol [194]; tetrahydro-3-hydroxy2H-pyran, 3,4-dihydro-2-methoxy-2H-pyran, 2-ethoxy-3,4-dihydro-2H-pyran, tetrahydro-2-(2-propynyloxy)2H-pyran [186]; *E*-/*Z*-rose oxide [66]
Halo-compounds	2-Halobutanes, 2-chloro-3-butene [186]; 1,2-dichloropropane [184]; 1,2-*E*-2,3-dichlorobutane, 2-bromo-1-chloropropane [186]; 1,2-dibromo-3-chloropropane [184]; exo-2-chloronorbornane, *1-chloro-2-propanol, 3-chloro-2-butanone, 2-chlorocyclopentanone, 2-chlorocyclohexanone, endo-/exo-3-chloro-2-norbornanone [186]; methyl 2-chloropropanoate [65]; ciprofibrate (ME) [190]; epihalohydrins, 2-chloro-/2-bromolactonitrile, 2-bromoalkanes, 1,2-/1,3-dibromobutane, exo-2-bromonorbornane, 2-bromo-1-phenylpropane, alkyl 2-bromopropanoates, *E*-/*Z*-2-alkyl 2-bromopropanoates, alkyl 2-bromobutanoates, *E*-/*Z*-2-bromo-4-pentanolide [186]

C) UNDILUTED CYCLODEXTRIN DERIVATIVES [42]	
Hydrocarbons	1,4-Dimethyltetralin [182]; endo-*E*-6-methyl-5-phenyl-2-norbornene [190]
Ketones	2-Methylcyclohexanone, endo-/exo-5-acetyl-2-norbornene [186]; menthone [66]; carvone [186]; α-ionone [68]; endo-/exo-5-benzoyl-2-norbornene [186]
Lactones	4-alkanolides [65]; 5-decen-4-olide[195]; *Z*-7-decen-4-olide = γ-jasmine lactone [64]; 5-alkanolides [65;66]; 2-decen-5-olide = massoi lactone [66]
Pesticides	α-1,2,3,4,5,6-Hexachlorocyclohexane = lindane, silvex (ME), phenyl mercuric lactate [184]
Miscellaneous	*4-(Hydroxymethyl)-2,2-dimethyl-1,3-dioxolane = solketal, 4-phenyl-1,3-dioxane, mandelonitrile, *2-/*3-methylpiperidine [186]
D) POLYSILOXANE-DISSOLVED CYCLODEXTRIN DERIVATIVES [35]	
CSP-71	Hexakis(2,3,6-tri-O-methyl)-α-cyclodextrin (10 % diss. in OV-1701)[7] (FS-CYCLODEX alpha-I/P) (#10 % diss. in SPB-35 [α-Dex])
Acetals	Cyclic acetals (e.g., frontalin, endo-/exo-brevicomin and related compounds) [103]
Acids	#2-Ethylhexanoic acid [196]; campholenic acid, fencholenic acid [197]
Alcohols	2-Methyl-1-pentanol, 4-methyl-2-pentanol [121]; #3-hexanol [196]; 1-hexen-3-ol, 2-cyclohexen-1-ol [103]; 5-methyl-2-hexanol [121]; 2-heptanol, 2-octanol [132]; 3-/4-octanol, 1-octen-3-ol, 1,7-octadien-3-ol, 3,5,5-trimethyl-1-hexanol [103]; #3-menthanols [196]; *Z-1-(2-hydroxyethyl)-2-isopropenyl-1-methylcyclobutane = grandisol, 2,7-dimethyl-1,7-octadien-3-ol [103]; campholenic alcohol and related compounds, fencholenic alcohol and related compounds [197]; 4-methyl-2-tert-butylcyclohexanol = rootanol [132]; 1-phenyl-1-alkanols [103]
Aldehydes	3-Methylhexanal [198]; 2-norbornene-5-carbaldehyde [121]; campholenic aldehyde, fencholenic aldehyde[197]; 2-phenylpropanal [103]
Diols	#1,3-/#E-2,3-Butanediol [196]
Epoxides	Glycidol and related compounds, Z-epoxyalkanes [103]; isobutyl 2-methylglycidate [121]; phenyloxiranes [7]
Esters	5-Cycloocten-*E*-1,2-diol diacetate, 2-tridecyl acetate, methyl lactate 103]; methyl cyclohexene-3-carboxylate [121]; methyl/ethyl mandelate [103]
Ethers	1-Alkoxy-2-butanols [121]
Furans & pyrans	#*E*-Tetrahydro-2,5-dimethylfuran [196]; *E*-/*Z*-tetrahydro-2-(hydroxymethyl)-5-methylfuran [103]; #*E*-tetrahydro-2,5-dimethoxyfuran, #*E*-tetrahydro-2,5-diethoxyfuran [196]; *E*-/*Z*-dihydrorose oxide [103]; *Z*-rose oxide [66]

D) POLYSILOXANE-DISSOLVED CYCLODEXTRIN DERIVATIVES [35]

Hydrocarbons	4-Vinylcyclohexene [121]; #limonene [196]; Z-3-(Z-1,3-butadienyl)-4-ethenyl-1-cyclopentene = viridiene, Z-3-(E-1,3-butadienyl)-4-ethenyl-1-cyclopentene [199]; anti-tricyclo[6.4.0.02,7]dodeca-3,11-diene, exo-tricyclo[6.2.2.02,7]dodeca-3,9-diene; 3-phenyl-1-butene [103]
Ketones	2-Methylcyclohexanone [103]; #3-methyl-2-heptanone 196]; 2-butylcyclohexanone [121]; 5,6-diacetyl-1-cyclooctene [132]
Lactones	3-Methyl-4-butanolide [132]
Miscellaneous	Alkylated 1,3-dioxanes, alkylated 1,3-dioxepines, methyl 4-(N,N-diethylamino)-E-2-pentenoate [103]; 5-cyano-2-norbornene [121]; 2-methyl-4-nitro-5-phenylcyclohexanone, 1,3-dichloropentane [103]; *E*-1,2-dichlorocyclopentane, 1,2-dichloro-1-phenylethane [121]; 2-chloro-3-methyl-1-butanol [103]; 2-bromohexane [121]; 2-mercapto-1-phenyl-1-propanone, alkyl-/phenylthiomorpholines [103]; #alkylarylsilane der. [86]
CSP-68	Heptakis(2,3,6-tri-O-methyl)-β-cyclodextrin (10 % diss. in OV-1701)[35] (FS-CYCLODEX beta-I/P; C-Dex B; CP-Cyclodextrin-β-2,3,6-M-19; Cydex B; FS-Hydrodex β-PM; etc.) (#10 % and ##20 % diss. in SPB-35, respectively [β-Dex 110 and β-Dex 120])
Acetals	Spiroacetals (e.g. *E-/Z*-chalcogran, frontalin, endo-/exo-brevicomin and related compounds) [87;103]
Acids	2-Methylalkanoic acids [200]; 2-ethylhexanoic acid [200;201]; 2-hydroxyalkanoic acids (dioxolanones via 1,1,3,3-tetrafluoro-1,3-dichloro-2-propanone) [202]; campholenic acid, fencholenic acid [197]; α-cyclogeranic acid [92]; ibuprofen [121]
Alcohols	2-Alkanols [203]; 1-cyclopropylethanol [52]; 2-methyl-1-alkanols [200]; #;##3,3-dimethyl-2-butanol [204]; 2-methyl-3-pentanol [89]; *E*-2-methylcyclopentanol, 3-methyl-2-cyclopenten-1-ol [35]; #3-hexanol [204]; 4-/5-methyl-3-hexanol [203]; *E*-2-methyl-1-cyclohexanol [205]; 1-methyl-2-cyclohexen-1-ol [206]; 3-methyl-2-cyclohexen-1-ol = seudenol [35]; 1-alken-3-ols [200]; 1-hepten-4-ol 52]; endo-/exo-2-norborneol [35;207]; 2-ethyl-1-hexanol [200]; 6-methyl-5-hepten-2-ol = sulcatol [208]; 2,3-/3,4-dimethylcyclohexanols [209]; Z-1-hydroxymethyl-3-isopropenyl-2,2-dimethylcyclobutane [87]; *Z-1-(2-hydroxyethyl)-2-isopropenyl-1-methylcyclobutane = grandisol [103]; 3-menthanols [87]; isopulegol, sabinene hydrate [66]; α-terpineol, 4-terpinenol [103]; iso-/borneol [35;207]; 6,7-dihydrolinalool [210]; linalool, Z-verbenol [52]; campholenic alcohol and related compounds, fencholenic alcohol and related compounds [197]; patchenol [52]; methylisoborneol [66]; homolinalool [210]; geosmin [211]; *E-/Z*-α-ionol [66]; *E-/Z*-nerolidol [121]; 1-phenyl-1-alkanols [41]; 1-(4-hydroxyphenyl)ethanol [121]; 1-(2-methoxyphenyl) ethanol [103]; 2-phenyl-1-propanol [132]; 1-phenyl-2-propanol [52]; 2-phenyl-2-butanol [205]; 4-phenyl-2-butanol [103]; 3-methyl-1-phenyl-3-pentanol [89]; *E*-2-phenylcyclohexanol [203]; *indanol [131]; 1-tetralinol [203]; 5,5,9-trimethylbicyclo[4.4.0]deca-6,8,10-trien-3-ol [52]; 1,1'-naphthylethanol [212]; 1,2'-naphthylethanol [203]

D) POLYSILOXANE-DISSOLVED CYCLODEXTRIN DERIVATIVES [35]

Aldehydes	3-Acetoxyhexanal [52]; 3-methylhexanal [103]; endo-2-norbornene-5-carbaldehyde [87]; 2-propylhexanal [103]
Amines	1-PEA [124]
Amino acids	*E*-/*Z*-Methyl isoleucinate [213]; ala, pro, ser (N-TFA/iPE) [203]
Diols & polyols	1,2-Alkanediols, 1,3-butanediol, *E*-2,3-butanediol [202]; 2-methyl-1,2-butanediol, 2-methyl-3-buten-1,2-diol [203]; 2,3-pentanediol [214]; *E*-2,4-pentanediol [202]; 2-methyl-2,4-pentanediol [203]; *E*-2,5-hexanediol [202]; *E*-1,2-cyclohexanediol [201]; *arabitol [215]; *E*-1,2-cyclooctanediol [121]; *E*-sobrerol [52]
Epoxides	3,3-Dimethylglycidol [161]; glycidyl isopropylether [121]; 1,4-epoxy-p-2-menthanone [119]; styrene oxide, alkylated phenyloxiranes [7]
Esters	2-Alkyl acetates [216]; *E*-/*Z*-3-hydroxy-2-butyl acetate, *E*-2,3-butanediyl diacetate [52]; alkanediol diacetates [103]; 3-oxo-2-butyl acetate [66]; 3-/4-/5-alkyl acetates, 2-methyl-1-alkyl acetates [103]; 1-alken-3-yl acetates [200]; 1,2-/1,3-*E*-cyclohexanediol diacetate [103]; *E*-1,3-bis(hydroxymethyl)cyclohexane diacetate [121]; *Z*-4-hepten-2-yl alkanoates [216;217]; 6-methyl-5-hepten-2-yl acetate = sulcatyl acetate[52]; 3-menthyl acetates [218]; α-cyclogeranyl acetate, nopyl acetate [52]; 1-(4-acetoxyhydroxyphenyl)ethyl acetate [121]; methyl lactate 103]; ethyl lactate, 2-hexyl crotonate, methyl/ethyl 2-hydroxybutanoate [52]; *ethyl 2-hydroxybutanoate [132]; methyl/ethyl 3-acetoxybutanoate 219]; **E*-dimethyl tartrate [131]; *E*-diisopropyl tartrate [121]; methyl/ethyl 2-methylbutanoate [219]; dimethyl/diethyl *E*-cyclobutane-1,2-dicarboxylate [103]; ethyl 3,4-hexadienoate [7]; ethyl 2-hydroxy hexanoate [52]; methyl 3-hydroxyhexanoate [219]; methyl 3-acetoxyhexanoate [52]; ethyl 3-hydroxyhexanoate/-octanoate [200]; ethyl 5-acetoxyhexanoate [219]; methyl bicyclo-[2.2.1]hept-5-en-2-carboxylate [87]; methyl 1,2-dihydroxycyclohexanecarboxylate [220]; dimethyl *E*-1,2-cyclohexanedicarboxylat[103]; endo-methyl 5-methyl-2-norbornene-5-carboxylate [205]; methyl/ethyl *E*-2-(2-oxopropyl)cyclopentanecarboxylate [103]; methyl *E*-2-(4-methyl-2-oxopentyl)cyclopentanecarboxylate, methyl 2-(2-oxohexyl)cyclohexanecarboxylate, methyl 2-(4-methyl-2-oxopentyl)cyclohexanecarboxylate [121]; methyl 3-hydroxytetradecanoate [221]; *E*-methyl jasmonate [92]; methyl mandelate [201]; methyl 2-phenylpropanoate, tert-butyl 3-hydroxy-2-methylene-3-phenylpropanoate [103]
Ethers	1-Ethoxy-7-butyl-2,4-cycloheptadiene [121]; 2-methoxy-5,5,9-trimethylbicyclo[4.4.0]deca-6,8,10-triene [52]
Furans & pyrans	*E*-Tetrahydro-2,5-dimethylfuran, *E*-tetrahydro-2,5-dimethoxyfuran, *E*-2,5-diethoxytetrahydrofuran [35]; tetrahydro-2-(hydroxymethyl)-5-methylfuran [198]; 5-methyl-2-(2,6,6-trimethyl-2-cyclohexen-1-yl) furan [103]; *E,E*-/*E,Z*-/*Z,E*-/*Z,Z*-tetrahydro-2,5-dimethoxyfuran-3-carbaldehyde [202]; 4-hydroxy2,5-dimethyl-3(2H)-furanone = furaneol, homofuraneols, 4-methoxy-2,5-dimethyl-3(2H)-furanone = methoxyfuraneol, *E*-/*Z*-linalool furanoxide, 2-cyclohexyl-5-methyl-3(2H)-furanone, 5-cyclohexyl-2-methyl-3(2H)-furanone [222]

D) POLYSILOXANE-DISSOLVED CYCLODEXTRIN DERIVATIVES [35]

Furans & pyrans	4-butanoyloxy-2,5-dimethyl-3(2H)-furanone = furaneol monobutanoate, tetrahydro-1-(2-hydroxy-1-isopropyl)-5-methyl-5-vinylfuran = lilac alcohol, *E*-/*Z*-theaspirane, *E*-/Z-3-theaspirene, *E*-/*Z*-8-theaspirone [52]; *E*-/*Z*-2-benzyloxytetrahydro-5-methylfuran [185]; *E*-/*Z*-2,5-divinyltetrahydro-2H-pyrane [223]; 3-/6-methyl-2,5-divinyltetrahydro-2H-pyrane [121]; neroloxide [205]
Halo-compounds	2,2,2-trifluoro-1-phenylethanol [85]; 1-(pentafluorophenyl) ethanol [203]; 4-chloroalkanes [103]; 1,4-/1,5-dichlorohexane [121]; 2-chloro-4-phenylbutane, Z-1-butyl-2-chlorol-cyclohexanol [103]; *E*-/*Z*-3-chloro-2-butyl acetate [52]; #methyl 2-chloropropanoate [204]; 4-chloromethyl-2,2-dimethyl-1,3-dioxolane [87]; 2-bromoalkanes [121;203;214]; 1,3-/*E*-2,3-dibromobutane [203]; 3-(4-bromophenyl)-1-butene [121]; alkyl 2-bromopropanoates [7]; 2-iodobutane [203]
Hydrocarbons	*E*-1,2-/*E*-1,3-Dimethylcyclohexanes, *E*/*Z*-1-alkyl-2-methylcyclohexanes, 1,1,3-trimethylcyclohexane, *E*-1,1,3,5-tetramethylcyclohexane, *E*-/*Z*-pinane [224]; limonene [35]; sabinene, ß-phellandrene [225]; camphene [219]; α-pinene [35]; ß-pinene [219]; fenchene [225]; 4-isopropylen-1,2,4-trimethylcyclohexene [121]; multifidene and related compounds [226]; *E*-viridiene [121]; 2,4,4,8-trimethylcyclo[3.3.3.0^{2,5}]undec-2-ene = modhephene [7]; endo-/exo-tricyclo[6.2.2.0^{2,7}]dodeca-3,9-diene [103]; 3-phenyl-1-butene [132]; exo-2-phenylnorbornane [205]; 6,6'-dimethyl-1,1'-biphenyls (substituted in 2,2' positions) [228].
Ketones	3-Hydroxy-2-butanone = acetoin, 3-/*4-hydroxy-2-pentanone, 2-hydroxy-3-pentanone [200]; 3-ethoxy-2-butanone [52]; bicyclo[3.1.0]hex-3-en-2-one [205]; *E*-2,5-dimethylcyclopentanone [103]; 5-norbornene-2-one [205]; bicyclo[3.2.0]hept-2-en-6-one [35]; #;##3-/#4-methyl-2-heptanone [204]; 5-methyl-5-norbornen2-one [87]; 3,5-dimethyl-2-cyclohexenone [52]; 5-methyl-*E*-2-hepten-4-one = filbertone, 5-methyl-*Z*-2-hepten-4-one [229]; 3,3,5-trimethylcyclohexanone [35]; 3-methyl-*E*-5-octen-4-one [87]; isomenthone [66]; pulegone [87]; fenchone 220]; 6-methylbicyclo-[4.4.0]dec-1-en-3-one, 1-methylbicyclo[4.4.0]dec-6-en-2,8-dione [203]; α-damascone [230]; α-ionone [231]; 2-(*E*-2,7-octadien-1-yl) cyclohexanone [103]; *E*-α-irone [66]; 4-acetyl-5,6,8,8-tetramethylbicyclo[4.4.0]dec-1-ene [232]
Lactones	4-Alkanolides [52]; 4-ethoxycarbonyl-4-butanolide [230]; pantolactone [230]; 2-hydroxy-3-methyl-2-penten-4-olide = sotolone [66]; solerone [230]; 2-acetyl-2-methyl-4-butanolide [52]; 2,3,3-trimethyl-4-butanolide, 3,3-dimethyl-4-pentanolide [205]; 2-hydroxy-3-methyl-2-hexen-4-olide = abhexon [66]; 3-alkyl-2-hydroxy-2-alken-4-olides, 3-alkyl-2-methoxy-2-alken4-olides [85]; *E*-/*Z*-whisky lactone [131]; 2-ethylidene-E-5-hepten-4-olide [132]; 4-phenyl-4-butanolide [161]; 5-alkanolides, 2-decen-5-olide (after reduction and methylation as 5-alkyl-5-lactol methyl ethers [results in *E*,*Z*-isomers]) [233]; 2-oxabicyclo[3.2.1]oct-6-en-3-one [205]; 2-oxabicyclo[3.3.0]oct-7-en-3-one [234]
Pesticides	α-1,2,3,4,5,6-Hexachlorocyclohexane = lindane [168]; mecoprop, dichlorprop (ME) [58]

D) POLYSILOXANE-DISSOLVED CYCLODEXTRIN DERIVATIVES [35]

Pharmaceuticals	3,3-Diethyl-5-methylpiperidine-2,4-dione = methyprylon 46]; hexobarbital, methylphenobarbital [202]; ethosuximide [235]; *ibuprofen [210]
Miscellaneous	Alkylated 1,3-dioxolanes, 2,2-dimethyl-4-phenyl-1,3-dioxolane, 4,7,11-trioxapentacyclo[6.3.0.02,6.03,10.05,9]undecane, alkyl-/phenyl-1,3-dioxanes [87;121]; 2-(hydroxyethyl)-2-methyl-1,3-dioxolane [200]; *4-(hydroxymethyl)-2,2-dimethyl-1,3-dioxolane = solketal [131]; alkyldihydro-1,3-dioxines [103]; 2-ethyl-5,6-dihydro-6-methyl-4-pyrone [87]; 4-isopropyl-1,3-dioxolan-2-one [121]; 4-phenyl-1,3-dioxolan-2-one [132]; 3-phenyl-5-vinyl-1,3-dioxolan-2-one [132]; *Z*-lactide [87]; 4-cyanooctane [103]; O-acetylmandelonitrile [208]; N,N-dimethyl(4-methylphenyl)acetic acid amide [201]; 1-isopropyl-3,3-dimethyldiaziridine, *E*-1,2,3,4-tetramethyldiaziridine [55]; methyl 4-(1-pyrrolidyl)-E-2-pentenoate, methyl 4-(1-piperidyl)-*E*-2-pentenoate [103]; 1-(4-pyridyl)ethanol [203]; nicotine, N-butanoylnornicotine [131]; 2,5-dihydro-2-isopropyl-3,6-dimethoxypyrazine [87]; 1,5-bis(tert-butoxycarbonyl)-2-pyrrolidinone [52]; 4-alkyl-1,3-oxazolidin-2-ones [103]; 1-(3-nitrophenyl)ethanol [121]; 2-methyl-4-nitro-5-phenylcyclohexanone [103]; 2,2-dimethyl-4-(4-nitrophenyl)-1,3-dioxolane [220]; tert-butyl 3-hydroxyalkylthioethers [180]; 3-(methylthio)1-hexanol l[236]; 2-mercapto-3-pentanone [66]; *E*-/*Z*-8-mercaptomenthone [237]; 2-mercaptopropanoic acid [52]; *E*-/*Z*-2-methyl4-propyl-1,3-oxathiane [100]; pyrrolidono[1,2-e]-4H-2,4-dimethyl-1,3,5-dithiazine [238]; #alkylarylsilane der.[86]; 2-(tert-butyldimethylsilyl)cyclopentanone [103]
CSP-69	Octakis(2,3,6-tri-O-methyl)-γ-cyclodextrin (10 % diss. in OV-1701) [7] (FS-CYCLODEX gamma-I/P) (#10 % diss. in SPB-35 [γ-Dex])
Acetals	Spiroacetals (e.g., *E*-2,3-dimethyl-1,4-dioxaspiro[4.4]nonane, *E*-/*Z*-chalcogran and related compounds) [7]
Acids	Fencholenic acid [197]; ibuprofen [132]
Alcohols	Menthol, campholenic alcohol and related compounds, fencholenic alcohol and related compounds [197]
Diols	#1,2-Alkanediols [10]
Esters	*E*-Diisopropyl tartrate [103]
Ketones	5,6-diacetylcyclooctene [132]
Pharmaceuticals	2-[4-(2-Methyl-1-propyl)phenyl]propanoic acid = ibuprofen [135]
Pyrans	Z-Rose oxide [66]
Miscellaneous	#Alkylarylsilane der. [86]
CSP-74	Heptakis(2,6-di-O-methyl-3-O-TFA)-β-cyclodextrin (10 % diss. in OV-1701) [224]
Acetals	Cyclic acetals (e.g. E-/Z-chalcogran and related compounds) [87]

D) POLYSILOXANE-DISSOLVED CYCLODEXTRIN DERIVATIVES [35]

Alcohols	Norbornanol der., *norbornanol der. [239]; Z-1-hydroxymethyl-3-isopropenyl-2,2-dimethylcyclobutane, 3-menthanols [87]; 1-phenyl-3,3-dimethyl-1-butanol [239]
Aldehydes	Bicyclo[2.2.1]hept-2-en-5-carbaldehyde [87]
Epoxides	1,2-Epoxy-3-methyl-3-butene [87]; norbornane der. [239]; styrene oxide [214]; methylphenyloxiranes [87;112;214]
Esters	Methyl/ethyl 2-methylbutanoate [240]; ethyl mandelate (no or TFA) [239]
Ethers	Phenyl (1-phenylethyl)ether, 1,2-benzo-3-cyclobutyl 2-butyl ether [239]
Furans & pyrans	*E*-Tetrahydro-2,5-dimethoxyfuran, *E*-/*Z*-rose oxide [112]
Halo-compounds	2-tert-Butyl-5-bromomethylene-1,3-dioxane [131]
Hydrocarbons	Norbornane der.[239]
Ketones	Bicyclo[3.1.0]hex-3-en-2-one [87]; bicyclo[2.2.1]heptan2-one [87]; 2-methylcyclohexanone, *E*-2,6-dimethylcyclohexanone [112]; 5-methyl-*E*-2-hepten-4-one = filbertone [241]; 1-methylbicyclo[2.2.1]hept-5-en-2-one [87]; 3,3,5-trimethylcyclohexanone [112]; 3-methyl-5-octen-4-one [87]; norbornanone der. [239]; pulegone, carvone [87]; α-ionone [52]
Lactones	4-Alkanolides [7;87]; 4-phenyl-4-butanolide [131]; 2-oxabicyclo[3.3.0]oct-7-en-3-one [112]; 5-alkanolides [53;74]; polycyclic lactones (e.g. strigol, epi-4'-strigol [242]
Miscellaneous	Alkylated 1,3-dioxolanes, 2,2,4-trimethyl-1,3-dioxane, 4,7,11-trioxapentacyclo[6.3.0.02,6.03,10.05,9]undecane, 2-ethyl-5,6-dihydro-6-methyl-4-pyrone [87]; cyano-nitro-norbornane der. [239]

E) POLYSILOXANE-DISSOLVED CYCLODEXTRIN DERIVATES [223]
(CD-concentrations of newly manufactured phases: 0.07 molar)

CSP-97	Hexakis(2,3,6-tri-O-pentyl)-α-cyclodextrin (10 % diss. in OV-1701) [41] (FS-CYCLODEX alpha-II/P)
Alcohols	2-Methyl-1-alkanols, 2-methyl-3-alkanols, 4-methyl-2-pentanol, 5-methyl-3-hexanol, 2-ethyl-1-hexanol, 1-octen-3-ol [103]
Epoxides	1,2-Epoxypropane, *E*-2,3-epoxybutane, 1,2-epoxy-3-butene [103]; 1,2:3,4-diepoxybutane [132]; glycidyl methylether [103]
Esters	Methyl lactate [103]
Halo-compounds	2-Halobutanes, 2-chloro-3-pentanone[103]; 2-chloro/2-bromocyclohexanone [41]; epichlorohydrin, epibromohydrin[103]; methyl 2-bromopropanoate [132]
Ketones	2-Methylcyclohexanone [41]

E) POLYSILOXANE-DISSOLVED CYCLODEXTRIN DERIVATES [223]
(CD-concentrations of newly manufactured phases: 0.07 molar)

Pyrans	Dihydrorose oxide [132]
Miscellaneous	2-azido-3-pentanone, 2-nitrobutane [103]
CSP-98	Heptakis(2,3,6-tri-O-pentyl)-β-cyclodextrin (10 % diss. in OV-1701) [41] (FS-CYCLODEX beta-II/P)
Acids	2-Ethylhexanoic acid [208]
Alcohols	2-Ethyl-1-hexanol, 6-methyl-5-hepten-2-ol = sulcatol, linalool [208]
Diols	*E*-1,2-Cyclohexanediol [208]
Epoxides	1,2-Epoxy-3-butene [103]
Esters	Methyl lactate [41]
Halo-compounds	2-Halobutanes [103]; methyl 2-bromopropanoate [41]
Miscellaneous	O-Acetylmandelonitrile [208]; 2-nitrobutane [103]
CSP-99	Hexakis(3-O-acetyl-2,6-di-O-methyl)-α-cyclodextrin (10 % diss. in OV-1701) [132] (FS-CYCLODEX alpha-III/P)
Diols	**E*-2,3-Butanediol [121]
Lactones	3-Butanolide, 4-alkanolides, pantolactone [121]
Miscellaneous	4-Alkyl-1,3-dioxolan-2-ones, *2-(2-aminopropyl)1,3-dioxolane [121]
CSP-100	Heptakis(3-O-acetyl-2,6-di-O-methyl)-β-cyclodextrin (10 % diss. in OV-1701) [103] (FS-CYCLODEX beta-III/P)
Aldehydes	2-(Oxocycloalkyl)acetaldehydes [103]
Ketones	*E*-2,6-Dimethylcycloalkanone [103]; 2-methylcycloalkanones [121]; camphor, 2-(*E*-2,7-octadien-1-yl)cyclopentanone [103]
Halo-compounds	1,1,2-Trichloropropane [121]; 1,3-/1,4-dichloropentane, 1,4-dichlorohexane [103]
Lactones	3-Butanolide, 4-alkanolides, pantolactone [103]; 12-tridecanolide, phoracantholide [121]
Miscellaneous	4-Alkyl-/4-phenyl-1,3-dioxolan-2-ones, 3-cyanooctane, 2-azido-3-pentanone, 7-azido-4-methyl-3-heptanone, 6-alkyl-2-piperidones [103]
CSP-101	Octakis(3-O-acetyl-2,6-di-O-methyl)-γ-cyclodextrin(10 % diss.in OV-1701) [132] (FS-CYCLODEX gamma-III/P)
Lactones	5-Nonanolide [121]

F) POLYSILOXANE-DISSOLVED CYCLODEXTRIN DERIVATES [77]

CSP-71	Hexakis(2,3,6-tri-O-methyl)-α-cyclodextrin (30 % diss. in OV-1701) [77] (Megadex 2; SCHB)
Alcohols	3-Octanol, menthol, patchenol [77]
Aldehydes	2-Phenylpropanal [77]
Esters	Z-3-Hexen-1-yl 2-methylbutanoate [77]
Hydrocarbons	Limonene [77]
Ketones	Carvone [77]
Lactones	4-Heptanolide, 5-alkanolides [77]
Pyrans	Z-Rose oxide [77]
Miscellaneous	4-Alkoxycarbonyl-2,2-dialkyl-1,3-dioxolanes [243]; 2-hydroxy-/2-butanoyloxy-7,9-dioxabicyclo[4.3.0]nonanes, 2-hydroxy-/2-butanoyloxy-6-methyl-7,9-dioxabicyclo[4.3.0]nonane [244]
CSP-68	Heptakis(2,3,6-tri-O-methyl)-β-cyclodextrin (30 % diss. in OV-1701) [77] (Megadex 1; SCHA)
Acetals	Spiroacetals (e.g. endo-/exo-brevicomin and related compounds) [245]
Alcohols	2-Butanol [93]; menthol, 4-terpinenol, a-terpineol, isopinocampheol, tetrahydrolinalool [77]; linalool [246]; patchenol, *E*-/*Z*-nerolidol, 1-phenylethanol [77]
Aldehydes	3-(4-Isopropylphenyl)-2-methylpropanal [77]
Epoxides	1,4-epoxy-1,4-dimethyl-2-cyclohexanol, 1,4-epoxy-p-2-menthanone, 1,8-epoxy-p-3-menthanone, 1,8-epoxy-9-(3-methyl-2-butenyl)-p-menthane [77]
Esters	Menthyl acetate, Z-3-hexen-1-yl 2-methylbutanoate, ethyl 3-hydroxyhexanoate, 1-phenylethyl acetate [77]
Hydrocarbons	Limonene [93]; a-pinene [247]
Ketones	α-Ionone [93]; *E*-α-irone, Z-γ-irone [77]
Lactones	4-Alkanolides [248]; *E*-5-decen-4-olide [249]; Z-6-dodecen-4-olide [250]
Miscellaneous	4-Alkoxycarbonyl-2,2-dialkyl-1,3-dioxolanes [243]; 4-(hydroxymethyl)-2,2-dimethyl-1,3-dioxolane [251]; 2-hydroxy-/2-butanoyloxy-7,9-dioxabicyclo[4.3.0]nonanes, 2-hydroxy-/2-butanoyloxy-6-methyl-7,9-dioxabicyclo[4.3.0]nonanes [244]
CSP-69	Octakis(2,3,6-tri-O-methyl)-γ-cyclodextrin(30 % diss. in OV-1701) [77] (Megadex 3; SCHC)

F) POLYSILOXANE-DISSOLVED CYCLODEXTRIN DERIVATIVES [77]

Alcohols	6-Methyl-2-(4-methylphenyl)-2-heptanol [77]
Lactones	2-Decen-5-olide = massoi lactone [77]
Pyrans	Z-Rose oxide [77]
Miscellaneous	4-Alkoxycarbonyl-2,2-dialkyl-1,3-dioxolanes [243]; 2-hydroxy-/2-butanoyloxy-7,9-dioxabicyclo[4.3.0]nonanes, 2-hydroxy-/2-butanoyloxy-6-methyl-7,9-dioxabicyclo[4.3.0]nonanes [244]
CSP-75	Octakis(2,6-di-O-methyl-3-O-TFA)-γ-cyclodextrin (10 % diss. in OV-1701) [7; 252] (Megadex 4; SCHD) (#with a different cyclodextrin concentration)
Lactones	4-/5-Alkanolides [#7; 252]
CSP-88	Heptakis(2,6-di-O-methyl-3-O-pentyl)-β-cyclodextrin (30 % diss. in OV-1701) [50;93] Megadex 5 (also available diss. in OV-1 and OV-225) (#30 % diss. in OV-1701, ##20 % diss. in OV-1701, for all other separations 50 % diss. in OV-1701)
Acids	2-Alkyl-/2-arylalkanoic acids [253]; 2-methylbutanoic acid [68]; 2-/3-hydroxyalkanoic acids, malic acid, 2-methoxysuccinic acid, 2-methyl-3-oxobutanoic acid, 2-methylsuccinic acid, *E*-2-hydroxy-3-methylsuccinic acid, citramalic acid, 2-methylglutaric acid, isocitric acid, 3-methyl-4-oxohexanoic acid, *E*-cycloalkanedicarboxylic acids (ME) [253];dimethylallenedicarboxylate, 3-ethyl-4-methylpentanoic acid, 2-methyl-/2-pentylcyclohexanone-2-carboxylic acid (ME) [254]; 2-cyclohexylpropanoic acid, *E-/Z*-dihydrojasmonic acid, *E-/Z*-jasmonic acid, mandelic acid, 2-/3-phenyllactic acid, abscisic acid (ME) [253]; 3-(3-phenylcyclobutylidene)propanoic acid (ME) [254]
Alcohols	#2-Butanol, #1-penten-3-ol [255]; 2-methyl-1-penten-3-ol, 3-heptanol, 2,4-dimethoxy-1-nonanol [254]; #α-terpineol, #linalool [255]; fenchol, *borneol, isoborneol, 3-menthanols, *menthol, *E-/Z*-sabinene hydrate, *E-/Z*-carveol; myrtenol, grandisol, *E-/Z*-verbenol, ispsenol, hotrienol [254]; lavandulol, 4-terpineol, α-terpineol, linalool [50]; #patchenol [80]; geosmin, dehydrogeosmin [51]; *E-/Z*-nerolidol [50]; *E-/Z*-α-bisabolol, *1-phenylethanol, 1-phenyl-3-butanol, 1-(4-methoxyphenyl)ethanol, 2-(3-phenylcyclobutylidene)ethanol, 1-(2,6-dimethylphenoxy)-2-propanol [254]; #6-methyl2-(4-methylphenyl)-2-heptanol [255]; 1-indanol [254]
Aldehyde	Citronellal [254]
Amino acids	Ala (TFA/ME) [51]; val, as, met, phe (TFA/ME) [254]
Diols	E-Sobrerol [157]
Epoxides	1,2-Epoxyalkanes [256]; #1,4-epoxy-p-2-menthanone, #1,8-epoxyp-3-menthanone [80]; 1,2-epoxy-1-methylcyclohexane [256]; 2,3-epoxy-1-decanol, 3-(*E*-1-heptenyl)glycidol, 2-nonylglycidol [254]

F) POLYSILOXANE-DISSOLVED CYCLODEXTRIN DERIVATIVES [77]

Epoxides	caudoxirene [256]; #1,8-epoxy-9-(3-methyl-2-butenyl)-p-menthane [255]; Z-3,4-epoxy-Z,Z-6,9-nonadecadiene and related compounds, methyl 10,11-epoxy-3,7,11-trimethyl-2,6-dodecadienoate and related compounds (juvenile hormones I-III), styrene oxide [256]; alkylated phenyloxiranes, *E*-1,2-diphenyloxirane, benzyl glycidylether, 1,2-epoxytetralin [254]
Esters	Lavandulyl acetate [254]; #bornyl acetate, #isobornyl 2-methylpropanoate[80]; #methyl/ethyl 3-hydroxybutanoat, #methyl 3-acetoxybutanoate, #ethyl 2-methylbutanoate, #methyl 3-hydroxy-2-methylenepentanoate, #ethyl 3-hydroxyhexanoate [68]; ethyl 2-cyclohexylpropanoate, ethyl 2-benzyloxypropanoate [254]; #ethyl 2-phenylbutanoate [255]
Ethers	1-Methoxy-2,6-dimethyl-2,7-octadiene [254]
Furans & pyrans	#*E*-/*Z*-Tetrahydro-5-(1-hydroxy-1-methylethyl)-2-methyl-2-furanacetaldehyde, #*E*-/*Z*-theaspirane, #*E*-/*Z*-vitispirane, #*E*-/*Z*-theaspirone, #*E*-/*Z*-1,3,7,7-tetramethyl-2-oxabicyclo[4.4.0]deca-5,9-diene = edulane [68]
Halo-compounds	2-Chlorosuccinic acid (ME) [253]; methyl 2-bromopropanoate [51]
Hydrocarbons	1-tert-Butylallenes, 1,2-cycloalkanedienes [257]; limonene [51]; #limonene [255]; α-/β-pinene, camphene, sabinene, α-thujene [254]; ectocarpene, desmarestene, dictyotene [254]; 1-tert-butyl-1,2-cyclooctadiene [257]; α-copaene, α-/β-bisabolene [254]; ##α-/β-bisabolene, ##α-curcumene [258]; 2-phenylbutane [254]
Ketones	2-Hydroxy-3-hexanone, 3-hydroxy-2-hexanone [254]; #3-methylcyclohexanone, #carvone [255]; #pulegone, #3-oxo-a-ionone, #3,4,7,8-tetra-hydro--7-hydroxy-4,4,7-trimethyl-6H-naphthalen-2-one [68]; 3-methyl-pentadecanone = muscone [50]
Lactones	#4-Alkanolides [80;255]; #2-hydroxy-3-methyl-2-penten-4-olide = sotolone [68]; 4-(2-methoxy-3-hydroxypropyl)-4-butanolide, 5-heptanolide [254]
Pesticides	Mecoprop, dichlorprop, fenoprop (ME) [253]; heptachlor [254]
Pharmaceuticals	Ibuprofen, fenoprofen, flurbiprofen (ME) [253]; hexobarbital, phenobarbital, ketoprofen (ME), ethosuximide, phensuximide, 3-ethyl-5-phenylhydantoin = ethotoin, 4-ethyl-1-methyl-4-phenylhydantoin = mephenytoin, methyprylon, ifosfamide, 3-dechloroethyl-ifosfamide, prilocain, *ketamine, mefenorex, viloxazin [157]; glutethimide, *methoxyphenamine, prilocain [254]
Miscellaneous	#1-p-Menthen-8-thiol, #E-8-mercaptomenthone [68]; *E*-dicyanoalkene-dithiacrownether, 5-acetoxy-4-methyl-1-trimethylsilyl-2,3-pentadiene [51]
CSP-102	Octakis(2,6-di-O-methyl-3-O-pentyl)-γ-cyclodextrin (30 % diss. in OV-1701) [50] Megadex 7 (also available diss. in OV-1 and OV-225) [all described separations on 50 % diss. in OV-1701]
Acetals	Spiroacetals (e.g., 7-methyl-2,6-dioxaspiro[4.5]decane) [254]

F) POLYSILOXANE-DISSOLVED CYCLODEXTRIN DERIVATIVES [77]	
Acids	2-Alkyl-/2-arylalkanoic acids, 2-methylglutaric acid, *E*-cycloalkanecarboxylic acid (ME) [253]; 5,6,6-trimethyl-3,4-hexadienoic acid (ME) [254]; jasmonic acid, mandelic acids (ME) [253]; 2-phenylbutanoic acid [254]; 2-phenylsuccinic acid, 3-phenyllactic acid, abscisic acid (ME) [253]
Alcohols	2-/3-Octanol, 1-octen-3-ol, grandisol, 3-menthanols, 4-terpinenol, α-terpineol, *E*-verbenol, myrtenol, linalool, lavandulol, hotrienol, dehydrogeosmin, *1-phenylethanol [254]; *E*-/*Z*-α-bisabolol [50]
Aldehydes	Citronellal [258]
Amines	*2-Aminoalkanes [254]
Amino acids	Ala (TFA/ME) [51]; pro, asp, glu, phe (TFA/ME) [254]
Epoxides	1,2-Epoxyalkanes, Z-9,10-epoxy-Z,Z-3,6-heptadecadiene [254]
Esters	Citronellyl actate [258]; linalyl acetate, lavandulyl acetate, α-terpinyl acetate, menthyl acetate [254]
Halo-compounds	2-Chloroheptane, 2-chloropropyl 1,3-dichloroisopropylether, 1,3-dichloroisopropyl 2,3-dichloropropylether [254]; 2-haloalkanoic acids (ME) [253]; polychlorinated biphenyls = PCBs [259]; 1,2-dibromoalkanes [254]
Hydrocarbons	1,2-Cycloalkanedienes [257]; limonene [51]; desmarestene [254]
Ketones	Iso-/menthone, carvone, pulegone [254]
Lactones	4-/5-Hexanolide [254]
Pesticides	1,1,1-Trichloro-2-(2-chlorophenyl)-2-(4-chlorophenyl)ethane = o,p'-DDT [260]; dinoseb [157]; *E*-/*Z*-chlordane, heptachlor, ifosfamide and related compounds [254]; *E*-/*Z*-permethrinic acid (ME) [254]
Pharmaceuticals	2-(2-Chloroethyl)-2,3-dihydro-1,3-benzoxazin-4-one = chlorthenoxazine [260]; *3,4-oximethylenemethamphetamine, *methoxyphenamine, *methylphenidate [157]
Miscellaneous	5-Alkylhydantoins and related compounds [261]; *2,2'-dipyrrolidine, 5-alkyl-/5-arylhydantoins, 5-(methoxycarbonylmethyl) hydantoin, 5-(methylthioethyl)hydantoin, ethyl methylsulphoxide, 1,1,3,3,6,6-hexamethyl-1-sila-4-cycloheptene [254]; silylated ylidenic β-lactam [51]

G) POLYSILOXANE-DISSOLVED CYCLODEXTRIN DERIVATES [101]	
CSP-72	Heptakis(2,3-di-O-acetyl-6-O-tert-butyldimethylsilyl)-β-cyclodextrin (50 % diss. in OV-1701) [100]
Alcohols	2-Alkanols (no or TFA) [262]; 2-methyl-1-alkanols [100]; 3-hexanol [262]; 2-ethyl-1-hexanol, 3-octanol [100]; *3-octanol [243]; 4-octanol [117]; 1-octen-3-ol [100]; lavandulol [263]; 3-menthanols [264]; 4-terpinenol [263]; α-terpineol [265]; citronellol, linalool [100]

G) POLYSILOXANE-DISSOLVED CYCLODEXTRIN DERIVATIVES [101]

Esters	2-Acetoxypropanal, 2-alkyl acetates, 3-hexyl acetate, 5-norbornen-2-yl acetate [264]; linalyl acetate [263]
Ethers	2-Alkyl methylethers [262]
Furans & pyrans	*E*-/*Z*-linalool furanoxide [263]; *E*-/*Z*-rose oxide [100]
Halo-compounds	Sec. haloalkanes [117]
Ketones	Iso-/menthone [265]; carvone [100]; camphor [263]
Lactones	4-/5-Alkanolides [100]
Miscellaneous	4-Alkoxycarbonyl-2,2-dialkyl-1,3-dioxolanes[117]; *E*-/*Z*-2-methyl-4-propyl-1,3-oxathiane [100]
CSP-77	Octakis(2,3-di-O-acetyl-6-O-tert-butyldimethylsilyl)-γ-cyclodextrin (50 % diss. in OV-1701) [101]
Acids	2-Ethylhexanoic acid [101]
Alcohols	2-/3-Alkanols, 4-octanol [117]
Esters	tert-Butyl 3-hydroxy-4-pentenoate, tert-butyl 3-acetoxy-4-pentenoate [266]
Ketones	Pulegone [267]
Lactones	4-/5-Alkanolides [101]
Miscellaneous	4-Alkoxycarbonyl-2,2-dialkyl-1,3-dioxolanes [117]; 3-mercaptohexyl alkanoates, 3-methylthiohexyl acetates [268]; *E*-/*Z*-8-mercaptomenthone, *E*-/*Z*-8-(acetylmercapto)menthone [267]

H) POLYSILOXANE-ANCHORED CYCLODEXTRIN DERIVATES [107]

CSP-82	Hexakis(2,3,6-tri-O-methyl)-mono[2,3-di-O-methyl-6-(7-octen-1yl)]-β-cyclodextrin (anchored on polymethylsiloxane; obilized) [89;107;108] (Chirasil-Dex)
Acids	2-Ethylhexanoic acid, 2-phenylpropanoic acid [203]
Alcohols	2-Alkanols, 2-alkyl-1-alkanols, 3-/4-methyl-2-pentanol, 2-methyl-3-pentanol, 5-methyl-2-hexanol, 4-/5-methyl-3-hexanol [203]; 3-methyl-2-cyclohexen-1-ol [89]; exo-norborneol, 6-methyl-2-heptanol, 4-methyl-3-heptanol, 2-methyl-4-heptanol, 2,6-dimethyl-3-heptanol, 6-methyl-5-hepten-2-ol = sulcatol, 3-octanol, 3-methyl-3-nonanol [203]; 3-menthanols [109;203]; α-terpineol, isoborneol, fenchol, linalool, 1-phenyl-1-alkanols, 2-/3-phenyl-2-butanol, 3-methyl-5-phenyl-3-pentanol, *E*-2-phenyl-1-cyclohexanol, 1-tetralinol, 1,2'-naphthylethanol [203]
Alkanolamines	1-Isopropylamino-3-[2-(2-propen-1-yl)phenoxy]-2-propanol (oxazolidinone via phosgene)

H) POLYSILOXANE-ANCHORED CYCLODEXTRIN DERIVATES [107]	
Amino acids	Ala, pro, phe, ser, cys (N-TFA/iPE) [203]; pro (N-TFA/ME or N-ac./ME or as methylthiohydantoin der.) [269]
Diols	1,2-/1,3-Alkanediols, 2-methyl-1,2-butanediol, 2-methyl-3-buten-1,2-diol, 1,4-/*E*-2,4-pentanediol, 2-methyl2,4-pentandiol, *E*-1,2-cyclohexanol [203]; 3-(2-methoxyphenoxy)1,2-propanediol (cyclic carbonate via phosgene) [111]
Esters	Methyl 3-hydroxy-2-methylpropanoate, dimethyl malate, methyl 3-(1-methyl-2-oxo-1-cyclohexyl)propanoate [203]; ethyl mandelate [109]
Ethers	*E*-1,4-Dimethoxy-2,3-butanediol [203]
Furans	1-(2-Furyl)ethanol [203]; *E*-tetrahydro-2,5-diethoxyfuran [89]
Halo-compounds	1-phenyl-2,2,2-trifluoroethanol, 1-(pentafluorophenyl) ethanol [203]; 1-chloro-1-phenylethane [109]; 2-bromoalkanes, 1,3-/*E*-2,3-bromobutane, 2-iodobutane [203]
Hydrocarbons	*E*-/*Z*-1-Ethyl-2-methylcyclohexane, 1,1,3-trimethylcyclohexane, *E*-1,1,3,5-tetramethylcyclohexane, Z-2-methyl-3-propylcyclohexane [203]; limonene [203]; *E*-/*Z*-pinane [109;203]; α-pinene [203]
Ketones	3-Hydroxy-2-butanone = acetoin, 4-hydroxy-2-pentanone, 2-hydroxycyclohexanone, 5-methyl-E-2-hepten-4-one = filbertone [203]; 3,3,5-trimethylcyclohexanone [89]; pulegone [109]; fenchone, verbenone, 1-methylbcyclo[4.4.0]dec-6-en-2,8-dione, 6-methylbicyclo[4.4.0]dec-1-en-3-one [203]
Lactones	4-Alkanolides, pantolactone, 3,3-dimethyl-4-pentanolide [203]; 4-phenyl-4-butanolide [111]; 2-benzyl-3,3-dimethyl-4-butanolide [203]
Pesticides	*E*-/*Z*-Pyrethroid acids and related compounds, dichlorprop (ME) [269]
Pharmaceuticals	Hexobarbital [270]; glutethimide, guafenesine, deprenyl [269]
Miscellaneous	4-Hydroxymethyl-2,2-dimethyl-1,3-dioxolane, 4-phenyl-2,2-dimethyl-1,3-dioxolane [89]; methyl 2,2,5-trimethyl1,3-dioxolane-4-carboxylate, 1-(4-pyridyl)ethanol, 1-indanol, 1-Z-2-tert-butyl-3-methyl-4-imidazolidinone, 1-BOC-2-tertbutyl-3-methyl-4-imidazolidinone [203]
CSP-83	Hexakis(2,6-di-O-methyl-3-O-TFA)-mono[2-O-methyl-3-O-TFA-6-(7-octen-1-yl)]-β-cyclodextrin (anchored on polymethylsiloxane; immobilized) [112] (Chirasil-Dex-TFA)
Acetals	Cyclic acetals (e.g. *E*-/*Z*-chalcogran and related compounds) [112]
Alcohols	3-Menthanols [112]
Epoxides	Styrene oxide, methylphenyloxiranes [112]
Furans & pyrans	*E*-tetrahydro-2,5-dimethoxyfuran, *E*-tetrahydro-2,5-diethoxyfuran, *E*-/*Z*-rose oxide [112]

H) POLYSILOXANE-ANCHORED CYCLODEXTRIN DERIVATES [107]	
Ketones	2-Methylcyclohexanone, bicyclo[3.1.0]hex-3-en-2-one, *E*-2,6-dimethylcyclohexanone, 5-methyl-*E*-2-hepten-4-one = filbertone, 3,3,5-trimethylcyclohexanone, pulegone [112]
Lactones	4-Alkanolides, 2-oxabicyclo[3.3.0]oct-7-en-3-one [112]
Miscellaneous	2-Ethyl-5,6-dihydro-6-methyl-4-pyrone, 4,7,11-trioxapentacyclo-$[6.3.0.0^{2,6}.0^{3,10}.0^{5,9}]$undecane [112]

α-AA = amino acids; ac. = acetyl; BE = butyl ester; BOC = tert-butyloxycarbonyl; der. = derivative; EE = ethyl ester; HFB = heptafluorobutanoyl; iBE = isobutyl ester; iPA = isopropylamide; iPC = isopropylcarbamate (= isopropylurethane); iPE = isopropyl ester; iPU = isopropyl ureido; MCA = monochloroacetyl; ME = methyl ester; NEA = 1,1'-naphthylethylamine; no = underivatized; (R!) = enantiomeric stationary phase; PE = pentyl ester; PEA = 1-phenylethylamine; PFP = pentafluoropropanoyl; PrA = propyl amide; PrE = propyl ester; prod. = product; tBA = tert-butylamide; tBC = tert-butylcarbamate (= tert-butylurethane); tBE = tert-butyl ester; tBU = tert-butyl ureido; TCA = trichloroacetyl; TFA = trifluoroacetyl; TFM = trifluoromethyl; TMS = trimethylsilylala = alanine; arg = arginine; asn = asparagine; asp = aspartic acid; cys = cysteine; gln = glutamine; glu = glutamic acid; gly = glycine; his = histidine; ile = isoleucine; leu = leucine; lys = lysine; met = methionine; orn = ornithine; phe = phenylalanine; pro = proline; ser = serine; thr = threonine; trp = tryptophan; tyr = tyrosine; val = valineall = allose; ara = arabinose; frc = fructose; fuc = fucose; gal = galactose; glc = glucose; lyx = lyxose; man = mannose; rha = rhamnose; rib = ribose; sor = sorbose ; tal = talose; xyl = xylose.

Table 3.5.8 Stationary achiral phases (to dissolve cyclodextrins) and their manufacturers

Squalane:	2,6,10,.15,19,23-Hexamethyltetracosane
OV-1 and SE-30:	Polydimethylsiloxane (high grade of polymerization)
OV-101 and PS-347.5:	Polydimethylsiloxane (medium grade of polymerization)
PS-255 and PS-268:	Polydimethylsiloxane
PS-537:	Polydimethylsiloxane (low grade of polymerization)
SDE-52 and SPB-5:	Polyphenylmethylsiloxane (5% phenyl)
PS-086:	Polyphenylmethylsiloxane (12 or 15% phenyl)
OV-7 and SPB-20:	Polyphenylmethylsiloxane (20% phenyl)
CP-Sil 13 and DC-550:	Polyphenylmethylsiloxane (25% phenyl)
OV-61:	Polyphenylmethylsiloxane (33% phenyl)
SPB-35:	Polyphenylmethylsiloxane (35% phenyl)
DC-710 and OV-17:	Polyphenylmethylsiloxane (50% phenyl)
DB-5 and SE-54:	Polyphenylvinylmethylsiloxane (5% phenyl), 1% vinyl)
XE-60:	Polycyanoethylmethyldimethylsiloxane (25% cyanoethyl)
BPX-BPX-70:	Polydicyanopropyldimethylsiloxane (70% cyanopropyl)
CP-Sil 19, OV-1701* and BP-10:	Polycyanopropylphenylvinylmethylsiloxane (5% cyanopropyl, 7% phenyl) (6% cyanopropyl, 6% phenyl) (7% cyanopropyl, 7% phenyl) (7.5% cyanopropyl, 7.5% phenyl) (7% cyanopropyl, 7% phenyl, 1% vinyl)
OV-225:	Polycyanopropylmethylphenylmethylsiloxane (25% cyanopropyl, 25% phenyl)
Ucon 50-HB-5100:	Poly(oxyethylenoxypropylene) (MG = 5100)
PEG 400:	Polyethyleneglycol (MG 400)
CW-20M:	Carbowax (polyethyleneglycol) (MG = 20000)

Table 3.5.8 Stationary achiral phases (to dissolve cyclodextrins) and their manufacturers

DB-Wax:	Bonded Carbowax
Celite:	Diatomite (Kieselgur)
Chromosorb W:	Treated with calcined flux diatomite (kieselgur)

* The composition of OV-1701 depends on the manufacturer!

BP and BPX	SGE
Celite and Chromosorb	Johns-Manville, Denver (CO, USA)
CP-Sil	Chrompack, Middelburg (The Netherlands)
CW and Ucon	Union Carbide, New York (NY, USA)
D XX	J & W Scientific, Folsom (CA, USA
DC	Dow Corning, Midland (MI, USA)
HP	Hewlett Packard, Palo Alto (CA, USA)
OV	Ohio Valley Specialty Chem., Marietta (OH, USA)
SE and XE	General Electric, Fairfield (CT, USA)
SPB	Supelco, Bellefonte (PA, USA)
PS	Petrarch Systems, Bristol (PA, USA)

References

[1] Saenger W. *Angew. Chem.* 1980, *92*, 343
[2] Freudenberg K., Cramer F. Z. *Naturforsch.* 1948, *3b*, 464
[3] Bender M.L., Komiyama M. *Cyclodextrin Chemistry*, Springer: Berlin, 1978
[4] Szejtli J. *Kontakte (Darmstadt)* 1988, *(1)*, 31
[5] Pagington J.S. *Chem. Britain* 1987, 455
[6] Jones S.P., Grant D.J.W., Hadgraft J., Parr G.D. *Acta Pharm. Technol.* 1984, *30*, 213
[7] Schurig V., Nowotny H.-P. *Angew. Chem.* 1990, *102*, 969
[8] Cramer F. *Angew. Chem.* 1952, *64*, 136
[9] Cramer F., Hettler H. *Naturwissenschaften* 1967, *54*, 625
[10] Mani V., Woolley C.L., Feibush B. Poster-Abstract, Fa. Supelco, Bad Homburg, presented in 1991
[11] Cramer F., Dietsche W. *Chem. Ber.* 1959, *92*, 378
[12] MacNicol D.D., Rycroft D.S. *Tetrahedron Lett.* 1977, *25*, 2173
[13] Hamilton J.A., Chen L. *J. Am. Chem. Soc.* 1988, *110*, 5833
[14] Harada A., Furue M., Nozakura S.-I. *J. Polym. Sci. Polym. Chem. Ed.* 1978, *16*, 189
[15] Debowski J., Sybilska D., Jurczak J. *J. Chromatogr.* 1982, *237*, 303
[16] Ko]cielski T., Sybilska D., Jurczak J. *J. Chromatogr.* 1983, *280*, 31
[17] Armstrong D.W., DeMond W. *J. Chromatogr. Sci.* 1984, *22*, 411
[18] Terabe S., Ozaki H., Otsuka K., Ando T. *J. Chromatogr.* 1985, *322*, 211
[19] Alak A., Armstrong D.W. *Anal. Chem.* 1986, *58*, 582
[20] Macaudiére P., Caude M., Rosset R., Tambuté A. *J. Chromatogr.* 1987, *405*, 135
[21] Snopek J., Jelínek I., Smolková-Keulemansová E. *J. Chromatogr.* 1988, *438*, 211
[22] Armstrong D.W., He F.-Y., Han S.M. *J. Chromatogr.* 1988, *448*, 345
[23] Gutman A., Paulus A., Cohen A.S., Grinberg N., Karger B.L. *J. Chromatogr.* 1988, *448*, 41

[24] Macaudiére P., Caude M., Rosset R., Tambuté A. *J. Chromatogr. Sci.* 1989, *27*, 383
[25] Vigh G., Quintero G., Farkas G. *J. Chromatogr.* 1990, *506*, 481
[26] Mayer S., Schurig V. *J. High Resol. Chromatogr. Chromatogr. Commun.* 1992, *15*, 129
[27] Sand D.M, Schlenk H. *Anal. Chem.* 1961, *33*, 1624
[28] Alexander G., Juvancz Z., Szejtli J. *J. High Resol. Chromatogr. Chromatogr.Commun.* 1988, *11*, 110
[29] Casu B., Reggiani M., Sanderson G.R. *Carbohydr. Res.* 1979, *76*, 59
[30] Mizobuchi Y., Tanaka M., Shono T. *J. Chromatogr.* 1980, *194*, 153
[31] Tanaka M., Mizobuchi Y., Kuroda T., Shono T. *J. Chromatogr.* 1981, *219*, 108
[32] Tanaka M., Kawano S., Shono T. *Fresenius Z. Anal. Chem.* 1983, *316*, 54
[33] Smolková E., Králová H., Krysl S., Feltl L. *J. Chromatogr.* 1982, *241*, 3
[34] Smolková-Keulemansová E., Neumannová E., Feltl L. *J. Chromatogr.* 1986, *365*, 279
[35] Schurig V., Nowotny H.-P. *J. Chromatogr.* 1988, *441*, 155
[36] Juvancz Z., Alexander G., Szejtli J. *J. High Resol. Chromatogr. Chromatogr. Commun.* 1987, *10*, 105
[37] Szejtli J. *Stärke* 1987, *39*, 357
[38] Venema A., Tolsma P.J.A. in: *9th Int. Symp. on Capillary Chromatography, Monterey 1988,* Sandra P. (Ed.), Hüthig: Heidelberg 1988, p. 167
[39] Knoll A., Böhme M., Merker S., Fabian G., Häfner B., Theil F. *Chimia* 1991, *45*, 195
[40] König W.A., Lutz S., Mischnick-Lübbecke P., Brassat B., Wenz G. *J. Chromatogr.* 1988, *447*, 193
[41] Köhnes A., Römer H. *CLB Chem. Lab. Biotech.* 1990, *41*, 70
[42] Armstrong D.W., Li W., Pitha J. *Anal. Chem.* 1990, *62*, 214
[43] Ciucanu I., Luca C. *Rev. Roum. Chim.* 1989, *34*, 1829
[44] Schurig V., Nowotny H.-P., Schleimer M., Schmalzing D. *J. High Resol. Chromatogr. Chromatogr. Commun.* 1989, *12*, 549
[45] Schurig V., Zhu J., Muschalek V. *Chromatographia* 1993, *35*, 237
[46] König W.A. *Gas Chromatographic Enantiomer Separation with Modified Cyclodextrins*, Hüthig: Heidelberg, 1992
[47] König W.A., Krebber R., Evers S., Bruhn G. *J. High Resol. Chromatogr. Chromatogr. Commun.* 1990, *13*, 328
[48] König W.A. *Nachr. Chem. Tech. Lab.* 1989, *37*, 471
[49] König W.A., Icheln D., Runge T., Pforr I., Krebs A. *J. High Resol. Chromatogr. Chromatogr. Commun.* 1990, *13*, 702
[50] König W.A., Gehrcke B., Icheln D., Evers P., Dönnecke J., Wang W. *J. High Resol. Chromatogr. Chromatogr. Commun.* 1992, *15*, 367
[51] Icheln D., Gehrcke B., Runge T., König W.A. in: *15th Int. Symp. on Capillary Chromatography, Riva del Garda 1993*, Sandra P. (Ed.), Hüthig: Heidelberg 1993, p. 278 and poster presentation
[52] Bernreuther A. *Doctoral Thesis*, University of Würzburg, Germany 1992
[53] Bernreuther A., Bank J., Krammer G., Schreier P. *Phytochem. Anal.* 1991, *2*, 43
[54] Schubert V., Dietrich A., Ulrich T., Mosandl A. *Z. Naturforsch.* 1992, *47c*, 304
[55] Jung M., Schurig V. *J. Am. Chem. Soc.* 1992, *114*, 529

[56] Juvancz Z., Grolimund K., Francotte E., Schurig V. in: *15th Int. Symp. on Capillary Chromatography, Riva del Garda 1993*, Sandra P. (Ed.), Hüthig: Heidelberg, 1993, poster presentation

[57] König W.A., Krebber R., Mischnick P. *J. High Resol. Chromatogr. Chromatogr. Commun.* 1989, *12*, 732

[58] König W.A., Icheln D., Runge T., Pfaffenberger B., Ludwig P., Hühnerfuß H. *J. High Resol. Chromatogr. Chromatogr. Commun.* 1991, *14*, 530

[59] König W.A., Icheln D., Hardt I. *J. High Resol. Chromatogr. Chromatogr. Commun.* 1991, *14*, 694

[60] Kobor F., Angermund K., Schomburg G. in: *15th Int. Symp. on Capillary Chromatography, Riva del Garda 1993*, Sandra P. (Ed.), Hüthig: Heidelberg, 1993, poster presentation

[61] Schmarr H.-G., Maas B., Mosandl A., Bihler S., Neukom H.-P., Grob K. *J. High Resol. Chromatogr. Chromatogr. Commun.* 1991, *14*, 317

[62] Armstrong D.W., Li W., Stalcup A.M., Secor H.V., Izac R.R., Seeman J.I. *Anal. Chim. Acta* 1990, *234*, 365

[63] Armstrong D.W., Li W., Chang C.-D., Pitha J. *Anal. Chem.* 1990, *62*, 914

[64] Fischer N., Hammerschmidt F.-J., Brunke E.-J. in: *Progress in Flavour Precursor Studies*, Schreier P., Winterhalter P. (Eds.), Allured: Carol Stream, 1993, p. 287

[65] Armstrong D.W., Chang C.-D., Li W.Y. *J. Agric. Food Chem.* 1990, *38*, 1674

[66] Werkhoff P., Brennecke S., Bretschneider W. *Chem. Mikrobiol. Technol. Lebensm.* 1991, *13*, 129

[67] Werkhoff P., Brennecke S., Bretschneider W., Güntert M., Hopp R., Surburg H. *Z. Lebensm. Unters. Forsch.* 1993, *196*, 307

[68] Full G. Doctoral Thesis, University of Würzburg, Germany 1993

[69] Armstrong D.W., Jin H.L. *J. Chromatogr.* 1990, *502*, 154

[70] Takeichi T., Shimura S., Toriyama H., Takayama Y., Morikawa M. *Chem. Lett.* 1992, 106

[71] Stoev G., Gancheva M. in: *15th Int. Symp. on Capillary Chromatography, Riva del Garda 1993*, Sandra P. (Ed.), Hüthig: Heidelberg, 1993, p. 290

[72] Sandra P., Jaques K., Courselle P. in: *12th Int. Symp. on Capillary Chromatography, Kobe, Japan 1990*, Jinno K., Sandra P. (Eds.), Ind.Publ.Consult.: Tokyo 1990, p. 828

[73] Haase-Aschoff I., personal communication 1990

[74] Haase-Aschoff I., Haase-Aschoff K., Patz C.-D. in: *15th Int. Symp. on Capillary Chromatography, Riva del Garda 1993*, Sandra P. (Ed.), Hüthig: Heidelberg, 1993, poster presentation

[75] Galli M., personal communication 1993

[76] Fellous R., Lizzani-Cuvelier L., Loiseau A.-M. *J. High Resol. Chromatogr. Chromatogr. Commun.* 1990, *13*, 785

[77] Bicchi C., Artuffo G., D'Amato A., Nano G.M., Galli A., Galli M. *J. High Resol. Chromatogr. Chromatogr. Commun.* 1991, *14*, 301

[78] Schurig V., Schleimer M., Jung M., Grosenick H., Fluck M. in: *15th Int. Symp. on Capillary Chromatography, Riva del Garda 1993*, Sandra P. (Ed.), Hüthig: Heidelberg, 1993, poster presentation

[79] Maas B., Dietrich A., Karl V., Kaunzinger A., Lehmann D., Köpke T., Mosandl A. in: *15th Int. Symp. on Capillary Chromatography, Riva del Garda 1993*, Sandra P. (Ed.), Hüthig: Heidelberg, 1993, poster presentation

[80] Bicchi C., Artuffo G., D'Amato A., Manzin V., Galli A., Galli M. *J. High Resol. Chromatogr. Chromatogr. Commun.* 1993, *16*, 209
[81] Armstrong D.W., Tang Y., Ward T., Nichols M. *Anal. Chem.* 1993, *65*, 1114
[82] Duvekot J., Buyten J.C., van der Poel W.P.C., Mussche P., Schmalzing D., Jung M., Schleimer M., Schurig V. in: *13th Int. Symp. on Capillary Chromatography, Riva del Garda 1991*, Sandra P. (Ed.), Hüthig: Heidelberg, 1991, p. 197
[83] Wan H., Wang Y., Ou Q., Yu W. *J. Chromatogr.* 1993, *644*, 202
[84] Blum W., Aichholz R. *J. High Resol. Chromatogr. Chromatogr. Commun.* 1990, *13*, 515
[85] Dietrich A, Mass B., Messer W., Bruche G., Karl V., Kaunzinger A., Mosandl A. *J. High Resol. Chromatogr. Chromatogr. Commun.* 1992, *15*, 590
[86] Feibush B., Woolley C.L., Mani V. *Anal. Chem.* 1993, *65*, 1130
[87] Nowotny H.-P., Schmalzing D., Wistuba D., Schurig V. *J. High Resol. Chromatogr. Chromatogr. Commun.* 1989, 12, 383
[88] Buda W.M., Jaques K., Venema A., Sandra P. in: *15th Int. Symp. on Capillary Chromatography, Riva del Garda 1993*, Sandra P. (Ed.), Hüthig: Heidelberg, 1993, poster presentation
[89] Jung M., Schurig V. *J. Microcol. Sep.* 1993, *5*, 11
[90] Schmarr H.-G., Mosandl A., Neukom H.-P., Grob K. *J. High Resol. Chromatogr. Chromatogr. Commun.* 1991, *14*, 207
[91] Jung M., Schmalzing D., Schurig V. *J. Chromatogr.* 1991, *552*, 43
[92] Lindstrom M. *J. High Resol. Chromatogr. Chromatogr. Commun.* 1991, *14*, 765
[93] Bicchi C., Artuffo G., D'Amato A., Galli A., Galli M. *J. High Resol. Chromatogr. Chromatogr. Commun.* 1992, *15*, 655
[94] Hardt I., König W.A. *J. Microcol. Sep.* 1993, *5*, 35
[95] Askari C., Hener U., Schmarr H.-G., Rapp A., Mosandl A. *Fresenius Z. Anal. Chem.* 1991, *340*, 768
[96] Rettinger K., Karl V., Schmarr H.-G., Dettmar F., Hener U., Mosandl A. *Phytochem. Anal.* 1991, 2, 184
[97] Nehrings A., personal communication 1993
[98] Römer H., personal communication 1993
[99] Schmarr H.-G., Kaunzinger A., Mosandl A. in: *13th Int. Symp. on Capillary Chromatography, Riva del Garda 1991*, Sandra P. (Ed.), Hüthig: Heidelberg, 1991, p. 188
[100] Dietrich A, Mass B., Karl V., Kreis P., Lehmann D., Weber B., Mosandl A. *J. High Resol. Chromatogr. Chromatogr. Commun.* 1992, *15*, 176
[101] Schmarr H.-G., Mosandl A., Kaunzinger K. *J. Microcol. Sep.* 1991, *3*, 395
[102] Dietrich A., Maas B., Brand G., Karl V., Kaunzinger A., Mosandl A. *J. High Resol. Chromatogr. Chromatogr. Commun.* 1992, *15*, 769
[103] Keim W., Köhnes A., Meltzow W., Römer H. *J. High Resol. Chromatogr. Chromatogr. Commun.* 1991, *14*, 507
[104] König W.A., Krüger A., Icheln D., Runge T. *J. High Resol. Chromamatogr. Chromatogr. Commun.* 1992, *15*, 184
[105] Runge T., Lange M., König W.A. in: *15th Int. Symp. on Capillary Chromatography, Riva del Garda 1993*, Sandra P. (Ed.), Hüthig: Heidelberg, 1993, p. 279 and poster presentation
[106] Fischer P., Aichholz R., Bölz U., Juza M., Krimmer S. *Angew. Chem.* 1990, *102*, 439

[107] Schurig V., Schmalzing D., Mühleck U., Jung M., Schleimer M., Mussche P., Duvekot C., Buyten J.C. *J. High Resol. Chromatogr. Chromatogr. Commun.* 1990, *13*, 713

[108] Schurig V., Juvancz Z., Nicholson G.J., Schmalzing D. *J. High Resol. Chromatogr. Chromatogr. Commun.* 1991, *14*, 58

[109] Schurig V., Schmalzing D., Schleimer M. *Angew. Chem.* 1991, *103*, 94

[110] Schurig V., Schleimer M., Jung M., Mayer S., Glausch A. in: *Progress in Flavour Precursor Studies*, Schreier P., Winterhalter P. (Eds.), Allured: Carol Stream, 1993, p. 63

[111] Juvancz Z., Grolimund K., Schurig V. *J. High Resol. Chromatogr. Chromatogr. Commun.* 1993, *16*, 202

[112] Jung M., Schurig V. *J. High Resol. Chromatogr. Chromatogr. Commun.* 1993, *16*, 282

[113] Bradshaw J.S., Yi G., Rossiter B.E., Reese S.L., Petersson P., Markides K.E., Lee M.L. *Tetrahedron Lett.* 1993, *34*, 79

[114] Yi G., Bradshaw J.S., Rossiter B.E., Reese S.L., Petersson P., Markides K.E., Lee M.L. *J. Org. Chem.* 1993, *58*, 2561

[115] Berthod A., Li W., Armstrong D.W. *Anal. Chem.* 1992, *64*, 873

[116] Venema A., Henderiks H., van Geest R. *13th Int. Symp. on Capillary Chromatography, Riva del Garda 1991*, Sandra P. (Ed.), Hüthig: Heidelberg, 1991, p. 273

[117] Jaques K., Buda W.M., Venema A., Sandra P. in: *15th Int. Symp. on Capillary Chromatography, Riva del Garda 1993*, Sandra P. (Ed.), Hüthig: Heidelberg, 1993, poster presentation

[118] Grob K., Neukom H.-P., Schmarr H.-G., Mosandl A. *J. High Resol. Chromatogr. Chromatogr. Commun.* 1990, *13*, 433

[119] Bicchi C., Artuffo G., D'Amato A., Galli A., Galli M. *Chirality* 1992, *4*, 125

[120] Hardt I.H., König W.A. in: *15th Int. Symp. on Capillary Chromatography, Riva del Garda 1993*, Sandra P. (Ed.), Hüthig: Heidelberg, 1993, p. 229 and poster presentation

[121] CS-Chromatographie Service Catalog Nr. D VI-6000-03.92-prdr, Langerwehe, Germany 1992

[122] Gyllenhaal O., Gustavsson K., Vessman J. in: *15th Int. Symp. on Capillary Chromatography, Riva del Garda 1993*, Sandra P. (Ed.), Hüthig: Heidelberg, 1993, p. 311

[123] Anonymous *Chrompack News* 1990, *(3)*, 16

[124] Mayer S., Schmalzing D., Jung M., Schleimer M. *LC-GC Intern.* 1992, *5(4)*, 58

[125] Mayer S., Schmalzing D., Jung M., Schleimer M. *LC-GC Mag. Separ. Sci.* 1992, *10*, 782

[126] König W.A. DE 38 10 737 (Cl.C08B37/16), 12 Oct 1989

[127] König W.A., Lutz S., Wenz G. *Angew. Chem.* 1988, *100*, 989

[128] König W.A., Mischnick-Lübbecke P., Brassat B., Lutz S., Wenz G. *Carbohydr. Res.* 1988, *183*, 11

[129] Macherey-Nagel Catalog (Chiral e1/10/0/2.89 PD), Düren, Germany 1989

[130] Reiher T., Hamann H.-J. *J. High Resol. Chromatogr. Chromatogr. Commun.* 1992, *15*, 346

[131] Macherey-Nagel Catalog (GC Katalog d2/5/5/8.92 PD), Düren, Germany 1992

[132] CS-Chromatographie Service Catalog, Langerwehe, Germany 1990

[133] König W.A. *Carbohydr. Res.* 1989, *192*, 51

[134] König W.A. in: *New Trends in Cyclodextrins and Derivatives*, Duchêne D. (Ed.), Editions de Santé: Paris, 1991, p. 551

[135] Römer H., Meltzow W., Köhnes A. *Abstract from CS-Chromatographie Service*, Langerwehe 1990

[136] König W.A., Lutz S. in: *Chirality and Biological Activity. Proc. Int. Symp., Tübingen 1988*, Holmstedt B., Frank H., Testa B. (Eds.), Liss: New York, 1990, p. 55

[137] König W.A., Lutz S., Wenz G., Görgen G., Neumann C., Gäbler A., Boland W. *Angew. Chem.* 1989, *101*, 180

[138] König W.A., Lutz S., Mischnick-Lübbecke P., Brassat B., von der Bey E., Wenz G. *Stärke* 1988, *40*, 472

[139] Meinwald J., Thompson W.R., Pearson D.L., König W.A., Runge T., Francke W. *Science* 1991, *251*, 560

[140] König W.A., Lutz S., Wenz G. in: *Proc. of the 4th Int. Symp. on Cyclodextrins, München 1988*, Huber O., Szejtli J. (Eds.), Kluwer: Dordrecht, 1988, p. 465

[141] König W.A. in: *Drug Stereochemistry: Analytical Methods and Pharmacology*, Wainer I.W. (Ed.), 2nd Ed., Dekker: New York, 1992, p. 107

[142] Wan H., Ou Q. Fenxi Huaxue *Chin. J. Anal. Chem.* 1992, *20*, 394

[143] König W.A., Lutz S., Hagen M., Krebber R., Wenz G., Baldenius K., Ehlers J., tom Dieck H. *J. High Resol. Chromatogr. Chromatogr. Commun.* 1989, *12*, 35

[144] Wan H., Dong Y., Ou Q. *Sepu* 1991, *9*, 214

[145] Blank I., Grosch W., Eisenreich W., Bacher A., Firl J. *Helv. Chim. Acta* 1990, *73*, 1250

[146] König W.A., Krebber R., Wenz G. in: *10th Int. Symp. on Capillary Chromatography, Riva del Garda 1989*, Sandra P. (Ed.), Hüthig: Heidelberg, 1989, p. 12

[147] König W.A., Krebber R., Wenz G. *J. High Resol. Chromatogr. Chromatogr. Commun.* 1989, *12*, 790

[148] Boland W., König W.A., Krebber R., Müller D.G. *Helv. Chim. Acta* 1989, *72*, 1288

[149] Brunner H., Lautenschläger H.-J., König W.A., Krebber R. *Chem. Ber.* 1990, *123*, 847

[150] Kappes D., Gerlach H., Zbinden P., Dobler M., König W.A., Krebber R., Wenz G. *Angew. Chem.* 1989, *101*, 1744

[151] König W.A. *Kontakte (Darmstadt)* 1990, *(2)*, 3

[152] König W.A., Lutz S., Colberg C., Schmidt N., Wenz G., von der Bey E., Mosandl A., Günther C., Kustermann A. *J. High Resol. Chromatogr. Chromatogr. Commun.* 1988, *11*, 621

[153] Wörner M., Schreier P. *Z. Lebensm. Unters. Forsch.* 1990, *190*, 425

[154] Bernreuther A., Lander V., Huffer M., Schreier P. *Flav. Fragr. J.* 1990, *5*, 71

[155] Guichard E., Mosandl A., Hollnagel A., Latrasse A., Henry R. *Z. Lebensm. Unters. Forsch.* 1991, *193*, 26

[156] Wirth D., Fischer-Lui I., Boland W., Icheln D., Runge T., König W.A., Phillips J., Clayton M. *Helv. Chim. Acta* 1992, *75*, 734

[157] König W.A., Gehrcke B. in: *15th Int. Symp. on Capillary Chromatography, Riva del Garda 1993*, Sandra P. (Ed.), Hüthig: Heidelberg, 1993, p. 277 and poster presentation

[158] Wörner M., Schreier P. *Phytochem. Anal.* 1991, *2*, 260

[159] König W.A., Lutz S., Wenz G., von der Bey E. *J. High Resol. Chromatogr. Chromatogr. Commun.* 1988, *11*, 506

[160] Mosandl A., Schubert V. *J. Ess. Oil Res.* 1990, *2*, 121

[161] Macherey-Nagel Catalog, Düren, Germany 1991

[162] Mosandl A., Kustermann A. *Z. Lebensm. Unters. Forsch.* 1989, *189*, 212

[163] Wörner M., Schreier P. *Z. Lebensm. Unters. Forsch.* 1991, *193*, 317

[164] Hollnagel A., Menzel E.-M., Mosandl A. *Z. Lebensm. Unters. Forsch.* 1991, *193*, 234

[165] Mosandl A., Hollnagel A. *Chirality* 1989, *1*, 293
[166] Wörner M., Schreier P. *Z. Lebensm. Unters. Forsch.* 1991, *193*, 21
[167] Guichard E., Kustermann A., Mosandl A. *J. Chromatogr.* 1990, *498*, 396
[168] Mössner S., Spraker T.R., Becker P.R., Ballschmiter K. *Chemosphere* 1992, *24*, 1171
[169] König W.A., Lutz S., Evers P., Knabe J. *J. Chromatogr.* 1990, *503*, 256
[170] de Vries N.K., Coussens B., Meier R.J., Heemels G. *J. High Resol. Chromatogr. Chromatogr. Commun.* 1992, *15*, 499
[171] Schubert V., Mosandl A. *Z. Lebensm. Unters. Forsch.* 1992, *194*, 552
[172] Schulz S., Toft S. *Science* 1993, *260*, 1635
[173] Bornscheuer U., Schapöhler S., Scheper T., Schügerl K., König W.A. *J. Chromatogr.* 1992, *606*, 288
[174] Klärner F.-G., Band R., Glock V., König W.A. *Chem. Ber.* 1992, *125*, 197
[175] Von dem Bussche-Hünnefeld, Cescato C., Seebach D. *Chem. Ber.* 1992, *125*, 2795
[176] König W.A., Nippe K.-S., Mischnick P. *Tetrahedron Lett.* 1990, *31*, 6867
[177] Valterová I., Unelius C.R., Vrko0 J., Norin T. *Phytochem.* 1992, *31*, 3121
[178] Borg-Karlson A.-K., Lindstrøm, Norin T., Persson M., Valterová I. *Acta Chem. Scand.* 1993, *47*, 138
[179] Grégoire J.-C., Baisier M., Drumont A., Dahlsten D.L., Meyer H., Francke W. *J. Chem. Ecol.* 1991, *17*, 2003
[180] Haase B., Schneider M.P. *Tetrahedron Asymm.* 1993, *4*, 1017
[181] Berthod A., Li W.Y., Armstrong D.W. *Carbohydr. Res.* 1990, *201*, 175
[182] Armstrong D.W., Tang Y., Zukowski J. *Anal. Chem.* 1991, *63*, 2858
[183] Armstrong D.W. US 4,948,395 (Cl.B01D15/08), 14 Aug 1990
[184] Armstrong D.W., Reid III, G.L., Hilton M.L., Chang C.-D. *Environ. Poll.* 1993, *79*, 51
[185] Mosandl A., Palm U. *Z. Lebensm. Unters. Forsch.* 1991, *193*, 109
[186] Li W.-Y., Jin H.L., Armstrong D.W. *J. Chromatogr.* 1990, *509*, 303
[187] Alltech Catalog 250, Deerfield (IL), USA 1991
[188] Jin H.L., Armstrong D.W. *Sepu* 1991, *9*, 148
[189] Frykman H., Yhrner N., Norin T., Hult K. *Tetrahedron Lett.* 1993, *34*, 1367
[190] ICT Catalog (04/91), Frankfurt/M., Germany 1991
[191] Kouzi S.A., Nelson S.D. *J. Org. Chem.* 1993, *58*, 771 (Correction: *J. Org. Chem.* 1993, *58*, 3222)
[192] Smith I.D., Simpson C.F. *Anal. Proc.* 1992, *29*, 245
[193] Smith I.D., Simpson C.F. *J. High Resol. Chromatogr. Chromatogr. Commun.* 1992, *15*, 800
[194] Fischer N., Hammerschmidt F.-J. *Chem. Mikrobiol. Technol. Lebensm.* 1992, *14*, 141
[195] Leal W.S. *Naturwissenschaften.* 1991, *78*, 521
[196] Mani V., Feibush B., Woolley C. in: *15th Int. Symp. on Capillary Chromatography, Riva del Garda 1993*, Sandra P. (Ed.), Hüthig: Heidelberg, 1993, p. 300
[197] Reinhardt R., Steinborn A., Engewald W., Schulze K., Beutmann K. in: *15th Int. Symp. on Capillary Chromatography, Riva del Garda 1993,* Sandra P. (Ed.), Hüthig: Heidelberg, 1993, p. 337
[198] Römer H., Meltzow W., Köhnes A. *GIT Fachz. Lab. Spezial Chromatographie* 1991, *(1)*, 33
[199] Wirth D., Boland W., Müller D.G. *Helv. Chim. Acta* 1992, *75*, 751
[200] Mosandl A., Rettinger K., Fischer K., Schubert V., Schmarr H.-G., Maas B. *J. High Resol. Chromatogr. Chromatogr. Commun.* 1990, *13*, 382

[201] Aichholz R., Bölz U., Fischer P. *J. High Resol. Chromatogr. Chromatogr. Commun.* 1990, *13*, 234
[202] Duvekot J., Buyten J.C., Peene J.A., Lips J., Schurig V. in: *11th Int. Symp. on Capillary Chromatography, Monterey 1990*, Sandra P., Redant G. (Eds.), Hüthig: Heidelberg, 1990, p. 147
[203] Schleimer D., Jung M., Mayer S., Rickert J., Schurig V. *J. High Resol. Chromatogr. Chromatogr. Commun.* 1992, *15*, 723
[204] Anonymous Supelco - *Biotext* 1992, *5(2)*, 6
[205] Chrompack Catalog, Middelburg, The Netherlands 1990
[206] Lindgren B.S., Gries G., Pierce Jr. H.D., Mori K. *J. Chem. Ecol.* 1992, *18*, 1201
[207] Kreis P., Juchelka D., Motz C., Mosandl A. *Dtsch. Apoth. Ztg.* 1991, *131*, 1984
[208] Meier-Augenstein W., Burger B.V., Spies H.S.C., Burger W.J.G. *Z. Naturwiss.* 1992, *47b*, 877
[209] Steinborn A., Reinhardt R., Engewald W. in: *15th Int. Symp. on Capillary Chromatography, Riva del Garda 1993*, Sandra P. (Ed.), Hüthig: Heidelberg, 1993, p. 892
[210] J & W Scientific/Fisons Catalog, Mainz-Kastel, Germany 1990
[211] Korth W., Bowmer K.H., Ellis J. *J. High Resol. Chromatogr. Chromatogr. Commun.* 1991, *14*, 704
[212] Kobor F., Angermund K., Schomburg G. *J. High Resol. Chromatogr. Chromatogr. Commun.* 1993, *16*, 299
[213] Leal W.S., Matsuyama S., Kuwahara Y., Wakamura S., Hasegawa M. *Naturwiss.* 1992, *79*, 184
[214] Schurig V. Jung M., Schmalzing D., Schleimer M., Duvekot J., Buyten J.C., Peene J.A., Lips J. in: *11th Int. Symp. on Capillary Chromatography, Monterey 1990*, Sandra P., Redant G. (Eds.), Hüthig: Heidelberg, 1990, p. 14
[215] Roboz J., Yu Q., Holland J.F. *J. Microbiol. Meth.* 1992, *15*, 207
[216] Mosandl A., Fischer K., Hener U., Kreis P., Rettinger K., Schubert V., Schmarr H.-G. *J. Agric. Food Chem.* 1991, *39*, 1131
[217] Schubert V., Diener R., Mosandl A. *Z. Naturforsch.* 1991, *46c*, 33
[218] Kreis P., Mosandl A., Schmarr H.-G. *Dtsch. Apoth. Ztg.* 1990, *130*, 2579
[219] Takeoka G., Flath R.A., Mon T.R., Buttery R.G., Teranishi R., Güntert M., Lautamo R., Szejtli J. *J. High Resol. Chromatogr. Chromatogr. Commun.* 1990, *13*, 202
[220] Spilkin A. *The J & W Separation Times* 1991, *5(2)*, 6
[221] Küsters E., Spöndlin C., Volken C., Eder C. *Chromatographia* 1992, *33*, 159
[222] Mosandl A., Bruche G., Askari C., Schmarr H.-G. *J. High Resol. Chromatogr. Chromatogr. Commun.* 1990, *13*, 660
[223] Keim W., Meltzow W., Köhnes A., Röthel T. *J. Chem. Soc. Chem. Commun.* 1989, 1151
[224] Schurig V., Nowotny H.-P., Schmalzing D. *Angew. Chem.* 1989, *101*, 785
[225] Cool L.G., Zavarin E. *Biochem. Syst. Ecol.* 1992, *20*, 133
[226] Kramp P., Helmchen G., Holmes A.B. *J. Chem. Soc. Chem. Commun.* 1993, 551
[227] Engewald W., Reinhardt R., Haufe G. *J. prakt. Chem.* 1992, *334*, 41
[228] Walther W., Cereghetti M. *J. High Resol. Chromatogr. Chromatogr. Commun.* 1991, *14*, 57
[229] Güntert M., Emberger R., Hopp R., Köpsel M., Silberzahn W., Werkhoff P. in: *Flavour Science and Technology*, Bessière Y., Thomas A.F. (Eds.), Wiley: Chichester, 1990, p. 29
[230] Guichard E., Hollnagel A., Mosandl A., Schmarr H.-G. *J. High Resol. Chromatogr. Chromatogr. Commun.* 1990, *13*, 299
[231] Werkhoff P., Bretschneider W., Güntert M., Hopp R., Surburg H. in: *Flavour Science and Technology*, Bessière Y., Thomas A.F. (Eds.), Wiley: Chichester, 1990, p. 33

[232] Neuner-Jehle N., Etzweiler F. in: *Perfumes - Art, Science and Technology,* Müller P.M., Lamparsky D. (Eds.), Elsevier: Amsterdam, 1991, p. 153
[233] Nago H., Matsumoto M. *Biosci. Biotech. Biochem.* 1993, *57,* 422
[234] Schurig V., Jung M. in: *Recent Advances in Chiral Separations,* Stevenson D., Wilson I.D. (Eds.), Plenum: New York, 1991, p. 117
[235] SGE Catalog (Publ.No.5000259), Weiterstadt, Germany 1992
[236] Werkhoff P., Brennecke S., Bretschneider W. *Contact (Holzminden)* 1991, *(50),* 3
[237] Köpke T., Mosandl A. *Z. Lebensm. Unters. Forsch.* 1992, *194,* 372
[238] Kubota K., Nakamoto A., Moriguchi M., Kobayashi A., Ishii H. *J. Agric. Food Chem.* 1991, *39,* 1127
[239] Stoev G. *J. Chromatogr.* 1992, *589,* 257
[240] Haase-Aschoff K., Haase-Aschoff I., Laub H. *Lebensmittelchemie* 1991, *45,* 107
[241] Jauch J., Schmalzing D., Schurig V., Emberger R., Hopp R., Köpsel M., Silberzahn W., Werkhoff P. *Angew. Chem.* 1989, *101,* 1039
[242] Samson E., Frischmuth K., Berlage U., Heinz U., Hobert K., Welzel P. *Tetrahedron* 1991, *47,* 1411
[243] Jaques K., Buda W.M., Pottie M., van der Eycken J., Vandewalle M., Venema A., Sandra P. in: *15th Int. Symp. on Capillary Chromatography, Riva del Garda 1993,* Sandra P. (Ed.), Hüthig: Heidelberg, 1993, p. 319
[244] Jaques K., Buda W.M., Dumortier L., van der Eycken J., Venema A., Sandra P. in: *15th Int. Symp. on Capillary Chromatography, Riva del Garda 1993,* Sandra P. (Ed.), Hüthig: Heidelberg, 1993, p. 327
[245] Pedrocchi-Fantoni G., Servi S. *J. Chem. Soc. Perkin Trans. I* 1991, 1764
[246] Cotroneo A., Stagno d'Alcontres I., Trozzi A. *Flav. Fragr. J.* 1992, *7,* 15
[247] Kotzias D., Spartà C., Duane M. *Naturwissenschaften* 1992, *79,* 24
[248] Cardillo R., Fronza G., Fuganti C., Grasselli P., Mele A., Pizzi D., Allegrone G., Barbeni M., Pisciotta A. *J. Org. Chem.* 1991, *56,* 5237
[249] Ercoli B., Fuganti C., Grasselli P., Servi S., Allegrone G., Barbeni M., Pisciotta A. *Biotechnol. Lett.* 1992, *14,* 665
[250] Allegrone G., Barbeni M., Cardillo R., Fuganti C., Grasselli P., Miele A., Pisciotta A. *Biotechnol. Lett.* 1991, *13,* 765
[251] Carlo Erba Reagenti Catalog, Milano, Italy 1993
[252] Bicchi C., Artuffo G., D'Amato A., Pellegrino G., Galli A., Galli M. *J. High Resol. Chromatogr. Chromatogr. Commun.* 1991, *14,* 701
[253] König W.A., Gehrcke B. *J. High Resol. Chromatogr. Chromatogr. Commun.* 1993, *16,* 175
[254] Anonymus *J. High Resol. Chromatogr. Chromatogr. Commun.* 1993, *16,* 338
[255] Bicchi C., Artuffo G., D'Amato A., Manzin V., Galli A., Galli M. *J. High Resol. Chromatogr. Chromatogr. Commun.* **1992**, *15,* 710
[256] König W.A., Gehrcke B., Peter M.G., Prestwich G.D. *Tetrahedron Asymmetry* 1993, *4,* 165
[257] Pietruszka J., Hochmuth D.H., Gehrcke B., Icheln D., Runge T., König W.A. *Tetrahedron Asymmetry* 1992, *3,* 661
[258] Gehrcke B., Riek A., Krüger A., König W.A. in: *15th Int. Symp. on Capillary Chromatography, Riva del Garda 1993,* Sandra P. (Ed.), Hüthig: Heidelberg, 1993, p. 1294 and poster presentation

[259] König W.A., Gehrcke B., Runge T., Wolf C. *J. High Resol. Chromatogr. Chromatogr. Commun.* 1993, *16*, 376

[260] König W.A. *Trends Anal. Chem.* 1993, *12*, 130

[261] Lickefett H., Krohn K., König W.A., Gehrcke B., Syldatk C. *Tetrahedron Asymmetry* 1993, *4*, 1129

[262] Krupcik J., Benicka E., Majek P., Skacani I., Sandra P. in: *15th Int. Symp. on Capillary Chromatography, Riva del Garda 1993*, Sandra P. (Ed.), Hüthig: Heidelberg, 1993, p. 286

[263] Kreis P., Mosandl A. *Flav. Fragr. J.* 1992, *7*, 187

[264] Buda W.M., Jaques K., Venema A., Sandra P. in: *15th Int. Symp. on Capillary Chromatography, Riva del Garda 1993*, Sandra P. (Ed.), Hüthig: Heidelberg, 1993, p. 230

[265] Kreis P., Mosandl A. *Flav. Fragr. J.* 1993, *8*, 161

[266] Jaques K., Buda W.M., Vrielinck S., Vandewalle M., Venema A., Sandra P. in: *15th Int. Symp. on Capillary Chromatography, Riva del Garda 1993*, Sandra P. (Ed.), Hüthig: Heidelberg, 1993, poster presentation

[267] Köpke T., Schmarr H.-G., Mosandl A. *Flav. Fragr. J.* 1992, *7*, 205

[268] Weber B., Haag H.-P., Mosandl A. Z. *Lebensm. Unters. Forsch.* 1992, *195*, 426

[269] Juvancz Z., Grolimund K., Schurig V.E in: *15th Int. Symp. on Capillary Chromatography, Riva del Garda 1993*, Sandra P. (Ed.), Hüthig: Heidelberg, 1993, p. 698 and poster presentation

[270] Jung M., Schurig V. *J. High Resol. Chromatogr. Chromatogr. Commun.* 1993, *16*, 215

3.5.2.4 Other chiral phases

Besides the above mentioned main groups of chiral phases, i.e. amide phases, metal complex phases as well as cyclodextrin phases, a few chiral phases exist which cannot be divided into these groups. Triglyceride phases, like (-)-1,2,3-tris-(2-benzyloxypropanoyloxy)propane (CSP-103), as described by Houche et al. [1], showed only low separation power. The point of interest in this work is the positive effect of an electric field on the separation of enantiomers. In 1979, Berrod et al. [2] and Binet et al. [3] used tartaric acid esters and cyclic acetals of tartaric acid esters (1,3-dioxolane-derivatives) as CSPs, achieving only a partial resolution of the enantiomers of 1-phenylethanol and some other alkanols. Oi and co-workers [4] introduced bis[(-)-menthyl] (+)-tartrate (CSP-104) and bis[(+/-)-menthyl](+)-tartrate (CSP-105) as stationary phases. The first-mentioned phase was suitable for the separation of N-TFA-amino acid isopropyl esters and several secondary N-TFA-amines. However, broad peaks and often very long retention times (e.g. approximately 170 min. for the resolution of N-TFA-1-phenylpropylamine) have been observed. The separation was explained via diastereomeric association complexes with hydroxyl groups.

Krimse and co-workers [5,6] synthesized phases with optically active polypropylene glycol (CSP-106). This CSPs are less useful for application in GC, because the analysis times are very long (4-14 hours [!] with a column length of approximately 120 - 150 m). Nevertheless, the first separation of enantiomers using multidimensional gas chromatography (MDGC) was carried out with such a column; underivatized bicyclic norterpene alcohols were separated [7]. It should be mentioned that as early as 1966 Gil-Av et al. [8] had unsuccessfully attempted separations with optically active propylene glycol [8].

In 1987, Aggarwal et al. [9] developed several amide-free, polysiloxane-bonded stationary phases such as *R*-2-phenylbutanoic acid 4-(4-methoxyphenyl) phenylester (bonded on OV-101; CSP-107), which are also suitable for SFC. As a more recent development, the application of chiral polysilphenyl siloxanes have been described (structure type: $[\text{-Si(CH}_3)_2\text{-C}_6\text{H}_5\text{-Si(CH}_3)_2\text{-O-}]_n$; [10]]. On this phases amino acid enantiomers were successfully separated.

Most recently, Betts [11] reported the baseline separation of linalool enantiomers on a phase coated with 10 % cholesteryl acetate on Chromosorb W AW (CSP-108). Other chiral compound could not be separated (e.g., carvone, citronellal, limonene or α-pinene). It has to be mentioned that the previous attempts of Kirk and Shaw [12] to use several cholesteryl esters as stationary phases for the separation of racemic steroids failed.

References

[1] Houche J., Moreau M., Longeray R., Dreux J. *Chromatographia* 1974, *7*, 306

[2] Berrod B., Bourdon J., Dreux J., Longeray R., Moreau M., Schifter P. *Chromatographia* 1979, *12*, 150

[3] Binet S., Dreux J., Longeray R. *Chromatographia* 1984, *18*, 294

[4] Ti N., Kitahara H., Doi T. *J.Chromatogr.* 1981, *207*, 252

[5] Krimse W., Siegfried R. *J.Am.Chem.Soc.* 1983, *105*, 950

[6] Krimse W., Siegfried R., Streu J. *J.Am.Chem.Soc.* 1984, *106*, 2465

[7] Scheidt F., Raskopf H. *LaborPraxis* 1984, *8*, 332

[8] Gil-Av E., Feibush B., Charles-Sigler R. in: *6th Int. Symp. on Gas Chromatography and Associated Techniques, Rome 1966*, Littlewood A.B. (Ed.), Inst. of Petroleum: London, 1967, p. 227

[9] Aggarwal S.K., Bradshaw J.S., Eguchi M., Parry S. *Tetrahedron* 1987, *43*, 451

[10] Abe I., Wasa T. *Chem.Express* 1989, *4*, 329-332 (C.A. 1990, 112:48012p)

[11] Betts T.J. *J.Chromatogr.* 1992, *600*, 337

[12] Kirk D.N., Shaw P.M. *J.Chem.Soc.C* 1971, 3979

3.5.2.5 Further developments

In the future, various types of stationary phases can be expected. One possibility is the use of chiral crown ethers as stationary phases in GC. Similar to cyclodextrins, crown ethers are able to form inclusion compounds. For HPLC, chiral crown ethers have been used for nearly 20 years [1]; for GC as well these compounds were tentatively used as stationary phases - but these crown ethers were not chiral [2,3]. If the use of chiral molecules of crown-ether structure in GC is successful (cf. chiral

HPLC phases), then even non-liquid phases could be used for GC, if they can be dissolved in polysiloxanes, for example in analogy to certain cyclodextrin phases [4]. One advantage might be that in contrast to the cyclodextrins, crown-ethers with quite different ring sizes are possible.

Also calixarenes, which are able to form inclusion complexes [5] might be interesting molecules to use as CSPs. Achiral calixarenes have already been used in gas solid chromatography (GSC) [6]. Other classes of promising compounds are catenanes and molecular knots, that exist as topological enantiomers [7], as well as cyclodextrin-based rotaxanes [8] and steroid-derived macrocycles (cholaphanes) [9], etc. Recently, even tubes made from cyclodextrin units were synthesized [10]. Imaginable are also cyclic polyethers, made from chiral 1,2-propanediol or similar molecules.

There are many possible, suitable compounds which have not been investigated to date as chromatographic stationary phases. But according to the present state of research, also in future, enantiomer separations without any optically active auxiliary hardly will not be possible [4].

References

[1] Helgeson R.C., Koga K., Timko J.M., Cram D.J. *J.Am.Chem.Soc.* 1973, *95*, 3021

[2] Ayyangar N.R., Tambe A.S., Biswas S.S. *J.Chromatogr.* 1991, *543*, 179

[3] Zeng Z.-R., Wu C.-Y., Yan H., Huang Z.-F., Wang Y.-T. *Chromatographia* 1992, *34*, 85

[4] Schurig V. *Angew.Chem.* 1984, *96*, 733

[5] Vicens J., Böhmer V. *Calixarenes: A Versatile Class of Macrocyclic Compounds,* Kluwer: Dordrecht, 1991

[6] Mangia A., Pochini A., Ungaro R., Andreetti G.D. *Anal .Lett.* 1983, *16*, 1027

[7] Sauvage J.-P. *Acc.Chem.Res.* 1990, *23*, 319

[8] Davis A.P. *Chem.Soc.Rev.* 1993, 243

[9] Isnin R., Kaifer A.E. *Pure & Appl.Chem.* 1993, *65*, 495

[10] Harada A., Li J., Kamachi M. *Nature* 1993, *364*, 516

3.6 Supercritical fluid chromatography (SFC)

Supercritical fluid chromatography occupies an intermediate position between the established methods of gas chromatography (GC) and high performance liquid chromatography (HPLC). The physical chemistry of SFC presents unique advantages in enantiomer separation [1,2]. The advantages and disadvantages of chiral analysis performed by SFC are outlined in Table 3.6.1.

Table 3.6.1 The advantages and disadvantages of chiral analysis performed by SFC.

Advantages	Disadvantages
Density program offers fine tuning of retention power without a polarity shift of the mobile phase	Column coating is rather difficult, since the CSP have moderate molecular mass and polarity
Stationary phase swelling makes it possible to use the CSP under its "melting point"	Analysis time is slower than in gas chromatography
Wide choice of detectors improves the selectivity of the system	Composition of the CSP needs careful tuning
Mild analysis conditions offer a high α-value and help to avoid the sample degradation and the racemization of the CSP	Solvent sphere from mobile phase caused decreased α values compared to gas chromatography
Solvent sphere from the mobile phase improves the peak shape	

The moderate working temperature allows improvements in various selectivity factors. Open-tubular columns can be used, which results in high efficiency as a consequence of the low-viscosity supercritical fluid mobile phases. The composition and density of the mobile phase and the analysis temperature can be selected and finetuned to improve the resolution of enantiomers. The analysis is much faster in SFC compared with LC. This is a result of the high self-diffusion coefficients of mobile phases in SFC. Packed-column SFC can complement LC separations by providing faster analysis, even on a semi-preparative scale. The mobile phase has a solvating power with tunable polarity and enhanced selectivity features in both the mobile and the stationary phases [2]; this is similar to LC. Open-tubular column SFC can complement capillary GC separations by providing higher selectivity. Until now, this newer technique has found only limited application. However, recent developments in instrumentation, advances in column technology, and an increased awareness of SFC will lead to more applications in the field of enantiomer separation.

3.6.1 *Properties of the mobile phase used in SFC*

The self-diffusion rate of the mobile phase can be changed over the working density range. It is limited by a gas-like state with high efficiency, low density, and high diffusion, and by a liquid-like state with less efficiency, high solvating power, and low diffusion. The solvating power of a pure mobile phase can be increased using a density gradient without causing a considerable change in polarity and selectivity. An entire homologous series of enantiomers can be eluted with the same resolution, because only the strength of the mobile phase increases without a noticeable change in temperature of polarity. Close to the critical point, the solvating power of the mobile phase as a function of pressure will form an S-shaped curve. This explains why for a pure mobile phase a density programme, showing a liner relationship between the density and solvating power, is preferred over a pressure programme. The solvating power of the mobile phase can also be increased by the addition of polar modifiers. This will alter the strength, polarity, and selectivity of the mobile phase. The selectivity of the eluent can be tuned using different modifiers in various amounts.

It is common practice to add polar organic modifiers to carbon dioxide for packed-column chiral SFC. The modifier gives the mobile phase increased solvating strength and selectivity and it also blocks the high-energy sites of the stationary phase. There is no general rule, however, that correlates the α–value with the type and concentration of modifier used. Which influence a modifier has depends on the type of modifier, the type of stationary phase, and the solutes [1,3,4]. An example is shown in Table 3.6.2.

Table 3.6.2 Dependence of the separation factor α on the composition of the mobile phase (alcohol/CO_2). (Solute: p-toluic acid amide derivative of 2-aminoheptane; column: 25 cm x 4.6 mm, ChiralCel OB [1].

Alcohol	Alcohol content [10^{-3} mol/g]	Selectivity [α]
methanol	0	1.90
	0.1	1.98
	0.2	2.12
	0.3	2.10
	0.5	2.00
	1.0	1.85
ethanol	0	1.90
	0.1	2.12
	0.2	2.15
	0.3	2.10
	0.5	1.90
	1.0	1.87
2-propanol	0	1.90
	0.1	2.05
	0.2	2.20
	0.3	2.25
	0.5	2.15
	1.0	2.10
1-butanol	0	1.90

Alcohol	Alcohol content [10^{-3} mol/g]	Selectivity [α]
	0.1	2.00
	0.2	2.10
	0.3	2.05
	0.5	1.98
	1.0	1.85
2-butanol	0	1.90
	0.1	2.15
	0.2	2.18
	0.3	2.19
	0.5	2.15
	1.0	2.10

The α-value reaches a maximum value as a function of the concentration of alcohol in the mobile phase, using ChiraCel OB as the stationary phase. The modifier regulates the polarity and selectivity of the mobile phase and competes with the solute molecules for the polar, optically active sites of the stationary phase. These opposing effects produce a maximum separation factor α. In SFC the separation mechanism differs from the mechanism recognized in LC. The alcohols used as modifiers can be divided into two groups - linear alcohols (in the example in Table 3.6.2, 1-butanol) and branched alcohols (in the example, 2-butanol) - depending on the effect they have on the separation. The linear alcohols provide less selectivity because they hinder the solutes forming hydrogen bonds. The solvation properties of supercritical fluids influence the capacity factor, and produce selectivities different from those measured under GC or LC conditions using the same column. In SFC, the α-value of an enantiomer is also different from the value measured in GC or LC. Several authors have reported lower α-values in SFC compared with GC at same operating temperature.

3.6.2 *Influence of various separation parameters*

3.6.2.1 Temperature

It is well known from the practice of chromatography that a decrease in analysis temperature usually results in larger α-values for enantiomer pairs. The Gibbs-Helmholtz equation (cf. Section 3.5) shows that the enantioselectivity of a system decreases with increasing analysis temperature. Only one practical exception to this rule has been found: ligand-exchange chromatography works in the entropy region in which α improves with increasing temperature [5].

In SFC the solutes do not need to have vapour pressure during the analysis, which is in contrast to GC, and the partition processes are similar to those of LC. Typical analysis temperatures are between ambient and 90°C for applications in chiral separations. Frequently, chiral analysis is performed below the critical temperature. Because of the use of modifiers, the critical parameters of the mobile phase become much higher than when analysis is performed in pure carbon dioxide.

The characteristics of the mobile phase do not change sharply at the border between the subcritical and supercritical region. Therefore, the advantages of the super-critical state are, more or less, also valid for the subcritical state (SubFC). However, there is a loss of efficiency in the subcritical state which is caused by the slower mass transfer at the lower temperatures, but the larger α-values sometimes compensate for this and produce maximum resolution in the subcritical region.

3.6.2.2 Dimension of the columns

The commercial LC columns used for enantiomer separation can usually be adapted for packed-column SFC. Microbore columns (15 cm x 1.2 mm) have also been used for chiral analysis employing SFC. To date, however, no articles have been published describing the use of packed SFC columns longer than the common LC columns. The lower viscosity of the supercritical fluid should permit the use of longer columns in SFC than is possible in LC [6].

In SFC, open-tubular columns are commonly 5 - 20 m long, with a preferred inner diameter of 50 µm for enantiomer separation. The inner diameter of open-tubular columns has to be as small as possible for efficiency reasons because the effects of diameter are more pronounced in SFC than in GC [7], but column preparation becomes increasingly more difficult with decreasing diameter (< 100 µm i.d.) [8].

The capacity of a chiral stationary phase is lower for enantiomers than it is for achiral compounds, because only a small part of the interaction sites has chiral properties. The low substitution concentration is a consequence of the large space requirement of chiral substituents. An efficient chiral polymer stationary phase for open-tubular columns usually has only a 10-15% optically active substitution rate. Although applications of semi-preparative SFC enantiomer separations have been published [9], an SFC column has less capacity than a LC column because of the limited solvating power of the supercritical fluids studied.

The open-tubular columns used in SFC have a capacity of approximately 50-200 ng/solute. The maximum sample capacity decreases with a decrease in the column diameter. Alternatively, thick-film columns (as great as 1 µm) can be used in SFC without significant loss of efficiency because the mass transfer is slower in SFC than in GC.

3.6.2.3 Analysis time

Using the same chiral column, SFC takes approximately a third to a tenth of the analysis time of LC. In this test the chromatograms were adjusted to give the same resolution in LC and SubFC modes. The faster analysis times occur since solute diffusion in a supercritical fluid is 5 - 20 times faster than in a liquid. Quick, efficient analysis is the greatest advantage of packed-column SFC. In addition, better resolution is obtained with SFC than with LC for the same analysis time.

3.6.3 *Stationary phases used in SFC*

The chiral phases used in SFC are based on three points of interaction (see Section 3.4). These three points do not need to be real points in a geometrical sense; they can be replaced by a plane or an axis. The chiral phases based on collective interactions, such as serum albumin and α-acid glycoprotein, are not useful for SFC applications. These phases have a slow mass transfer in the stationary phase which is not compatible with the relatively fast diffusion properties of supercritical fluid mobile phases.

3.6.3.1 Packed columns

The first reported enantiomer separation using SFC was performed by Mourier et al. [10] in 1985, using a commercial Pirkle-type LC column. Phosphine oxides were separated with supercritical and subcritical mobile phases. Using various Pirkle phases it has been established that the separation mechanism of enantiomers is similar in LC and SubFC. The selectivity factors are similar in both chromatographic modes, but under SubFC conditions the analysis required less time as compared with normal-phase LC.

Gasparrini et al. [11] synthesized a π-acidic stationary phase in a packed microbore column for use under SubFC and SFC conditions. They separated sulfoxides, flurbiprofen, and alprenolol enantiomers. The oxazolidine derivatives of these amino alcohols yielded a good peak shape, and they protected amines from reacting with the carbon dioxide mobile phase.

Macaudière et al. [12] introduced β-cyclodextrin LC columns for SubFC analysis of phosphine oxides and various amides. Selectivity, efficiency, and time were gained using SubFC as compared with LC. They showed that the high polarity of cyclodextrin necessitates the use of a high modifier concentration. Hexane, a common eluent in normal-phase LC, is held more strongly in the cavity of a cyclodextrin as compared with carbon dioxide. According to the same authors, this results in smaller α-values because the solute molecules need more energy to displace the mobile phase from the cyclodextrin cavity. They also determined that the mechanisms of enantiomer separation are different in normal and reversed mode using cyclodextrin columns.

The first report on the application of a cellulose-derivative chiral column was made by Lienne et al. [13]. The cellulose-tribenzoate stationary phase has different chiral separation mechanisms for analyzing various amides and lactams under SubFC conditions as compared with LC conditions. The α-value showed a maximum as a function of increased alcohol content in the mobile phases. Such a maximum was not observed under LC conditions.

Nitta et al. [14] used cellulose tris(phenylcarbamate) as a stationary phase to separate stilbene oxide under SFC and SubFC conditions. They gained resolution as well as time using SFC instead of LC. This demonstrated that, compared with SubFC, SFC is more efficient and faster, although it is less selective. Dobashi et al. [4] introduced chiral amide-containing silica stationary phases for SFC. These phases are suitable for the separation of derivatized amino acid enantiomers. This analysis was also faster using SFC as compared with LC. The α-value was slightly lower using SFC (2.32), but the resolution value R_S (10.75) allowed the exact determination of enantiomer purity, even above 99%, for both enantiomers. The enantiomer selectivity of these phases is based on hydrogen-bond interactions.

Macaudière et al. [1] also introduced (+)-poly(triphenylmethyl) methacrylate ('ChiralPak OT (+)') as a stationary phase for chiral separation of α,α'-bi-β-naphthol and γ-lactones using SubFC. Again, the analysis time was shorter with SubFC than with LC. In this instance, the mass transfer in the stationary phase was too slow for SFC purposes. The retention, selectivity, nature of the modifiers, and analysis temperature were systematically studied. The chiral separation mechanisms of the stationary phase were found to be different in SubFC and LC. The chiral cavities between two adjacent chains appeared to play no role in the separation mechanism of SubFC.

The greatest shortcoming of packed-column chiral SFC is the influence of silica packing material on retention. The surface of silica is only partly covered by chiral functional groups. The solute molecules are adsorbed onto high-energy surface points of the silica particles as a competitive retention mechanism to chiral recognition of the stationary phase. This results in unwanted retention, but even more seriously it gives a mixed retention mechanism with lower selectivity for chiral separation. Polar modifiers can be used to coat dynamically the high-energy suface, as well as to regulate the selectivity of the mobile phse and to improve the peak shape. A mobile phase that contains a modifier, however, has higher critical parameters as compared with pure carbon dioxide. Consequently, when using considerable amounts of modifier, the analysis will usually be made under less efficient subcritical conditions. Both time and efficiency can be gained using inert columns because the inert sufaces require less modifier for deactivation than the surfaces of commercial chiral LC columns. In addition, the competition of the polar modifier and the chiral analytes for the enantiomer-selective sites on the stationary phase cannot be ignored. In conclusion, there will be a growing need for inert polymer-based enantiomer-selective SFC columns.

3.6.3.2 Open tubular columns

The first separation of chiral compounds in capillary SFC was reported by Röder et al. [15,16] in 1987. They also applied a Pirkle-type selector, which was anchored to an immobilizable polysiloxane, for the separation of several derivatized amino acids. The comparison of selectivities under SFC (capillary) and LC (packed) conditions, which is not unambiguous, showed smaller separation factors under supercritical conditions. This was explained by the higher temperatures used in SFC as well as by the higher polarity of CO_2 (which may itself undergo interactions with the chiral selector) as compared to the n-heptane and added modifier used in normal phase HPLC. Bradshaw et al. [17] synthesized polysiloxanes containing pendant chiral amide side chains. Rouse [18] used this derived from *S*-1-(1-naphthyl)-ethylamine for the separation of racemic proline with a different derivatization in capillary SFC.

'Chirasil-Val', a well-known CSP for the enantiomer separation of derivatized amino acids in GC [19], was used in SFC by Lai et al. [20] (column with 100 µm i.d.). Significantly lower enantioselectivities were found in SFC as compared with GC at the same temperature. The loss in enantioselectivity was ascribed to an inhibitory effect of solvated solute and selector during the diastereomeric interaction due to strong hydrogen bonding with the supercritical mobile phase carbon dioxide.

Petersson et al. [21] and Rossiter et al. [22] developed a new strategy for the synthesis of chiral polysiloxanes. They used block polymers, in which the polysiloxane units are connected via a chiral cyclohexylene-bis-benzamide derivative. Several diols were separated with high efficiency on a long (20 m x 50 µm) capillary column.

Schurig et al. used an immobilized (non-extractable) CSP, consisting of a permethylated monokis-(6-O-octamethylene)-β-cyclodextrin chemically bonded to a dimethylpolysiloxane ('Chirasil-Dex') in both GC [23] and SFC [24]. The enantiomer separations of underivatized pharmaceuticals such as syncumar, dihydrodiazepam, and ibuprofen were achieved on a short (2.5 m x 50 µm i.d.) non-deactivated fused silica capillary column.

Although the enantioselectivity as expressed by the separation factor α is lower than in most LC separations, the high resolution obtained in combination with the use of the universal FID brings certain advantages in the trace analysis of enantiomeric excess greater than 90%. A systematic comparison of enantiomer separation in inclusion GC and SFC [24] showed a dramatic loss of enantioselectivity under supercritical conditions at a given temperature. This was attributed to the blocking of the hydrophobic cyclodextrin cavity by nonpolar carbon dioxide molecules at high densities. The effect of temperature on resolution was also studied, showing the same typical maximum observed for the capacity factors of the enantiomers. Generally, a better resolution is obtained at a lower inlet pressure (density).

More recently, Schurig et al. [26] reported on the use of a polysiloxane-based CSP containing a chiral metal chelate (Chirasil-Metal) in capillary GC and SFC. In complexation chromatography enantiomer separation is achieved by donor-acceptor interactions between a chiral solute possessing lone electron pairs and a sterically and electronically unsaturated metal chelate bearing non-racemic terpene ketonate ligands such as nickel (II)-bis[(3-heptafluorobutanoyl)-(1*R*)-camphorate]. Thus, the enantioselectivity of this system is strongly affected by variations in either the ligand (e.g., derivatives of camphor, pulegone or menthone) or the Lewis acidity of the metal ion (M(II), e.g., Ni(II), Mn(II), Cu(II), Zn(II)). The standard system developed, is a polysiloxane-anchored Ni(II)-bis[(3-heptafluorobutanoyl)-(1*R*)-camphorate], 'Chirasil-Nickel'. This CSP shows no loss in enantioselectivity over the whole range of temperature and density applied, indicating that blocking of the chiral selector by supercritical carbon dioxide does not play a negative role. In some cases higher separation factors are observed in SFC than in GC. At constant resolution (e.g., R_S = 2.0, measured for 1-phenylethanol), the analysis temperature decreases with increasing pressure from 145°C in GC to 45°C in SFC. As pointed out before, the dramatic loss of efficiency in SFC as compared to GC can thus be compensated for by applying low-temperature SFC.

References

[1] Macaudière, P., Caude, M., Rosset, R., Tambuté, A. *J. Chromatogr. Sci.* (1989), *27*, 383

[2] Lee, M.L., Markides, K.E., Eds., *Analytical Supercritical Fluid chromatography and Extraction*, Chromatography Conferences, Brigham Young Univ. Press: Provo, 1990

[3] Macaudière, P., Caude, M., Rosset, R., Tambuté, A. *J. Chromatogr.* (1987), *405*, 135

[4] Dobashi, A., Dobashi, Y., Ono, T., Hara, S., Saito, M., Higashidate, S., Yamauchi, Y. *J. Chromatogr.* (1989), *461*, 121

[5] Gübitz, G. *J. Liq. Chromatogr.* (1986), *9*, 516

[6] Payne, K.M. Thesis, Brigham Young Univ: Provo, 1990

[7] Fields, S.M., Kong, R.C., Fjeldsted, J.C., Lee, M.L., Paeden, P.A. *J. High Resolut. Chromatogr. Chromatogr. Commun.* (1984), *7*, 312

[8] Sumpter, S.R., Woolley, C.L., Huang, E.C., Markides, K.E., Lee, M.L. *J. Chromatogr.* (1990), *517*, 503

[9] Pericles, N., Giorgetti, A., Dätwyler, P. in *SFC Applications Handbook*, Markides, K.E., Lee, M.L., Eds., Bringham Young Univ. Press, Provo, 1988, 44

[10] Mourier, P.A., Eliot, E., Caude, M.H., Rosset, R.H., Tambuté, A. *Anal. Chem.* (1985), 57, 2819

[11] Gasparrini, F., Misti, D., Villani, C. *J. High Resolut. Chromatogr.* (1990), *13*, 182.

[12] Macaudière, P., Caude, M., Rosset, R., Tambuté, A. *J. Chromatogr.* (1988), *450*, 255

[13] Lienne, M., Caude, M., Rosset, R., Tambuté, A. *J. Chromatogr.* (1988), *448*, 55

[14] Nitta, T., Yakushijin, Y., Kametani, T., Katajama, T. *Bull. Chem. Soc. Japan* (1990), *63*, 1365.

[15] Röder, W., Ruffing, F.-J., Schomburg, G., Pirkle, W.H. *J. High Resolut. Chromatogr.* (1987), *10*, 665.

[16] Ruffing, F.-J., Lux, A., Roeder, W., Schomburg, G. *Chromatographia* (1988), *26*, 19.

[17] Bradshaw, J.S., Aggarwal, S.K., Rouse, C.J., Tarbet, B.J., Markides, K.E., Lee, M.L. *J. Chromatog.* (1987), *405*, 169.

[18] Rouse, C. in *SFC Applications Handbook*, Markides, K.E., Lee, M.L., eds., Brigham Young Univ. Press: Provo, 1988, p. 38.

[19] Frank, H., Nicholson, G.J., Bayer, E. *J. Chromatogr.* (1978), *167*, 187.

[20] Lai, G., Nicholson, G.J., Mühleck, U., Bayer, E. *J. Chromatogr.* (1991), *540*, 217.

[21] Petersson, P., Markides, K.E., Deborah, F., Johnsson, D.F., Rossiter, B.E., Bradshaw, J.S., Lee, M.L. *J. Microcol. Sep.* (1992), *4*, 155.

[22] Rossiter, B.E., Petersson, P., Johnson, D.F., Eguchi, M., Bradshaw, J.S., Markides, K.E., Lee, M.L. *Tetrahedron Lett.* (1991), 32, 3609.

[23] Schurig, V., Schmalzing, D., Mühleck, U., Jung, M., Schleimer, M., Musche, P., Duvekot, D., Buyten, J.C. *J. High Resolut. Chromatogr.* (1990), *13*, 713.

[24] Schurig, V., Juvancz, Z., Nicholson, G., Schmalzing, D. *J. High Resolut. Chromatogr.* (1991), *14*, 58.

[25] Schmalzing, D., Nicholson, G., Jung, M., Schurig, V. *J. High Resolut. Chromatogr.* (1991), *14*, 58.

[26] Schurig, V., Schmalzing, D., Schleimer, M. *Angew. Chem. Int. Ed. Engl.* (1991), *8*, 30.

3.7 Electrophoresis

In recent years the number of publications on chiral separation by means of capillary electrophoresis has increased dramatically. Only less than 10 % of all the articles, reviews, books, etc. published appeared before 1990 (incl. paper electrophoresis). In the following section, only an overview of this subject can be given.

For the most part, no detailed information on the achiral buffer components has been provided. Technical details, e.g. voltage, electric field strength, capillary dimensions, etc., have also been given in only a few cases. In general, pH-values have been reported, which refer in some cases to the total pH, in other cases only to the pH of the added buffer. For theoretical considerations see ref. [1,2], for reviews see ref. [3-14].

In the literature many different terms and abbreviations are used. The most important ones are listed below:

ACE Affinity Capillary Electrophoresis
AEC Affinity Electrokinetic Chromatography
AGE Affinity Gel Electrophoresis
CAE Capillary Affinity Electrophoresis
CAGE Capillary Affinity Gel Electrophoresis
CAZE Capillary Affinity Zone Electrophoresis
CE Capillary Electrophoresis
CEC Capillary Electrochromatography
CES Capillary Electroseparation
CGE Capillary Gel Electrophoresis
CIEF Capillary Isoelectric Focussing
CITP Capillary IsoTachoPhoresis
CMEC Capillary Micellar Electrokinetic Chromatography
CZE Capillary Zone Electrophoresis
DE Displacement Electrophoresis
EC ElectroChromatography
ECC Electrokinetic Capillary Chromatography
EKC Electrokinetic Chromatography
FSCE Free Solution Capillary Electrophoresis
FZCE Free Zone Capillary Electrophoresis
FZE Free Zone Electrophoresis
GE Gel Electrophoresis
GZE Gel Zone Electrophoresis
HPCE High Performance Capillary Electrophoresis
HPCGE High Performance Capillary Gel Electrophoresis
HPE High Performance Electrophoresis
HPPE High Performance Paper Electrophoresis
HPZE High Performance Zone Electrophoresis
HVE High Voltage Electrophoresis

HVPE	High Voltage Paper Electrophoresis
IEF	Isoelectric Focussing
ITP	IsoTachoPhoresis
LEE	Ligand Exchange Electrophoresis
MCE	Microemulsion capillary electrophoresis
MEC	Micellar Electrokinetic Chromatography
MECC	Micellar Electrokinetic Capillary Chromatography
MEEKC	MicroEmulsion ElectroKinetic Chromatography
MEKC	Micellar ElectroKinetic Chromatography
PAGE	Polyacrylamide Gel Electrophoresis
PE	Paper Electrophoresis
TLE	Thin-Layer Electrophoresis
ZE	Zone Electrophoresis

Equations and terms used to describe the quality of an enantiomer separation for liquid chromatography (LC) and gas chromatography (GC), e.g., α-value, resolution, ee-value, etc., can be adopted for capillary electrophoretic separations as well. For details and the calculation of electrophoretic terms (e.g., electrophoretic mobility μ_{ep}) see ref. [15-18] and for considerations of quantitative enantiomer separation see ref. [19, 20].

3.7.1 Introduction

Electrophoresis is a separation technique based on differences of ion linear velocity in an electric field. In other words, charged molecules migrate in the presence of an electric field to the electrode with the opposite charge. The instrumentation for capillary electrophoresis is simple. It consists of a high-voltage power supply, two buffer reservoirs, a capillary, an injection device, a detector (e.g., modified HPLC detector) and a data recording system (e.g., computer) (Figure 3.7.1).

The dimensions of capillaries used for CE are very small. The range of the length is 10 to 100 cm with inner diameters of 5 to 100 μm (for ITP up to 800 μm). The analyzing times can be very short (less than 1 min, but they can also be as long as 2 h). The working temperature is mainly room temperature (0-40°C). The main advantages of CE are the very high numbers of theoretical plates (>1'000'000) and the extreme sensitivity (e.g., achieved with laser fluorescence detection). Different modes of capillary electrophoretic separation can be performed using a standard CE instrument [16]:

(i) Capillary zone electrophoresis (CZE)

CZE separation is based on differences in electrophoretic mobility resulting in different migration velocities of sample ions in the electrophoretic buffer contained in the capillary. The separation mechanism is mainly based on differences in solute size and charge at a given pH-value. When a constant electric field is applied, sample ions migrate with a linear velocity proportional to their mobility. When the mobilities differ, separate zones are formed. The carrier electrolyte prevents the formation of non-conductive gaps in between the zones. The zones are similar to chromatographic peaks. They represent the analyte distribution (caused by zone or peak spreading) around a sharply defined 'ideal' analyte position. Preferentially, the ion concentration in the sample is much lower than that of the background, so that local conductivity and electric field are not disturbed by the presence of the zone. Under these conditions, all the zones move at a constant, characteristic speed [17].

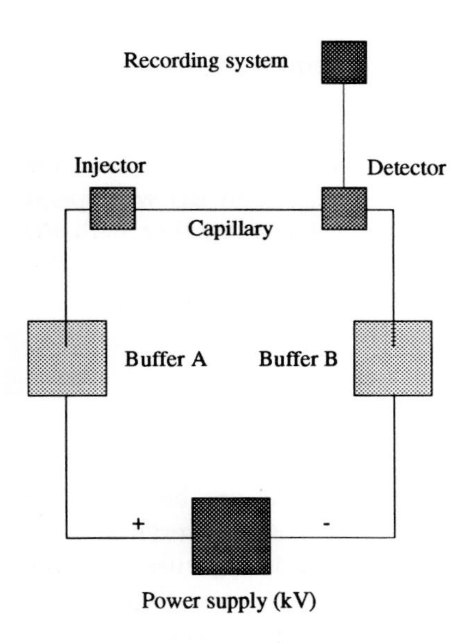

Figure 3.7.1 Schematic diagram of an instrumentation for capillary electrophoresis.

Most capillaries used for CE are made of fused silica, containing surface silanol groups. These silanol groups may become ionized in the presence of the electrophoretic medium. The interface between the fused silica tube wall and the electrophoretic buffer consists of three layers: the negatively charged silica surface (at pH >2), the diffuse layer of cations adjacent to the surface of the silica which tend to migrate towards the cathode, and the immobile layer. This migration of cations results in a concomitant migration of fluids through the capillary (electroosmotic flow = EOF) [16].

(ii) Micellar electrokinetic capillary chromatography (MECC or MEKC)

An important development in CE is the introduction of micellar electrokinetic capillary chromatography. The main separation principle is based on solute partitioning between the micellar phase and the solution phase. This technique makes possible the resolution of neutral molecules as well as charged molecules by CE. Micelles are formed in solution when a surfactant (e.g., sodium dodecyl sulphate, SDS) is added to water (buffer) in concentrations above its critical micelle concentration (CMC). Micelles consist of aggregates of surfactant molecules with typical lifetimes less than 10 μs. In the case of SDS, the micelles can be considered as small droplets of oil with a highly polar surface which is negatively charged. Even though these anionic micelles are attracted toward the anode, in an uncoated fused silica capillary they still migrate toward the cathode because of electroosmotic flow. However, the micelles move more slowly toward the cathode than the bulk of the liquid because of their attraction towards the anode. Neutral molecules partition in and out of the micelles based on the hydrophobicity of each analyte. Consequently, the micelles of MEKC are often referred to as a pseudo (or moving) stationary phase.

A very hydrophilic neutral molecule (e.g., methanol) will spend almost no time inside the micelle and will therefore migrate at the same rate as the bulk flow. On the other hand, a very hydrophobic neutral molecule (e.g., Sudan III) will spend nearly all its time inside the micelles and therefore elute later, together with the micelles. All other solutes with intermediate hydrophobicity will elute within this migration range [16].

(iii) Capillary gel electrophoresis (CGE)

The main separation principle in capillary gel electrophoresis (CGE) is based on differences in solute size as analytes migrate through the pores of the gel-filled column. Gels are potentially useful for electrophoretic separations because they permit separation based on 'molecular sieving'. Additionally, they serve as anti-convective media, they minimize solute diffusion, which is a cause of zone broadening, they prevent solute adsorption to the capillary walls and they help to eliminate electroosmosis. The gel must, however, possess certain characteristics, such as temperature stability and the appropriate range of pore size. Furthermore, the technique is subject to the limitation that neutral molecules do not migrate through the gel, since the electroosmotic flow is suppressed in this mode of operation. One of the most-used gels is polyacrylamide [16].

(iv) Capillary isoelectric focusing (CIEF)

Another separation method which can be performed using a CE instrument is iso-electric focusing (IEF), in which analytes are separated on the basis of their iso-electric points. The use of IEF is limited to the separation of amphoteric molecules (e.g., amino acids; proteins). The anodic end of the column is placed into an acidic solution (anolyte), the cathodic end in a basic solution (catholyte). Under the influence of an applied electric field, charged analytes migrate through the medium until they reside in a region of pH where they become electrically neutral and, therefore, stop migrating. After focusing, the zones can be migrated (mobilized) from the capillary by a pressurized flow. The resolving power of IEF can be expressed by the differences in the pI-values (ΔpI) of two sample components [16].

(v) Capillary isotachophoresis (CITP)

The main feature of isotachophoresis (ITP), also called displacement electrophoresis (DE), is that it is performed in a discontinuous buffer system. The sample mixture is applied between a leading electrolyte (LE) and a terminating electrolyte (TE), producing a steady-state migration configuration composed of consecutive sample zones [16]. Therefore, this mode of operation is different from other modes of capillary electrophoresis, such as CZE, which are normally carried out in a uniform carrier buffer and are characterized by sample peaks similar to those obtained in chromatographic separations, whereas in CITP the isotachopherogram obtained contains a series of steps, with each step representing an analyte zone. Unlike in other CE modes, where the amount of sample present can be determined, as in chromatography, from the area under the peak, quantitation of isotachopherograms is mainly based on the measured zone length, which is proportional to the amount of sample present [16].

(vi) Capillary electrochromatography (CEC)

In capillary electrochromatography (electrokinetic chromatography), the separation column is packed or coated with a chromatographic stationary phase which can retain solutes by the normal distribution equilibria upon which chromatography depends; it is, therefore, a special case of CE [16].

3.7.2 *Classical electrophoretic methods*

Only few publications deal with the separation of chiral compounds using classical electrophoretic methods. In the following some examples from paper and gel electrophoresis, as well as isoelectrical focusing, are represented.

3.7.2.1 Paper electrophoresis (PE)

After the first failing experiments in 1960 [21] and 1967 [22], in 1970, Yoneda and Miura [23] reported the first successful enantioseparation by means of paper electrophoresis. The authors used tartaric acid as the chiral counter-ion for chiral separation of tris(ethylenediamine)Co(III). The separation mechanism is based on formation of the diastereomeric salt [$Coen_3$]Cl-(-)-tartrate·$5H_2O$. Applying the same technique, Yoneda and Miura [24] resolved tris(ethylenediamine) metal complexes (e.g., Cr or Rh). The authors stressed the importance of the addition of aluminium chloride to the background solution and of the optimal pH (<7). Separations were obtained using arsenic potassium (+)-tartrate and other similar chiral counter-ions of metal complexes containing cobalt, ferrum and nickel [25]. Fanali et al. [26] used sodium (+)- or (-)-tartrate for tris(ethylenediamine)Co(III) to check the influence of the enantiomeric purity of the chiral counter-ion.

Separation of tris-oxalatoCo(III) and tris-oxalatoCr(III) with acetic acid solutions of brucine, quinine or cinchonine on Whatman No. 1 paper was reported by Cardaci and Ossicini [27]. Several other publications deal with the separation of metal complexes employing the same method [28-30]. In a recent publication a microcrystalline cellulose-type thin-layer electrophoresis is described for the enantioseparation of methyltryptophan derivatives [31].

3.7.2.2 Isoelectric focusing (IEF)

The separation of enantiomeric forms of dansylated amino acids (e.g., phenylalanine and tryptophan) by isoelectric focusing in immobilized pH gradients (IPG) was demonstrated by Righetti et al. [32]. Separations occurred in a pH 3.0-4.0 IPG interval, in the presence of 7 mol urea, 10 % methanol and 60 mmol β-cyclodextrin as chiral discriminator. Dansylated *S*-amino acids form weaker inclusion complexes with β-cyclodextrin than do the *R*-forms.

3.7.2.3 Gel zone electrophoresis (GZE)

Nishizawa et al. [33] used a polyacrylamide gel containing 70 mmol β-cyclodextrin for the enantioseparation of dansylated amino acids (e.g., asn, asp, cys, gln, glu, his, ile, leu, lys, met, ser, thr, tyr, val) at pH 7.9. With an acidic buffer (pH 5.6) no resolution was obtained. The different binding of (+)-anti- and (-)-anti-benzo[a]-pyrene-7,8-dihydrodiol-9,10-epoxide to DNA-oligomers was investigated via PAGE by Marsch et al. [34].

3.7.3 *Capillary electrophoretic methods*

Nowadays, capillary techniques are replacing more and more the classical electrophoretic techniques. More than 95 % of all applications on enantiomer separation described in the literature have been carried out with capillaries. In the following, several examples for the different electrophoretic techniques are presented.

3.7.3.1 Capillary gel electrophoresis (CGE)

With the incorporation of β-cyclodextrin within a polyacrylamide gel column the chiral resolution of dansylated amino acids for high-performance capillary gel electrophoresis (HPCGE) has been achieved [35]. The analysis was carried out with α-, β- or γ-cyclodextrins as buffer additives. Without incorporation of β-cyclodextrin in the gel no enantiomer separation was obtained. β-Cyclodextrin exhibited the highest enantioselectivity of all the cyclodextrins evaluated; α-cyclodextrin did not show any selectivity. In some cases the addition of 10 % methanol improved the enantioselectivity. Several dansylated amino acids (cf. Table 3.7.1) have been separated with a β-cyclodextrin-containing buffer solution. Stronger electric fields lead to smaller peaks and shorter times (e.g., by increasing the field from 700 V/cm to 1000 V/cm the migration time of several dansylated amino acids was halved) [35].

Cruzado and Vigh [36,37] synthesized polysubstituted allyl carbamoylated β-cyclodextrins copolymerized with acrylamine to form cross-linked cyclodextrin gel-filled capillaries for the enantioseparation of dansylated amino acids (e.g., asp, glu, leu, met, nleu, phe, ser, thr, val) and homatropine (at pH 8.3). Both chiral selectivity and peak resolution increased non-linearly with increasing cyclodextrin concentration. To avoid the known problems of cross-linked CE gels, i.e. bubble formation, poor reproducibility and short life-time, soft liquid gels were used, which do not contain external cross-linkers [37].

Several cyclodextrin derivatives have been used as chiral buffer additives by Schmitt and Engelhardt [38,39]. Hydroxypropylation, methylation or carboxymethylation of the cyclodextrin not only resulted in a better solubility of the cyclodextrin in aqueous solutions, but also favoured, via additional hydrogen bonding, the stabilization of cyclodextrin-analyte complexes [39]. The degree of derivatization of the cyclodextrins differs (no concrete derivatives were used). A reverse of the elution order for dansyl-phenylalanine was observed when the concentration of cyclodextrin increased [39]. In some cases, it was possible to carry out the separations with short capillaries (7 cm effective length) and short migration times (< 2 min) [38].

In Table 3.7.1 HPCGE enantiomer separations applying chiral additives like cyclodextrins are summarized. Due to the importance of the applied pH, the pH values of the buffers used are also given. With carboxymethylated β-cyclodextrin, a low pH has been applied. A pH >5 leads to deprotonation of the carboxylic groups of the derivatized cyclodextrin. Thus, the induced charge on the cyclodextrin molecule leads to mobility of the chiral selector; e.g., negatively charged enantiomers can no longer be separated because of electrostatic repulsion. The interaction of positively charged molecules with the charged cyclodextrin via ion-pairing is possible. Furthermore, uncharged enantiomers can be separated in a micellar-like system [39].

A different approach is the production of polyacrylamide-coated capillaries filled with an immobilized protein-dextran polymer network, where bovine serum albumine (BSA) is covalently linked to a high-molecular-mass dextran (M_r = 2'000'000) using cyanogen bromide [43]. The addition of such a dextran and BSA to the run buffer was reported by Sun et al. [44]. The resolution increased with increasing dextran concentration. Capillaries filled with gels consisting of bovine serum albumin cross-linked with glutaraldehyde have been used in affinity gel electrophoresis (AGE) [45] for the separation of tryptophan. Polyacrylamide-coated fused-silica capillaries with human serum albumin (HSA) have been produced by

Vespalec et al. [46]. The HSA has been heated for 30 min at pH 9 and 60°C to stabilize its enantioselectivity. After prolonged heating, e.g., for 24 h at 60° C, the albumin completely lost its enantioselectivity. Higher concentrations of HSA led to improved peak shape and better resolution. Coated capillaries were applied to reduce the electroosmotic flow of HSA.

Tanaka et al. [47] employed avidin, a basic protein isolated from egg white as chiral selector in capillary affinity gel electrophoresis (polyacrylamide gel) for the enantioseparation of some carboxylic acids (e.g., flurbiprofen, folinic acid, ibuprofen, ketoprofen, warfarin). Typical problems of these techniques are also outlined: (i) adsorption of the protein on the capillary wall, (ii) absorption of UV light at the shorter wavelength region and (iii) relatively low purity of the proteins.

Table 3.7.1 Applications of capillary gel electrophoresis (CGE) using polyacryl amide (PAA) gels and chiral gels or buffer additives.

Chiral selector	Type	Compounds	pH	Ref.
β-CD	G/B	DNS-amino acids (2-amino butanoic acid, asp, glu, leu, nleu, nval, met, phe, ser, thr, trp, val)	8.3	35
β-CD	G	DNS-amino acids (leu, ser, glu)	8.3	40
	B	Hexobarbital	8.3	39
AC-β-CD	G	DNS-amino acids (leu, phe)	8.3	37
HP-β-CD	B	DNS-phenylalanine	6.0	38
	B	Hexobarbital	8.3	39
	B	Naproxen	5.0	41
CM-β-CD	B	Doxylamine, ephedrine, dimetinden	2.5	38
	B	Propanolol, 2,2'-dihydroxy-1,1'-dinaphthyl	2.5	42
CE-β-CD	B	2,2'-Dihydroxy-1,1'-dinaphthyl, hexobarbital	5.5	42
HP-γ-CD	B	1,4-Dihydro-4-(3-nitrophenyl)pyridine 3,5-dicarboxylic acid 5-monobenzylester	6.0	38
BSA-dextran	G	Leucovorin	7.0	43
	B	DNS-amino acids (leu, nval), mandelic acid, ibuprofen, leucovorin	7.1	44
BSA	G	Trp	7.5	45
HSA	G	*E*-2,3-dibenzoyltartaric acid, N-2,4-dinitrophenyl-glutamic acid	8.0	46

AC = allyl carbamoyl; B = buffer additive; BSA = bovine serum albumin; CD = cyclodextrin; CE = carboxyethyl; CM = carboxymethyl; DNS = dansylated; G = gel additive; HSA = human serum albumin; HP = 2-hydroxypropyl

Terabe and Tanaka [48] used avidin and another egg white protein, ovomucoid (acidic protein), for enantioseparations. Avidin for acidic compounds such as atrolactic acid, mandelic acid and phenyllactic acid derivatives, etc.; ovomucoid for basic compounds such as bunitrolol, pindolol, verapamil, etc.

3.7.3.2 Capillary zone electrophoresis (CZE)

In contrast to capillary gel electrophoresis, for capillary zone electrophoresis uncoated capillaries are usually employed. Buffer additives reduce interactions between the silanol groups of the capillary wall and the buffer [49]. The main principle is maximization of the difference in electrophoretic mobility between two enantiomers; but the electroosmotic mobility is influenced as well [1]. In order to obtain high separation efficiencies, electromigration dispersion has to be minimized either by using very dilute samples in concentrated background electrolytes (BGE) or by matching the mobilities of the analyte and the co-ion of the BGE (e.g., zwitterionic compounds). By varying the co-ion mobilities the separation efficiency can be adjusted individually for each separation problem [50].

Chiral derivatizing agents (CDA)

In CZE the separation of enantiomers can be obtained via diastereomeric derivatives or by addition of chiral auxiliaries to the buffer. One possibility is the derivatization of amino acids with o-phthaldialdehyde (OPA) and N-acetylcysteine (ACC) to form diastereomeric isoindole derivatives (cf. Figure 3.7.2 [51] and Section 3.5). Dette et al. [51] applied this method in the opposite way by determining the optical purity of N-acetylcysteine via enantiomerically pure *S*-valine (pH 9.2) as demonstrated in Figure 3.7.3 [52]. Another possibility is the use of (+)-1-(9-fluorenyl)ethyl chloroformate to obtain fluorescent amino acid derivatives [53].

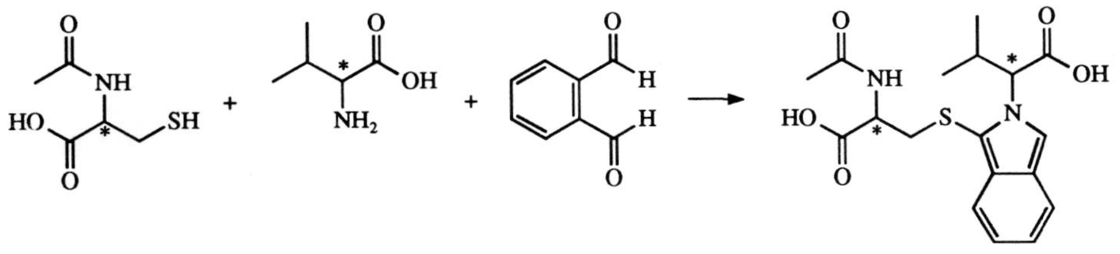

Figure 3.7.2 Derivatization of N-acetylcysteine with valine and o-phthaldialdehyde.

Amino acids can also be derivatized with L-(+)-diacetyl tartaric acid anhydride. The resulting diastereomeric diacetyl tartaric acids monoamides can be separated at pH 6.4 [54]. Polyvinylpolypyrrolidone (PVPP) was added to the buffer solution. PVPP is able to undergo hydrophobic as well as dipolar interactions with the sample components, thus influencing the effective mobility of the analytes [54]. Later, an analogue compound, L-(+)-dibenzoyl tartaric acid anhydride, has been used under similar conditions [56].

Ternary copper complexes

From 1970 to 1985 only paper electrophoresis was used for enantiomer separations. In 1985, Gassmann et al. [56] introduced capillary zone electrophoresis with a chiral support electrolyte for the separation of dansylated amino acids (e.g., 2-aminobutanoic acid, asp, cysteic aicd, cystine, glu, met, phe, tyr, val) with laser fluorescence detection. The analysis was fast (10 min) and sensitive (femtomole range) with a high separation efficiency. The authors used the copper(II) complex of *S*-histidine as an electrolyte (pH 7-8), which formed ternary complexes with the amino acids. This technique is similar to the already-mentioned ligand-exchange chromatography (LEC; cf. Section 3.4). Later, with better results, Cu(II)-aspartame was applied as a chiral electrolyte for the enantioseparation of several dansylated amino acids (pH 7-8) [57,58].

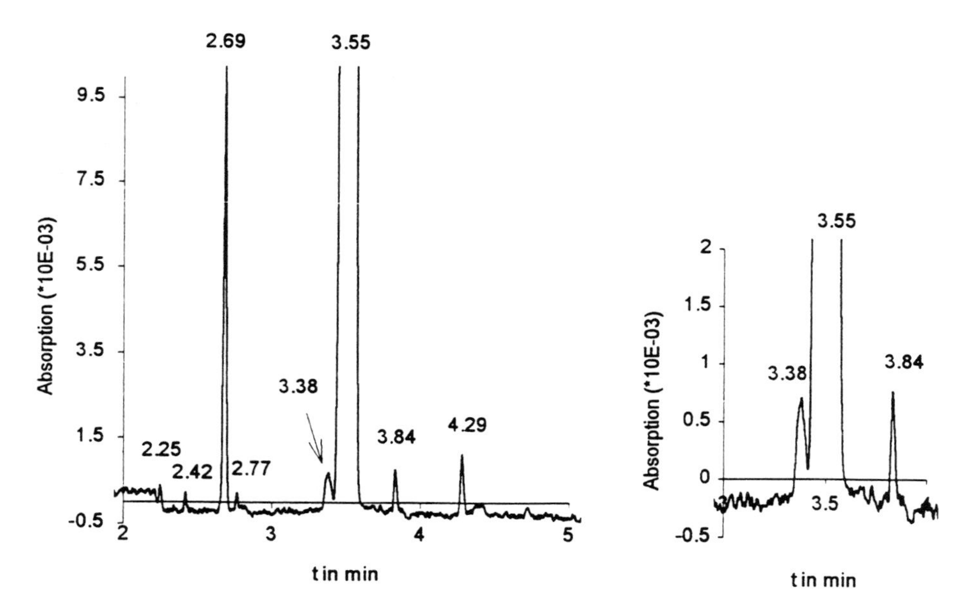

Figure 3.7.3 Determination of the optical purity of acetylcystein (ee = 98.8 %) after derivatization with valine and o-phthaldialdehyde. t_M = 3.38 min, L,D-; t_M = 3.55 min, L,L-product (30 cm x 50 µm i.d. fused silica capillary; borate buffer pH 10.4; 25°C; 25 kV) [52].

The influence of temperature was also tested (1 to 40°C); higher temperatures lead to shorter migration times, without any decrease in enantioselectivity. Recently, mandelic acid and analogues as well as phenyllactic acids could be separated into their enantiomers using Cu(II)-aspartame, Cu(II)-proline or Cu(II)-hydroxyproline as background electrolytes (BGE) [59].

Tartaric acid

In 1989, Fanali et al. [60] used (+)-tartaric acid (pH 5.3) for the resolution of tris(ethylenediamine)cobalt(III) and amino acid cobalt complexes. The separation principle was adopted from former paper electrophoretic applications of this working group [26].

Cyclodextrins and derivatives

Due to the high enantioselectivity of the cyclodextrins particular attention has been devoted to cyclodextrins as buffer additives for host-guest complexation capillary electrophoresis (inclusion capillary electrophoresis). Detailed information is provided in Sections 3.4 and 3.5. In Table 3.7.2 the scope of cyclodextrins is outlined.

Table 3.7.2 Use of cyclodextrins and derivatives in capillary electrophoresis.

Cyclodextrin	Compound	pH	Ref.
α-CD	Aminoglutethimide	2.5	61
	DNS-aspartic acid	9.0	62
	Bambuterol	?	63
	2-(4-Chloro-2-methylphenoxy)propanoic acid	4.8	64
	AHNS-/ANA-/AP-carbohydrates (ara, fuc, gal, lyx, man, rib, xyl)	9.2	65
	Dibenzooxepine-der., isochromane-der.	2.5	66
	2-(4,6-Dichloro-2-methylphenoxy)propanoic aicd	4.8	64
	2-(2,4-Dichlorophenoxy)propanoic acid	4.8	64
	1,1'-Dinaphthyl-2,2'-diyl hydrogenphosphate	9.0	65
	5-Methyltetrahydrofolate	7.1	67
	1-/5-/6-Methyl-trp, trp octyl ester	2.5	68
	Phenoxycarboxylic acid herbicides (MCPP and 2,4-DP)	4.8	69
	Phe, tyr	2.5	70
	Trp	2.2	71
2,6-DM-α-CD	DNS-phe	9.0	72
	Antiviral compound BCH189, clenbuterol	2.3	19
2,3,6-TM-α-CD	DNS-amino acids (leu, nleu, phe, val), pentobarbital, secobarbital	9.0	72
MG-α-CD	2,2'-Dihydroxy-1,1'-dinaphthyl	9.0	73
	1,1'-Dinaphthyl-2,2'-diyl hydrogenphosphate	9.0	73
β-CD	Ala-β-naphthylamide	2.0	74
	DNS-amino acids (asp, glu, leu, nleu, nval, val)	9.0	72
	DNS-amino acids (met, ser, thr)	9.5	62
	Brompheniramine, chlorpheniramine, pheniramine	2.5	75
	Carbinoxamine maleate (urea was added)	2.5	76

Cyclodextrin	Compound	pH	Ref.
	AHNS-/ANA-/AP-carbohydrates (all, ara, ery, fuc, gal, glu, ido, lyx, man, rib, tal, thr, xyl), ANA-glyceraldehyde	9.2	65
	Carvedilol, ephedrine, ketamine	3.3	77
	Z-Chloramphenicol	3.5	4
	2-(4-Chloro-2-methylphenoxy)propanoic acid	5.8	49
	Chlorpheniramine (urea was added)	3.0	78
	Clenbuterol, picumeterol	4.0	79
	1,1'-Dinaphthyl-2,2'-diol	9.0	73
	Fadrazole, glutethimide analogues	2.5	61
	Fenoprofen, ibuprofen (HEC and MES were added)	4.5	80
	Fluparoxan, dto. *N*-benzylated (urea was added)	2.7	81
	Hexobarbital, mephobarbital	7.0	61
	Homatropine (37°C; HEC was added)	6.3	82
	3-Hydroxy-*N*-methylmorphinan, 3-methoxy-*N*-methylmorphinan	9.1	83
	Indane-der., isochromane-der., dibenzooxepin-der.	2.5	66
	Ketamine	2.5	84
	Ketotifen (I) (MHEC was added)	3.5	85
	Ketotifen-intermediate (II)	2.5	85
	5-Methyltetrahydrofolate (urea was added)	7.1	67
	Phenoxycarboxylic acids (fenoxaprop, mecoprop)	4.5	86
	Pilocarpine	2.4	87
	Propranolol (urea was added), terbutaline	2.5	88
	Quinagolide	2.5	70
	Tröger's base	2.5	6
	Trp ethyl ester	2.5	68
	Zopiclone	3:3	89
α-/β-CD (1:1)	2,2'-Dihydroxy-1,1'-dinaphthyl	9.0	73
	1,1'-Dinaphthyl-2,2'-diyl hydrogenphosphate	9.0	73
2,3-DA-β-CD	Ala-β-naphthylamide	2.0	74
2,6-DM-β-CD	DNS-amino acids (asp, leu), hexobarbital	9.0	72
	Amphetamine, cathinone, cocaine, methamphetamine, methcathinone, norpseudoephedrine, pseudoephedrine	2.4	90
	Atenolol, pilocarpine, isopilocarpine	2.4	87
	Carvediol, bupivacaine, mepivacaine (MHEC, HTAB were added)	2.7	91
	Chloramphenicol	3.5	85
	Denopamine, trimetoquinol	2.2	92
	2'-Deoxy-3'-thiacytidine (= BCH 189)	2.3	93
	2-(2.4-Dichlorophenoxy)propanoic acid	4.8	64
	4,5-Dihydrodiazepam, Tröger's base, 1,1'-dinaphthyl-2,2'-diamine, 1,1'-dinaphthyl-2,2'-diyl hydrogenphosphate	7.0	94
	Ephedrine, epinephrine, isoproterenol	2.4	95
	Ergot alkaloids (meluol, nicergoline)	2.5	96
	Fadrozole, glutethimide analogues	2.5	61
	Ketamine, octopamine	2.5	84
	Ketotifen	3.5	4
	2-(2-Methyl-4-chlorophenoxy)propanoic acid	4.8	64
	Methylephedrine, pyrrolidine-der.	2.5	66
	Methylpseudoephedrine	2.5	52
	1-/5-/6-/7-Methyl-trp, 5-hydroxy-trp	2.5	68

Cyclodextrin	Compound	pH	Ref.
	Norephedrine, norepinephrine	2.4	95
	Picumeterol	2.3	97
	Pindolol (MHEC and HTAB was added)	3.0	98
	Propranolol	3.1	99
	Terbutaline	2.5	88
	Trp methyl ester, trp ethyl ester, trp butyl ester	2.5	68
2,3,6-TM-β-CD	DNS-amino acids (leu, phe), pentobarbital	9.0	72
	DNS-amino acids (nleu, nval, trp, val)	9.0	62
	Fluoxetine, verapamil (MHEC and HTAB were added)	2.8	91
	Hexahelicene-7,10-dicarboxylic acid, etodolac 1,1'-dinaphthyl-2,2'-diyl hydrogenphosphate	7.0	94
	Octopamine	2.5	84
	Phenoxycarboxylic acids (dichlorprop, fenoprop, mecoprop)	4.5	86
*Methyl.-β-CD	Atenolol, metoprolol, oxprenolol, propranolol	3.0	100
	5-Chlorowarfarin, warfarin	8.4	101
	Ephedrine, practolol	2.5	1
	Isoprenaline, methylephedrine, norephedrine	3.3	77
	Norfenefrine, salbutamol, synephrine	3.3	77
HE-β-CD	*N*-Desmethyldimetindene, 6-methoxydimetindene	3.3	77
	Propranolol	2.4	87
2-HP-β-CD	Ambucetamide, etilefrine, lofexidine, nomifensine	3.3	77
	Benzetimide (= R 4929)	4.0	102
	Clenbuterol	2.2	103
	N-Desmethyldimetindene, dimetindene	3.3	104
	N-Desmethyl-6-methoxydimetindene	3.3	104
	2,2'-Dihydroxy-1,1'-dinaphthyl	9.0	73
	Dimetindene-*N*-oxide, 6-methoxydimetindene	3.3	104
	1,1'-Dinaphthyl-2,2'-diyl hydrogenphosphate, DNS-amino acids (phe, trp)	9.0	73
	Epinephrine	2.5	105
	Fenoprofen, naproxen	4.5	106
	Hexobarbital	8.3	38
	Imafen, mefloquine, metomidate, mianserin, nefopam	3.3	77
	Mandelic acid, hydroxymandelic acids and related compounds	7.0	107
	Octopamine, pholedrine, sotalol, zopiclone	3.3	77
	2-/3-Phenyllactic acid, 4-hydroxymandelic acid	6.0	108
	Propranolol	2.4	87
	Tioconazole	4.3	109
CM-β-CD	3-Benzoyl-2-tert-butyl-1,3-oxazolidin-5-one	6.0	38
	DNS-phe	6.0	39
CME-β-CD	*E*-5,6-(3'-fluorobenzo)-2,3-di-(hydroxymethyl)-1,4-dioxane	12.4	81
6-MMA-β-CD	3,4-Dihydroxymandelic acid	5.0	108
	2-/3-Phenyllactic acid, 3-/4-hydroxymandelic acid	5.0	108
$6^A,6^D$-DMA-β-CD	3,4-Dihydroxymandelic acid	5.0	108

Cyclodextrin	Compound	pH	Ref.
	2-/3-Phenyllactic acid, 3-/4-hydroxymandelic acid	5.0	108
TS-β-CD	DNS-phe, 1,1'-dinaphthyl-2,2'-diol, hexobarbital, 4,5-dihydrodiazepam	7.0	94
	1,1'-Dinaphthyl-2,2'-diyl hydrogenphosphate	7.0	110
SBE-β-CD	Adrenaline, DOPA, ephedrine, noradrenaline, pseudoephedrine, tyr	2.5	111
	Amphetamine, cathinone, methamphetamine, methcathinone, norpseudoephedrine	2.4	90
	Clenbuterol, dimethindene, etilefrine, imafen, isoprenaline, lofexidine, mefloquine, metomidate, miansereine	3.1	112
	Methylephedrine, methylpseudoephedrine, norephedrine	10.0	52
SBE-/2,6-DM-β-CD	amphetamine, cathinone, cocaine, ephedrine, methamphetamine, methcathinone, norephedrine, norpseudoephedrine, propoxyphene	2.4	90
γ-CD	Ala-β-naphthylamide	2.0	74
	DNS-amino acids (asp, glu, leu, nleu, nval, val)	9.0	72
	DNS-amino acids (2-amino butanoic acid, met, thr)	9.0	62
	Aminoglutethimide	2.5	61
	AHNS-/ANA-/AP-carbohydrates (e.g. ara, gal, glu, man, rib, xyl)	9.2	65
	Ergot alkaloids (isolysergic acid, lisuride)	2.5	96
	Ergot alkaloids (meluol, nicergoline, terguride)	2.5	96
	Etodolac, 1,1'-dinaphthyl-2,2'-diyl hydrogenphosphate, 1,1'-dinaphthyl-2,2'-diol, 1,1'-dinaphthyl-2,2'-dicarboxylic acid	7.0	94
	Ketotifen (I) and its intermediates (II, III)	3.5	85
	Leucovorin, 5-methyltetrahydrofolate (urea was added)	7.1	67
	Mandelic acid	7.0	113
	Propranolol (TMA was added)	2.5	114
	Secobarbital	7.0	61
	Thioridazine	2.5	85
	Tocainide and related compounds (PVA was added)	3.0	115
	Trp butyl ester	2.5	68

* = degree of substitution for all three positions is 1.8; AHNS = 4-amino-5-hydroxy-2,7-naphthalene disulphonate; ANA = 5-amino-2-naphthalene sulphonate; AP = 2-aminopyridine; CD = cyclodextrin; CM = carboxymethyl; CME = carboxymethylethyl; DA = diacetyl; DM = dimethyl; DMA = dimethylamino; DNS = dansylated; HE = hydroxyethyl; HEC = hydroxyethylcellulose; HP = hydroxypropyl; HTAB = hexadecyltrimethylammonium bromide; MES = morpholine ethanesulphonic acid monohydrate; MG = monoglycosylated; MHEC = methylhydroxyethylcellulose; MMA = monomethylamino; PVA = poly(vinylalcohol); SBE = sulphobutyl ether; TM = trimethyl; TMA = tetramethylammonium; TS = mixture of mono, bis and tris(6-*O*-trimethylenesulphonic acid)

As shown in Table 3.7.2, many types of cyclodextrins and CD-derivatives have been used. For instance, carboxymethylated β-cyclodextrin [38] results not only in better solubility of the cyclodextrin in water, but also enables additional hydrogen bonding to favour the stabilization of a cyclodextrin-analyte complex. Carboxylated cyclodextrins can be used as charged molecules. For *E*-5,6-(3'-fluorobenzo)-2,3-di-(hydroxymethyl)-1,4-dioxane [81] a carboxymethylethyl-β-cyclodextrin was applied at pH 12.4 without the use of a micelle. An equilibrium model has been developed by Rawjee et al. [82] to describe the electrophoretic mobilities of the enantiomers of chiral weak bases and the resulting selectivities, as a function of pH and β-cyclodextrin concentration of the background electrolyte. Three different types of analytes have been described: (i) only the non-ionic forms interact differently with cyclodextrin (very low pH-values lead to maximal chiral selectivity and to faster migration times); (ii) only dissociated forms interact (optimizing of the separation by increasing the pH until the selectivity becomes sufficiently high); (iii) both the dissociated and the non-ionic forms interact (a detailed pH study is indispensable if the optimum enantioselectivity is to be determined; however, from a practical point of view, the use of low pH electrolyte seems the most promising approach). The validity of this model has been confirmed with homatropin as a model substance [82]. The influence of the degree of binding between the cyclodextrins and the solute on the enantioselectivity was discussed by St. Pierre and Sentell [116].

Using mandelic acid as model substance, Valkó et al. [113] investigated the importance of the nature of the electrolyte. Acceptable enantioseparation has only been achieved with a phosphate buffer. In some cases additives such as urea, isopropanol or methylhydroxyethylcellulose (MHEC) have been used to improve separations. For example, it was only possible to resolve tocainide and related compounds [115] with 0.05 % poly(vinylalcohol) (PVA) in the buffer. PVA also drastically increased the efficiency of non-chiral separations. Furthermore, a higher ionic strength improved separations significantly, e.g., in the case of the pharmaceuticals clenbuterol and picumeterol [79]. Kuhn et al. [70] studied the optimal separation temperature for quinagolide. The authors observed a decrease of enantiomer resolution at higher temperatures; the stability constants of cyclodextrin inclusion complexes in solution decreased with increasing temperature [84]. The influence of the capillary length was evaluated by McLaughlin et al. [76]; doubling the length nearly doubled the number of theoretical plates and increased the resolution tremendously. The influence of the inner diameter of the capillary has also been investigated [87]; the thinner the better the resolution. In the case of ephedrine, lower pH values and higher cyclodextrin concentrations improved separation as well. For tioconazole [109] the best separation has been obtained at a concentration of cyclodextrin that was equal to the reciprocal of the average binding constant (10 mmol cyclodextrin). Using a model system with (2-hydroxy)propyl-β-CD as chiral selector and propranolol as analyte the effect on enantioseparation of several parameters such as buffer pH, concentration of chiral selector, applied electric field and temperature was investigated in order to achieve maximum resolving power with minimal analysis time [117]. An improvement of enantioselectivity for warfarin enantiomers has been achieved by addition of sodium phytate to the buffer solution (containing 2,6-dimethyl-β-CD; pH 9.2) [118].

Significant influence of the degree of substitution of (2-hydroxy)propyl-β-CD on the enantioseparation of organic acids was reported by Valkó et al. [107]. Different positional isomers of methylated β-CD were studied by Yoshinaga and Tanaka [119] using dansylated amino acids. Very low enantioselectivity showed 2-monomethyl-β-

CD compared to underivatized β-CD, except for 2-aminobutanoic acid. The 6-monomethylated analogue exhibited excellent and the 3-monomethyl derivative medium enantioselectivity (with reversed elution orders). Using 3,6-dimethyl-β-CD no separation has been achieved, while the 2,3- and the 2,6-derivatives were found to be suitable phases, but less than the permethylated analogues, to separate some analytes. In comparison to underivatized β-CD, with all dimethyl- and the trimethyl-β-CD reversed elution orders have been observed [119]. Differences between cyclodextrin derivatives of different manufacturers were outlined by Nielen [66].

RP-18-coated capillaries have been used, but without any improvement of the resolution. In contrast, Belder and Schomburg [120] pointed out that the performance of chiral separations is strongly influenced by alteration of the surface coating. Adsorbed polybrene as a cationic coating and thermally immobilized PVA as a permanent hydroxylic non-ionic coating are well suited to improve the resolution for chiral organic acids (e.g., 2-phenylpropanoic acid, 3-phenyllactic acid, 3-phenylbutanoic acid, tocainide and analogues).

Soini et al. [91] analyzed the pharmaceuticals carvediol, bupivacaine, and mepivacaine. As buffer a mixture of tris(hydroxymethyl)aminomethane, methylhydroxyethylcellulose, and hexadecyltrimethylammonium bromide (HTAB), adjusted to pH 2.7, was employed. Although the authors used a micelle-forming agent with HTAB, capillary zone electrophoresis rather than MEKC appeared to be the mode of operation, as the concentration of HTAB was below the critical micelle concentration (CMC). In MEKC surfactant concentrations above the CMC are required (see below, Section 3.7.3.3). Short-chain tetraalkylammonium cations were applied together with native β-CD by Quang and Khaledi [121]. In some cases the presence of these tetraalkylammonium cations was essential for chiral separations.

In a more recent publication [65], even carbohydrates (e.g., ara, gal, glc, man, rib, xyl, and some others) could be separated into their enantiomers as fluorescent Schiff's base-derivatives of 2-aminopyridine (AP), 5-amino-2-naphthalene sulphonic acid (ANA) and 4-amino-5-hydroxy-2,7-naphthalene disulphonic acid (AHNS) using α-, β- or γ-CD with a borate buffer (pH 9.2). The use of several fluorescent derivatives for amino acid analysis was discussed by Ruyters and van der Wal [122]. The authors found 4-fluoro-7-nitrobenz-2,1,3-oxadiazol and naphthalene-2,3-dicarboxaldehyde to be the best derivatizing agents, because of their fast reaction times, their detectability by laser-induced fluorescence (LIF) and their capability for both CZE and MECC. A novel anionic cyclodextrin-type, the sodium salt of a randomly substituted β-CD sulphobutyl ether, was introduced as chiral selector by Tait et al. [111, 123]. The average degree of alkylation is approximately 4 (β-CD-SBE(IV)). The sulphonic group provides a derivative which is anionic over the entire pH range accessible to CE experiments and, with four methylene units serving as a spacer between the β-CD cavity and the sulphonate moiety, possesses inclusion properties similar to those of the parent β-CD. β-CD-SBE(IV) has dramatically improved aqueous solubility compared with that of β-CD with no buffer capacity and is transparent to both UV an fluorescence spectroscopy (for separation examples see Table 3.7.2). The difference of enantioselectivities for ephedrine analogues between 2,6-DM-β-CD and β-CD-SBE(IV) are shown in Figures 3.7.4 and 3.7.5 [52]. In some cases mixtures of β-CD-SBE(IV) with 2,6-DM-β-CD gave the best resolutions [90].

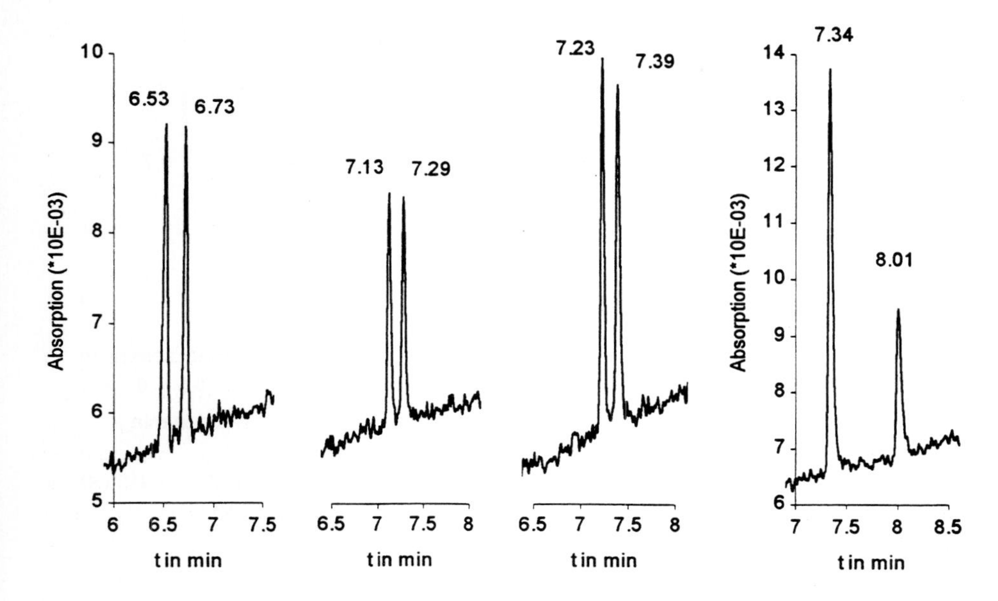

Figure 3.7.4 Enantioseparation ([-]- before [+]-isomer) of norephedrine, ephedrine, methyl ephedrine, and methylpseudoephedrine (from left to right) using 2,6-DM-β-CD. (30 cm x 50 μm i.d. fused silica capillary; phosphate buffer pH 2.5; 30°C; 15 kV; hydroxypropylcellulose, tetrabutyl ammonium bromide were added) [52].

Sun et al. [124] improved the chiral resolution of dansylated amino acids by adding a dextran (M_r = 2'000'000) to the β-CD-containing buffer solution (cyclodextrin-dextran polymer network). Optimal separation efficiency was obtained at 25°C. Water-soluble β-CD-polymers have been synthesized by Nishi et al. [125] using epichlorohydrin for condensations; the number of CD in a molecule varies from 3 to 50. Several pharmaceuticals were resolved (e.g., trimetoquinol, laudanosoline).

A carboxymethylated β-CD polymer was applied by Aturki and Fanali [126] which showed very good enantioselectivity for some pharmaceuticals, except ephedrine and norephedrine. An increase of the pH of the electrolyte caused a reversed elution order for propranolol and terbutaline due to strong complexation with the negatively charged CD-polymer.

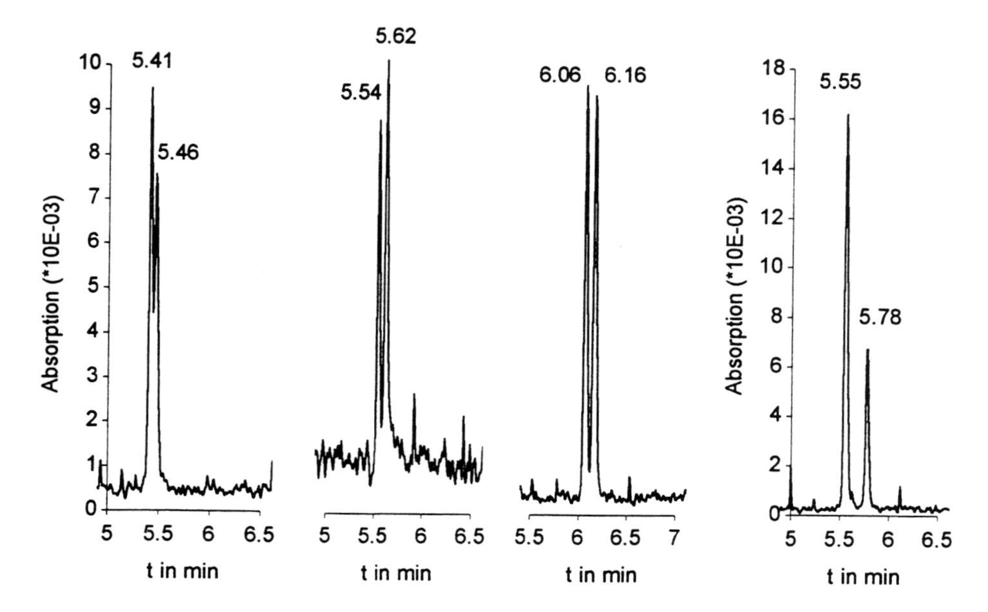

Figure 3.7.5 Enantioseparation of the same compounds as in Fig. 3.7.4 using β-CD-SBE(IV) (30 cm x 50 μm i.d. fused silica-capillary; borate buffer pH 10.0; 30°C; 15 kV) [52].

Carbohydrate polymers

There is little information [127] on the separation of enantiomers using maltodextrins, which are linear α-(1-4)-linked D-glucose polymers. Several pharmaceuticals have been analyzed like flurbiprofen, ibuprofen, 4-hydroxy-3-(3-oxo-1-phenylbutyl)coumarin (warfarin), 3-(α-acetonyl-p-chlorobenzyl)-4-hydroxycoumarin, chlorophenprocoumon, phenoprocoumon (all at pH 7.1), as well as cephalexin and cephadroxyl (at pH 7.5). In analogy to amylose, some maltodextrins display a balanced hydrophilic-hydrophobic surface, resulting from the helical conformation, which creates the steric environment necessary for chiral interactions [127]. Recently, some more chiral compounds have been resolved using long-chain dextrins, e.g., ketoprofen, simendan, and verapamil [128].

The separation of some carbohydrates with dextrin as chiral additive, e.g., arabinose, glucose, mannose, ribose, xylose, was reported by Stefansson and Novotny [129]. Nishi et al. [130] employed as chiral selector dextran sulphate, which is a mixture of a linear α-1,6-linked D-glucose polymer having a sulphate group in the molecule. Enantiomers of trimetoquinol and an analogue were separated at pH 5.5 using 3 to 5 % of dextran sulphate.

A two component polymer mixture consisting of polyethylene oxide and dextran together with 2,6-di-*O*-methyl-β-CD as chiral buffer additive was used for the resolution of several pharmaceuticals, among them some chiral ones such as disopyramide, nitrendipine, pindolol [131]. Heparin, which is a naturally occurring polydisperse polyanionic glycosaminoglycan was also investigated as chiral selector for CZE. Stalcup and Agyel [132] demonstrated its applicability for the separation of antimalarials (e.g., chloroquine, enpiroline, halofantrine, hydroxychloroquine, mefloquine, primaquine, quinacrine) and antihistamines (e.g., brompheniramine, carbinoxamine, chlorcyclyzine, chlorpheniramine, dimethindene, doxylamine, pheniramine, promethazine) as well as some other compounds (e.g., anabasine, bupivacaine, indapamide, nornicotine, tetramisole, trp, trp methyl ester).

Crown ethers

As with HPLC (see Section 3.4) crown ethers have also been employed as chiral auxiliaries for enantiomer separations. Kuhn et al. [70] were the first to use a chiral crown ether, i.e. 18-crown-6 tetracarboxylic acid for the separation of enantiomers of DOPA, phe, trp and tyr (at pH 2.2).

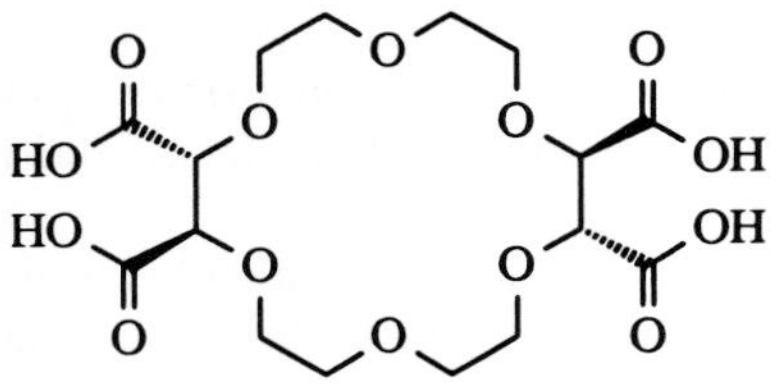

Later, further chiral compounds were analyzed, i.e. ala, p-amino-phe, *E-/Z*-2-amino-3-phenylbutanoic acid, 3-amino-3-phenylpropanoic acid, his, ile, leu, nval, thr and val, as well as noradrenaline, norephedrine, normetanephrine, phenylalaninol and 1,1'-naphthylethylamine (at pH 2.2) by Kuhn et al. [133]; phenylglycinol, methoxamine and octopamine (pH 2.1) by Höhne et al. [134]; decaline derivatives (pH 3.7), gly-phe and gly-trp (pH 2.5) by Kuhn et al. [135]; aminotetralin-, aminophenanthrene-, and aminodecaline derivatives (pH 2.2) by Walbroehl and Wagner [136]. In some cases, mixtures of cyclodextrins and 18-crown-6 tetracarboxylic acid were used, e.g., for trp [6], noradrenaline [137], as well as for DOPA and quinagolide [135]. The same authors also reported on the enantioseparation of some tyrosine analogues, as well as 4-amino-5-hexynoic acid and 4-amino-5,6-heptadienoic acid [138]. As possible separation mechanisms two main principles have been discussed [139]. Firstly, the crown ether substituents act as chiral barriers for the analyte and, secondly, lateral electrostatic interactions occur between host and guest.

Proteins

Bovine serum albumin (BSA) was first used as a chiral buffer additive for CZE enantioseparations by Barker et al. [140] studying leucovorin (pH 7.0). This particular type of CZE is also called affinity capillary electrophoresis (ACE) or capillary affinity zone electrophoresis (CAZE). A reversal of the elution order by modifying the buffer with 20 % polyethylene glycol PEG 8M-10 was observed. Additional applications comprise separations of tryptophan, benzoin, promethazine and warfarin (pH 6.8 with 1-propanol) by Busch et al. [141].

The importance of the selection of the appropriate buffer was shown by Arai et al. [142]. For instance, the separation of the antimicrobial agent ofloxacin could only be achieved using a phosphate buffer (pH 7.0), while no separation was observed employing a borate buffer (pH 7.0). The same compound was not resolved using human serum albumin (HSA) instead of BSA. HSA, which has a higher purity than BSA, was used by Vespalec et al. [46] for the enantioseparation of kynurenine, tryptophan, 3-indole lactic acid (pH 9.6) and *N*-2,4-dinitrophenylglutamic acid (pH 9.0). HSA was also employed for the separation of tryptophan and warfarin enantiomers [143]. Two different separation mechanisms were considered: the first based on HSA adsorbed to the capillary wall (for warfarin) and the second based on HSA in the running buffer (for tryptophan).

Busch et al. [141] have evaluated other proteins than BSA to use as additives for ACE. The enantioseparation of promethazine was achieved with the glycoprotein orosomucoid (M_r = 41'000) as an additive including *N,N*-dimethyloctylamine (pH 6.8). Other compounds such as tryptophan, benzoin, pindolol and warfarin could not be separated.

A glycoprotein, fungal cellulase isolated from *Aspergillus niger*, has also been investigated. Only the pindolol enantiomers were sufficiently separated (pH 6.8). The application of ovomucoid, a glycoprotein isolated from egg white, which is an excellent chiral stationary phase for HPLC, was not successful. [141]. In 1994, however, Ishihama et al. [144] reported some enantioseparations using ovomucoid as pseudo-stationary phase. In order to avoid the influence of protein adsorption on the wall polyethyleneglycol-coated capillaries were used. Successful results were obtained for benzoin, chlorpheniramine, eperisone and tolperisone by adding hydroxypropylcellulose (HPC) and organic solvents or zwitterions to the separation solution.

Enzymes like cellulase isolated from *Trichoderma reesei* (cellobiohydrolase I) were also used [145]. Good results have been obtained only with high concentrations of enzyme (40 mg/ml) in a buffer containing a high ionic strength (0.4 mol sodium phosphate; pH 5.1) and 2-propanol (which also improved the peak shapes). Chiral pharmaceuticals such as propranolol, pindolol, metoprolol, alprenolol and *E*-/*Z*-labetolol were successfully analyzed. As cellobiose, an inhibitor of the enzyme, strongly impairs the enantioselectivity, it is evident that the active site is involved in chiral recognition. It has been stressed, however, that it is not necessary that the protein and the enantiomers migrate in different directions to avoid disruption of the UV detection; it may be sufficient for them to have different migration velocities [145].

Antibiotics

Chu and Whitesides [146] used the glycopeptide antibiotic vancomycin as chiral buffer additive for ACE to separate the enantiomers of two peptides (*N*-FMOC-gly-ala-ala and *N*-FMOC-gly-ala-ala-ala) at pH 7.1. More recently, Armstrong et al. [147] introduced a new class of chiral selectors, i.e. macrocyclic antibiotics. Analytical details were given on rifamycin B which was added to phosphate buffer (pH 7.0) including 2-propanol. Several aminoalcohols such as alprenolol, atenolol, bamethan, ephedrine, epinephrine, isoproterenol, metanephrine, metaproterenol, metoprolol, norepinephrine, normetanephrine, norphenylephrine, octopamine, oxprenolol, pseudoephedrine, salbutamol, synephrine and terbutaline could be separated. Several conditions were studied, e.g., different concentrations of phosphate buffer, 2-propanol, and sodium chloride.

3.7.3.3 Micellar electrokinetic capillary chromatography (MECC)

Chiral derivatization agents (CDA)

The separation of diastereomeric derivatives of chiral compounds has also been applied to micellar electrokinetic capillary chromatography. However, the use of only few chiral derivatization agents (CDA) has been described in the literature. Nishi et al. [148] derivatized proteinogenic amino acids (except asp and glu), as well as cystine and phenylglycine with 2,3,4,6-tetra-O-acetyl-ß-glucopyranosyl isothiocyanate (GITC). MECC has been performed with sodium dodecyl sulphate (SDS) micelles (pH 9.0); methanol has been added to suppress tailing. Further separated GITC-compounds were amphetamine, methamphetamine, ephedrine, pseudoephedrine, norephedrine, and norpseudoephedrine [149]. Another carbohydrate CDA was used by Tivesten et al. [150]; amino acids were converted with o-phthaldialdehyde and 2,3,4,6-tetra-O-acetyl-1-thio-β-D-glucopyranose to fluorescent diastereomers to be detected by laser-induced fluorescence (LIF).

Leopold [151] and Tran et al. [152] used Marfey's reagent as CDA (for the structure see Section 3.8) to derivatize amino acids (e.g., ala, asp, glu, leu, phe, trp, val) and peptides (e.g., ala-ala, ala-ala-ala). As MECC conditions SDS micelles at pH 3.3 or 8.5 were employed. Kang and Buck [153] described the reaction of amino acids with o-phthaldialdehyde and N-acetyl-*S*-cysteine or N-tert-butyloxycarbonyl-*S*-cysteine to form diastereomeric isoindole derivatives. Using methanol and SDS-containing buffer (pH 9.6) several racemic mixtures of amino acids were resolved successfully, i.e. ala, arg, gln, glu, his, ile, leu, lys, met, phe, trp, tyr, and val.

Amino acid derivatives

In 1987, the use of mixed chiral micelles formed by N,N-didecyl-*S*-alanine and SDS (comicellar system) was reported by Cohen et al. [15]. Dansylated amino acids (e.g., leu, met, thr) were separated. The buffer solution contained Cu(II)-ions and glycerol. Another comicellar system consisting of equimolar amounts of sodium N-dodecanoyl-*S*-valinate and sodium dodecylsulphate (SDS) was used by Dobashi et al. [154]. N-3,5-Dinitrobenzoyl derivatives of amino acid isopropyl esters (e.g., ala, leu, phe, val) were resolved at pH 7.0. With the same chiral selector (sodium N-dodecanoyl-*S*-valinate), but without SDS N-3,5-dinitrobenzoyl-, N-4-nitrobenzoyl- and N-benzoyl-derivatives of amino acid isopropyl esters (e.g., ala, leu, phe, val) were separated at pH 7.0 [151]. The influence of methanol has been investigated; slightly better results have been obtained in a methanol-free system [151]. The use of N-dodecanoyl-*S*-alaninate instead of the analogue valinate led to insufficient separation of N-3,5-dinitrobenzoyl derivatives of amino acid isopropyl esters (e.g., ala, leu, phe, val) [151].

The separation of several phenylthiohydantoin (PTH) amino acids (e.g., nleu, nval, trp) has been described; a buffer (pH 7.0) containing methanol and urea was used (improving the separation efficiency) [156]. Further separations have been performed [157], e.g., α-hydroxy-α-phenylacetophenone (benzoin) and 4-hydroxy-3-(3-oxo-1-phenylbutyl)coumarin (warfarin) at pH 9.0, including methanol and urea in the buffer.

Sodium N-dodecanoyl-*S*-glutamate showed similar separation characteristics [158]. The system has been applied for phenylthiohydantoin-amino acids (e.g., 2-aminobutanoic acid, met, nleu, nval, trp, val). The buffer contained SDS, urea and methanol (pH 9.0). Later, the *S*-serine analogue was introduced to separate some PTH-amino acids [159]. The *S*-aspartic acid homologue as well as sodium N-tetradecanoyl-*S*-glutamate were also investigated, but satisfactory results were not obtained [159].

Recently, N-dodecyloxycarbonyl-*S*-valine was described as a novel chiral surfactant. It showed higher enantioselectivity for ephedrine and related substances compared to N-dodecanoyl-*S*-valine [160]. Furthermore, N-dodecyloxycarbonyl-*S*-valine exhibited significantly less background absorbance in the low UV.

A novel chiral micelle polymer, poly(sodium N-undecylenyl-*S*-valinate), has been synthesized by Wang and Warner [161]. It has been used for separations of 1,1'-dinaphthyl-2,2'-diol and laudanosine.

Bile salts

Chiral bile salts as micelles for MECC were introduced by Terabe [3]. The most promising bile salt seems to be sodium taurodeoxycholate (STDC). It has been used for many application (cf. Table 3.7.3). Possible bimolecular aggregates of the bile salts have been discussed [162].

The separation of 1,1'-dinaphthyl-2,2'-dicarboxylic acid has been achieved at pH 4.7; no enantiomer separation was possible at pH 10.3 [163]. In some cases, e.g., laudanosoline [164], STDC showed enantioselectivity only at neutral pH conditions. Further investigations have been performed with the sodium bile salts cholate, taurocholate, and deoxycholate [165] (cf. Table 3.7.3).

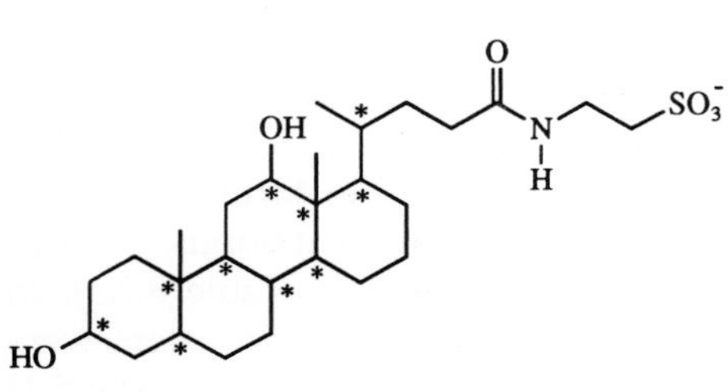

Taurodeoxycholate

Cyclodextrins and derivatives

Cyclodextrins were first applied by Nishi et al. [166]. These researchers used β-cyclodextrin for the separation of 2,2,2-trifluoro-1-(9-anthryl)ethanol and 1,1'-dinaphthyl-2,2'-diyl hydrogenphosphate with SDS micelles. Ueda et al. [167] derivatized amino acids with naphthalene-2,3-dialdehyde (NDA) in the presence of cyanide to form 1-cyano-2-substituted-benz[f]isoindoles (CBI) (cf. Table 3.7.3) for the separation, with β- or γ-cyclodextrin as additives. Additional achiral additives led to an improvement in resolution, e.g., of tetrabutylammonium salts [4], urea [78], and methanol [168]).

Heptakis(2,6-di-*O*-methyl)-β-CD as well as heptakis(2,3,6-tri-*O*-methyl)-β-CD were used by Nishi et al. [166], while mono(6-[2-aminoethylamino]-6-deoxy)-β-CD and the corresponding γ-CD were employed by Terabe [3] as well as Nishi and Terabe [169] (for applications see Table 3.7.3). α-CD, β-CD, heptakis(2,6-di-*O*-methyl)-β-CD, heptakis(2,3,6-tri-*O*-methyl)-β-CD and γ-CD were investigated by Nishi et al. [170]; γ-CD showed the highest enantioselectivity. Addition of methanol reduced the peak tailing of lipophilic compounds, but the enantioselectivity as well. It was only possible to separate the enantiomers of sodium thiopental and calcium pentobarbital by addition of (-)-camphor-10-sulphonate or (-)-menthoxyacetic acid to the micellar solution of SDS and γ-cyclodextrin. Tetraalkylammonium salts have been recommended for improving the separation of the pairs of amino acid isomers through ion pair formation between solutes and micelles [171].

A mixture of β- and γ-CD (molar ratio 5:1) together with SDS has been employed for the separation of dansylated amino acids, e.g., 2-aminobutanoic acid, asp, glu, leu, met, nleu, nval, phe, thr, and val [168]. In this case, the addition of 10 % methanol showed unfavourable results, i.e. broader peaks and higher migration times were observed.

With glucosyl-β-CD [168] almost identical results compared to those obtained with unbranched β-CD have been achieved. Linear maltoheptaose has also been used [168] for the separation of dansyl amino acids (e.g., leu, nleu, phe).

To overcome difficulties in reproducibility and, therefore, in peak assignment in chiral MECC, Sirén et al. [172] introduced the so-called migration indices for improved peak identification which are calculated in relation to two marker compounds, i.e. meso-2,3-diphenylsuccinic acid and triphenylacetic acid.

Mixtures of bile salts and cyclodextrins

A mixture of taurodeoxycholate and β-CD (molar ratio 5:2) to form a 'pseudo-stationary phase' was studied by Okafo et al. [173]. Several compounds, such as dansylated amino acids and pharmaceuticals were separated (cf. Table 3.7.3). In addition, CBI derivatives of amino acids and some phosphoric acid esters were resolved [174]. Finally, a taurodeoxycholate was mixed with γ-CD (molar ratio 5:1) [168] to separate dansylated amino acids.

Mixtures of cyclodextrins and other chiral additives

Not only bile salts or cyclodextrins have been used for chiral MECC. Nishi and Terabe [92] described the addition of (+)-camphor-10-sulphonate to a buffer mixture of SDS and γ-CD for the separation of barbiturates (cf. Table 3.7.3).

Table 3.7.3 Use of bile salts and cyclodextrin derivatives in micellar electrokinetic chromatography

Chiral auxiliary	Compounds	pH	Ref.
A) Bile salts			
SC	Carboline A and B, 2,2'-hydroxy-1,1'-dinaphthyl	7.0	165
	2,2'-Dihydroxy-1,1'-dinaphthyl	?	175
SDC	Carboline A and B, 2,2'-hydroxy-1,1'-dinaphthyl	7.0	165
	1,1'-Dinaphthyl-2,2'-diyl hydrogenphosphate	9.0	176
STC	Carboline A and B, 2,2'-hydroxy-1,1'-dinaphthyl	7.0	165
STDC	DNS-amino acids (met, nleu, nval, phe)	3.0	3
	Carboline A and B, 2,2'-dihydroxy-1,1'-dinaphthyl, diltiazem hydrochloride, 1,1'-naphthylethyl amine, trimetoquinol hydrochloride	7.0	165
	Deacetyl diltiazem	7.0	169
	1,1'-Dinaphthyl-2,2'-dicarboxylic acid	4.7	163
	Laudanosoline	7.0	164
	Diltiazem analogues, tetrahydropapaveroline	7.0	177
B) Cyclodextrins and derivatives			
α-CD/β-CD (1:2)	Alprenolol, atenolol, propranolol	*9.3	173
β-CD	CBI-amino acids (asp, ile, phe, thr, tyr)	*9.0	167
	DNS-amino acids (2-aminobutanoic acid, asp, glu, leu, met, nleu, nval, phe, ser, thr, trp, val)	*8.6	168
	Chlorpheniramine (TBA salt was added)	*7.5	4

Chiral auxiliary	Compounds	pH	Ref.
B) Cyclodextrins and derivatives			
	Chlorpheniramine (urea was added)	*3.0	78
	Cicletanine	*8.6	178
	1,1'-Dinaphthyl-2,2'-diyl hydrogenphosphate, 2,2,2-trifluoro-1-(9-anthryl)ethanol	*9.0	166
	4-Hydroxymephenytoin, 4-hydroxyphenytoin, mephenytoin	*9.1	179
	Juvenile hormone III bisepoxyde	*?	180
2,6-DM-β-CD	1,1'-Dinaphthyl-2,2'-diyl hydrogenphosphate	*9.0	166
	Diniconazole	*9.5	181
2,3,6-TM-β-CD	2,2'-Dihydroxy-1,1'-dinaphthyl[1], 1,1'-dinaphthyl-2,2'-diyl hydrogenphosphate[1]	*9.0	166
glucosyl-β-CD	DNS-amino acids (similar results as with β-CD)	*8.6	168
#maltoheptaose	DNS-amino acids (leu, nleu, phe)	*7.0	168
mono(6-AEAD-β-CD)	DNS-amino acids (glu, meth, nval, phe, ser, thr)	*7.0	3
γ-CD	CBI-amino acids (arg, leu, met, phe, ser, thr, tyr, val)	*9.0	167
	DNS-amino acids (2-amino butanoic acid, glu, leu, met, nleu, nval, phe, thr, val)	*8.3	182
	DNS-amino acids (ser, trp)[3]	*7.0	171
	Carprofen, flurbiprofen, tiaprofen	*9.0	184
	Cicletanine	*8.6	183
	Diniconazole, uniconazole and related compounds	*9.0	181
	Sodium thiopental[2], calcium pentobarbital[2]	*9.0	170
β-CD/γ-CD (5:1)	DNS-amino acids (2-aminobutanoic acid, asp, glu, leu, met, nleu, nval, phe, thr, val)	*8.6	168
C) Mixtures of bile salts and cyclodextrins			
STDC/β-CD (5:2)	DNS-amino acids (asp, glu, nval, thr, trp, val), mephenytoin, 4'-hydoxylated mephenytoin, fenoldopam, 4'-O-methylated fenoldopam	7.2	173
	DNS-amino acids (ala, arg, pro), CBI-amines and amino acids (arg, asp, glu, his, ile, met, phe, ser, trp, tyr, val), baclofen, phosphoric acid mono[3-amino-1-(4-chlorophenyl)] propyl ester, phosphoric acid mono(5-amino-3-phenyl)pentyl ester, phosphoric acid mono[6-amino-1-(4-trifluoromethylphenyl)]hexyl ester)	7.0	174
STDC/γ-CD (5:1)	DNS-amino acids (leu, met, nval, phe, thr, trp)	3.0	168
D) Mixtures cyclodextrins and (+)-camphor-10-sulphonate			
γ-CD (3:2)	Pentobarbital, thiopental	9.0	92

* = buffer contains SDS; # = linear molecule; 1 = buffer contains hydroxypropylcellulose (HEP); 2 = buffer contains (-)-camphor-10-sulphonate or (-)-menthoxyacetic acid; 3 = buffer contains tetraethylammonium bromide (TEAB); AEAD = mono(6-[2-aminoethylamino]-6-deoxy); CBI = 1-cyano-2-substituted-benz[f]isoindole; CD = cyclodextrin; DM = dimethyl; DNS = dansyl; SC = sodium cholate; SDC = sodium deoxycholate; STC = sodium taurocholate; STDC = sodium taurodeoxycholate. TBA = tetrabutylammonium; TM = trimethyl

Glucose derivatives

The application of MECC with in situ charged micelles like N-D-gluco-N-methyl-N-alkanamide boronates (alkanamide = octanoylamide, nonanoylamide and decanoylamide) or alkylglucoside boronates (alkyl = heptyl, octyl, nonyl and decyl) was recently reported [185,186]. With both types of surfactants dansylated amino acids were separated, using γ-cyclodextrin in the running buffer. The most recently introduced [187] similar chiral surfactants dodecyl β-D-glucopyranoside monophosphate and dodecyl β-D-glucopyranoside monosulphate were successfully used without cyclodextrin addition to resolve pharmaceuticals (e.g., cromakalin, ephedrine, fenoldopam, hexobarbital, hydroxymephenytoin, mephenytoin, metoprolol), dansylated amino acids, 1,1'-dinaphthyl-2,2'-diyl hydrogenphosphate, Tröger's base and some other compounds (phosphate/borate buffer, pH 8.0 at 25°C; 20 kV).

Digitonin

Otsuka and Terabe [188] introduced mixed chiral micelles formed by digitonin and SDS (at pH 3.0) for the resolution of several phenylthiohydantoin (PTH)-amino acids (e.g., 2-aminobutanoic acid, ala, nleu, nval, val). With SDS-free digitonin micelles (containing urea; pH 2.5) several PTH-amino acids have also been resolved (e.g,, aba, nval, val). Dansylated amino acids could not be separated with this micelle system [158].

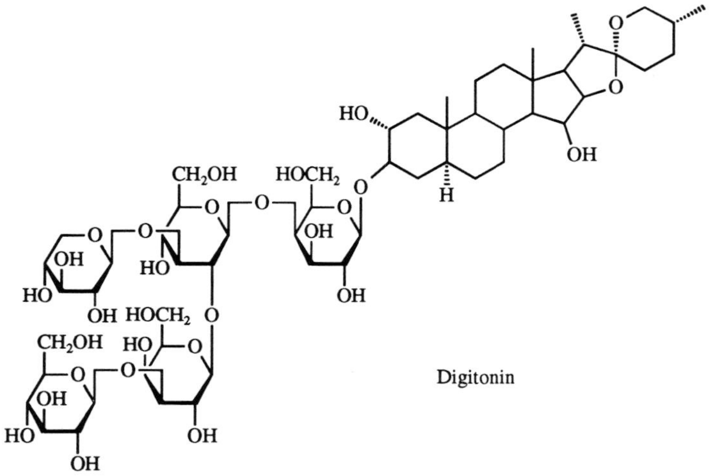

Digitonin

Tartaric acid esters

A different approach was taken by Aiken and Huie [189]. 2*R*,3*R*-Dibutyl tartrate was used together with SDS, tris(hydroxymethyl)aminomethane and 1-butanol (pH

8.1) to form microemulsions. The enantiomers of ephedrine were successfully separated. Microemulsions are clear, stable, monophasic mixtures of surfactant, cosurfactant, oil and water. The oil-in-water microemulsion systems show properties similar to micelles; hydrophobic compounds are solubilized, but with a much larger capacity. This high solubilization capacity permits the incorporation of a wide variety of hydrophobic additives, including lipophilic chiral selectors, into the core of the microemulsion oil droplet [189].

Saponins

Natural chiral triterpene glycosides (saponins) such as glycyrrhizic acid and β-escin have been applied in MECC by Ishihama and Terabe [190]. With a mixture of glycyrrhizic acid, octyl β-D-glucoside and SDS dansylated amino acids have been separated (e.g., leu, met, phe, ser, val) (pH 7.0). With a mixture of β-escin and SDS PTH-amino acids (e.g., ala, 2-aminobutanoic acid, met, nleu, nval, ser, thr, trp, val) have been resolved at pH 3.0.

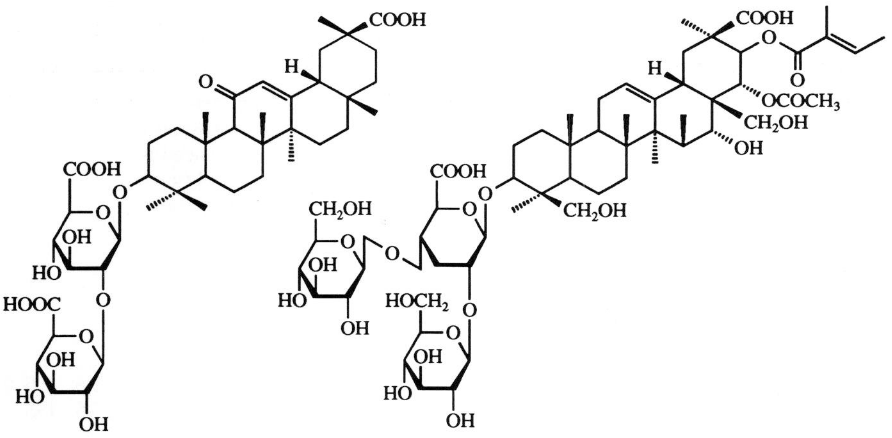

Glycyrrhizic acid

β-Escin

3.7.3.4 Capillary isotachophoresis (CITP)

In 1988, Snopek et al. [191] described for the first time a capillary isotachophoretic technique for resolving racemic mixtures of ephedrine alkaloids (cf. Table 4). Using a poly(tetrafluoroethylene) (PTFE) capillary and a conductivity detector, β-CD, heptakis(2,6-di-O-methyl)-β-CD or heptakis(2,3,6-tri-O-methyl)-β-CD was added to the leading electrolyte (LE). The LE consisted of 5 mmol sodium acetate and 0.2 % hydroxyethylcellulose (HEC) adjusted with acetic acid to pH 5.5. The terminating electrolyte (TE) contained 10 mmol β-alanine. The behaviour of cyclodextrin molecules is like that of a quasi-stationary phase, because they do not migrate during ITP. Underivatized cyclodextrins showed lower enantioselectivities [191].

Jelínek et al. [192] studied the influence of different counter-ions, e.g., acetate, pyridine-2-carboxylate and others. An increasing stability of the counter-ion inclusion complex resulted in a decreasing efficiency of the separation process. Acetate seemed to be the most promising counter-ion, because of its low affinity to β-CD and its methylated derivatives. Enantioseparations performed by means of CITP are summarized in Table 3.7.4. Snopek et al. [193] also applied a two-dimensional isotachophoresis system for the resolution of the enantiomers of 11-(3-dimethylaminopropyl)-6,11-dihydrodibenzo[b,e]thiepin (hydrothiadene).

Table 3.7.4 Use of cyclodextrins and derivatives in capillary isotachophoresis.

Cyclodextrin	Compounds	pH	Ref.
β-CD	*E*-2-Methylamino-1-phenyl-1-propanol (= pseudoephedrine), *E*-2-amino-1-phenyl-1-propanol (= norpseudoephedrine), *E*-1-acetoxy-2-methylamino-1-phenylpropan (= O-acetylpseudoephedrine), *E*-2-amino-1-(4-hydroxyphenyl)-1-propanol (= p-hydroxynorpseudoephedrine)	5.5	191
	9,10-Dihydro-4-(1-methyl-4-piperidylidene)-4*H*-benzo[4,5]cyclohepta-[1,2-b]thiophen-10-one (= ketotifen), 9,10-dihydro-4-hydroxy-10-methoxy-4-(1-methyl-4-piperidyl)-4*H*-benzo[4,5]cyclohepta[1,2-b]thiophene (II)	5.5	194
	Promethazine, cyamepromazine	5.6	195
2,6-DM-β-CD	Z-2-Methylamino-1-phenyl-1-propanol (= ephedrine), *E*-2-methylamino-1-phenyl-1-propanol (= pseudoephedrine), Z-2-amino-1-phenyl-1-propanol (= norephedrine), *E*-2-amino-1-phenyl-1-propanol (= norpseudoephedrine), *E*-1-acetoxy-2-methylamino-1-phenylpropan (= O-acetylpseudoephedrine), Z-2-amino-1-(4-hydroxyphenyl)-1-propanol (= p-hydroxynorephedrine), *E*-2-amino-1-(4-hydroxyphenyl)-1-propanol (= p-hydroxy-norpseudoephedrine)	5.5	191
2,3,6-TM-β-CD	*E*-2-Methylamino-1-phenyl-1-propanol (= pseudoephedrine), *E*-2-amino-1-phenyl-1-propanol (= norpseudoephedrine), *E*-2-amino-1-(4-hydroxyphenyl)-1-propanol (= p-hydroxynorpseudoephedrine)	5.5	191
γ-CD	Thioridazine	5.6	195
	11-(3-Dimethylaminopropyl)-6,11-dihydrodibenzo[b,e]thiepin (= hydrothiadene)	5.5	193

CD = cyclodextrin; DM = dimethyl; TM = trimethyl

3.7.3.5 Capillary electrochromatography (CEC)

In 1992, for the first time, Mayer and Schurig [196] introduced a wall-coated chiral stationary phase into electrochromatography, heptakis(2,3,6-tri-O-methyl)-β-cyclodextrin, which was covalently bonded to a polymethylsiloxane and immobilized to the capillary surface ('Chirasil-Dex'; for synthesis and structure cf. Section 3.5; for patent see ref. [197]). Racemic mixtures of 1-phenylethanol and 1,1'-dinaphthyl-2,2'-diyl hydrogenphosphate were separated at pH 7.0 using a 50 μm i.d. capillary. The separation was not dependent on the applied voltage. Since increasing voltage increases the number of theoretical plates, higher voltages have been preferentially used.

In comparison to electrophoretic methods which are applied by adding CDs to the buffer, the use of covalently bonded CDs shows important advantages; (i) no loss of CDs; and (ii) simultaneous masking of the silanol groups on the capillary surface by the immobilized polysiloxane backbone. Meanwhile, further separations have been published, i.e. cicloprofen and Tröger's base [198]; flurbiprofen and ibuprofen [199]; carprofen and etodolac [94]. The analog γ-CD phase has also been introduced [199], i.e. octakis(2,3,6-tri-O-methyl)-γ-CD covalently bonded to a polymethylsiloxane and immobilized to the capillary surface. Etodolac [199] and 1,1'-dinaphthyl-2,2'-diyl hydrogenphosphate [94] were successfully separated at pH 7.0.

Mayer et al. [110] applied 'Chirasil-Dex' with permethylated cyclodextrin containing buffers to investigate their influence on the enantioselectivity for hexobarbital. Peak inversion was observed when a trimethylenesulphonate derivatized β-CD as buffer additive was used for the separation of 1,1'-dinaphthyl-2,2'-diyl hydrogenphosphate. A dual chiral recognition system was proposed based on a so-called 'peak coalescence of the fourth kind' phenomena (see Section 3.3).

Most recently, Schurig et al. [200] demonstrated the so-called 'unified enantioselective chromatography' by the separation of hexobarbital enantiomers. A 'Chirasil-Dex' capillary column was successfully applied under four different chromatographic conditions: HRGC, HPLC, SFE, and CEC.

In 1993, Armstrong et al. [201] synthesized a different type of immobilized β-CD (for structure and details see Section 3.5) for the electrophoretic separation of chiral compounds like mephobarbitol (pH 7.8). With a γ-CD-coated fused-silica capillary epinephrine was resolved [202].

Li and Lloyd [203] used capillaries packed with α_1-acid glycoprotein (AGP) chiral stationary phase (see Section 3.4) for the separation of benzoin, hexobarbital, pentobarbital and some other pharmaceuticals (alprenolol, cyclophosphamide, diisopyramideifosfamide, metoprolol, oxprenolol). In addition, the influence of the pH-value, electrolyte concentration and different organic modifiers (methanol, ethanol, 1-propanol, 2-propanol, acetonitrile) was investigated. The same authors reported on capillaries packed with cyclodextrins, but only broad and tailing peaks were achieved [204]. In 1994, the same authors [205] employed a 50 μm i.d. fused-silica capillary packed with β-cyclodextrin with triethylammonium acetate in the running buffer. Several amino acid derivatives (e.g., 2,4-dinitrophenyl and dansyl) as well as benzoin and hexobarbital were resolved. Comparison to free-solution capillary electrophoresis (FSCE) showed that similar plate heights (HETP) could be achieved.

A new type of capillary coating, chitosan, was introduced by Sun et al. [206]. Chitosan, the partially deacetylated form of chitin, is a natural biopolymer composed primarily of D-glucosamine and N-acetyl-D-glucosamine. The cationic property of chitosan at pH below 6.5 created a reversed osmotic flow (EOF) in the capillary. It was applied for the separation of proteins, but also for chiral drugs like ephedrine and pindolol (methylated-β-cyclodextrin was added to the buffer).

3.7.3.6 Conclusions

In the following the influence of various parameters on the separation efficiencies in capillary electrophoretic techniques are summarized:
- pH (electroosmotic mobility of charged molecules is the lowest at the optimal pH; this pH value has to be determined empirically);
- ionic strength (in some cases, the higher the better);
- temperature (in most cases, the lower the better);
- composition and concentration of buffers;
- addition of achiral auxiliaries (e.g., methanol, urea, hydroxyethylcellulose, polyvinylalcohol, etc.; in many cases, separation was only achieved with one of these additives; some of them suppress electroosmosis);
- structure of chiral additives (e.g., alkaloids, bile salts, cyclodextrins, etc.);
- derivative type of chiral additives (e.g., native, methylated, hydroxypropylated, carboxymethylated cyclodextrin, etc.);
- quality of chiral additives (differences between the same products [e.g., cyclodextrin derivatives] of different manufacturers);
- concentration of chiral additives (e.g., for many cyclodextrin derivatives there is an optimal concentration; too high or too low concetrations lead to worse results);
- use of additional chiral auxiliaries (e.g., menthoxy acetic acid of camphor-10-sulphonic acid).

References

[1] Wren S.A.C. *J. Chromatogr.* (1993), *636*, 57

[2] Wren S.A.C., Rowe R.C., Payne R.S. *Electrophoresis* (1994), *15*, 774

[3] Terabe S. *Trends Anal. Chem.* (1989), *8*, 129

[4] Ong C.P., Ng C.L., Lee H.K., Li S.F.Y. *J. Chromatogr.* (1991), *588*, 335

[5] Grossman P.D., Colburn J.C. (Eds.) *Capillary Electrophoresis - Theory & Practice*, Academic Press: San Diego, London, 1992

[6] Kuhn R., Hoffstetter-Kuhn S. *Chromatographia* (1992), *34*, 505

[7] Snopek J., Jelínek I., Smolková-Keulemansová E. *J. Chromatogr.* (1992), *609*, 1

[8] Guzman N.A. (Ed.) *Chromatographic Science Series. Vol. 64. Capillary Electrophoresis Technology*, Dekker: New York, 1993

[9] Okafo G.N., Camilleri P. in: *Series: New Directions in Organic Biological Chemistry. Capillary Electrophoresis*, CRC Press: Boca Raton, 1993, p. 163

[10] Otsuka K., Terabe S. *Trends Anal. Chem.* (1993), *12*, 125

[11] Bereuter T.L. *LC-GC Int. (Eur. Ed.)* (1994), *7*, 78

[12] Isaaq H.J. *Instrumentation Sci. Technol.* (1994), *22*, 119

[13] Terabe S., Otsuka K., Nishi H. *J. Chromatogr. A* (1994), *666*, 295

[14] Vespalec R., Bocek P. *Electrophoresis* (1994), *15*, 755

[15] Cohen A.S., Paulus A., Karger B.L. *Chromatographia* (1987), *24*, 15

[16] Li S.F.Y. *Capillary Electrophoresis. J. Chromatogr.Libr. Vol. 52*, Elsevier: Amsterdam, 1992

[17] Vindevogel J., Sandra P. *Introduction to Micellar Electrokinetic Chromatography*, Hüthig: Heidelberg, 1992

[18] Weinberger R. *Practical Capillary Electrophoresis*, Academic Press: New York, 1993

[19] Altria K.D., Goodall D.M., Rogan M.M. *Electrophoresis* (1994), *15*, 824

[20] D'Hulst A., Verbeke N. *Electrophoresis* (1994), *15*, 854

[21] Ref. 1 taken from Ossicini, L., Celli C. *J. Chromatogr.* (1975), *115*, 655

[22] Mazzei M., Lederer M. *J. Chromatogr.* (1967), *31*, 196

[23] Yoneda H., Miura T. *Bull. Chem. Soc. Jap.* (1970), *43*, 574

[24] Yoneda H., Miura T. *Bull. Chem. Soc. Jap.* (1972), *45*, 2126

[25] Cardaci V., Ossicini L., Prosperi T. *Ann. Chimica* (1978), *68*, 713

[26] Fanali S., Lederer M., Masia P., Ossicini L. *J. Chromatogr.* (1988), *440*, 361

[27] Cardaci V., Ossicini L. *J. Chromatogr.* (1980), *198*, 76

[28] Fanali S., Cardaci V., Ossicini L. *J. Chromatogr.* (1983), *265*, 131

[29] Fanali S., Ossicini L., Prosperi T. *J. Chromatogr.* (1985), *318*, 440

[30] Fanali S., Masia P., Ossicini L. *J. Chromatogr.* (1987), *403*, 388

[31] Huynh T.K.X., Ossicini L., Polcaro C. *J. Chromatogr. A* (1994), *663*, 264

[32] Righetti P.G., Ettori C., Chafey P., Wahrmann J.P. *Electrophoresis* (1990), *11*, 1

[33] Nishizawa H., Nakajima K., Kobayashi M., Abe Y. *Anal. Sci.* (1991), *7*, 959

[34] Marsch G.A., Jankowiak R., Suh M., Small G.J. *Chem. Res. Toxicol.* (1994), *7*, 98

[35] Guttman A., Paulus A., Cohen A.S., Grinberg N., Karger B.L. *J. Chromatogr.* (1988), *448*, 41

[36] Cruzado I.D., Vigh G. *J. Chromatogr.* (1992), *608*, 421

[37] Cruzado I.D., Vigh G. in: *Minutes of the Sixth Int. Symp. on Cyclodextrins, Chicago 1992*, Hedges A.R. (Ed.), Editions de Santé: Paris, 1992, p. 594

[38] Schmitt T., Engelhardt H. in: *15th Int. Symp. on Capillary Chromatography, Riva del Garda 1993*, Sandra P. (Ed.), Hüthig: Heidelberg, 1993, p. 1506

[39] Schmitt T., Engelhardt H. *J. High Resol. Chromatogr.* (1993), *16*, 525

[40] Smith N. *Applied Biosystems Report* (1990), *8*, 1

[41] Guttman A., Cooke N. *J. Chromatogr. A* (1994), *685*, 155

[42] Schmitt T., Engelhardt H. *Chromatographia* (1993),*37*,475-481

[43] Sun P., Barker G.E., Hartwick R.A., Grinberg N., Kaliszan R. *J. Chromatogr.* (1993), *652*, 247

[44] Sun P., Wu N., Barker G., Hartwick R.A. *J. Chromatogr.* (1993), *648*, 475

[45] Birnbaum S., Nilsson S. *Anal. Chem.* (1992), *64*, 2872

[46] Vespalec R., Sustácek V., Bocek P. *J. Chromatogr.* (1993), *638*, 255

[47] Tanaka Y., Matsubara N., Terabe S. *Electrophoresis* (1994), *15*, 848

[48] Terabe S., Tanaka Y. in: *16th Int. Symp. on Capillary Chromatography, Riva del Garda 1994,* Sandra P., Devos G. (Eds.), Hüthig: Heidelberg, 1994, p. 1826

[49] Carpenter J., Goodall D.M., Robinson N.A., Wu Z., Wätzig H. *Personal communications,* 1992

[50] Rawjee Y.Y., Williams R.L., Vigh G. *Anal. Chem.* (1994), *66,* 3777

[51] Dette C., Wätzig H., Ebel S. *Scientia Pharm.* (1992), *60,* 188

[52] Dette C., Doctoral Thesis, University of Würzburg, 1994

[53] De Witt P., Deias R., Muck S., Galletti B., Meloni D., Celletti P., Marzo A. *J. Chromatogr. B* (1994), *657,* 67

[54] Schützner W., Fanali S., Rizzi A., Kenndler E. *J. Chromatogr.* (1993), *639,* 375

[55] Schützner W., Caponecchi G., Fanali S., Rizzi A., Kenndler E. *Electrophoresis* (1994), *15,* 769

[56] Gassmann E., Kuo J.E., Zare R.N. *Science* (1985), *230,* 813

[57] Gozel P., Gassmann E., Michelson H., Zare R.N. *Anal. Chem.* (1987), *59,* 44

[58] Gozel P., Zare R.N. *ASTM Spec. Tech. Publ. No. 1009,* (1988), 41

[59] Desiderio C., Aturki Z., Fanali S. *Electrophoresis* (1994), *15,* 864

[60] Fanali S., Ossicini L., Foret F., Bocek P. *J. Microcol. Separ.* (1989), *1,* 190

[61] Francotte E., Cherkaoui S., Faupel M. *Chirality* (1993), *5,* 516

[62] Tanaka M., Yoshinaga M., Asano S., Yamashoji Y., Kawaguchi Y. *Fresenius' Z. Anal. Chem.* (1992), *343,* 896

[63] Ref. 19 taken from [5]

[64] Nielen M.W.F. *J. Chromatogr.* (1993), *637,* 81

[65] Stefansson M., Novotny M. *J. Am. Chem. Soc.* (1993), *115,* 11573

[66] Nielen M.W.F. *Anal. Chem.* (1993), *65,* 885

[67] Shibukawa A., Lloyd D.K., Wainer I.W. *Chromatographia* (1993), *35,* 419

[68] Nardi A., Ossicini L., Fanali S. *Chirality* (1992), *4,* 56

[69] Nielen M.W.F. *Trends Anal. Chem.* (1993), *12,* 345

[70] Kuhn R., Stoecklin F., Erni F. *Chromatographia* (1992), *33,* 32

[71] Fanali S., Bocek P. *Electrophoresis* (1990), *11,* 757

[72] Tanaka M., Asano S., Yoshinago M., Kawaguchi Y., Tetsumi T., Shono T. *Fresenius' Z.. Anal. Chem.* (1991), *339,* 63

[73] Sepaniak M.J., Cole R.O., Clark B.K. *J. Liq. Chromatogr.* (1992), *15,* 1023

[74] Yamashoji Y., Ariga T., Asano S., Tanaka M. *Anal. Chim. Acta* (1992), *268,* 39

[75] Sydor W., Mularz E. presented at: *3rd Conference on HPCE, San Diego (CA), Febr. 1991*

[76] McLaughlin G.M., Nolan J.A., Lindahl J.L., Palmieri R.H., Anderson K.W., Morris S.C., Morrison J.A., Bronzert T.J. *J. Liq. Chromatogr.* (1992), *15,* 961

[77] Heuermann M., Blaschke G. *J. Chromatogr.* (1993), *648,* 267

[78] Otsuka K., Terabe S. *J. Liq. Chromatogr.* (1993), *16,* 945

[79] Altria K.D., Goodall D.M., Rogan M.M. *Chromatographia* (1992), *34,* 19

[80] Rawjee Y.Y., Staerk D.U., Vigh G. *J. Chromatogr.* (1993), *635*, 291

[81] Smith N.W. *J. Chromatogr.* (1993), *652*, 259

[82] Rawjee Y.Y., Williams R.L., Vigh G. *J. Chromatogr. A* (1993), *652*, 233

[83] Aumatell A., Wells R.J. *J. Chromatogr. Sci.* (1993), *31*, 502

[84] Schützner W., Fanali S. *Electrophoresis* (1992), *13*, 687

[85] Snopek J., Soini H., Novotnu M., Smolková-Keulemansová E., Jelínek I. *J. Chromatogr.* (1991), *559*, 215

[86] Schmitt P., Kettrup A. *GIT Fachz. Lab.* (1994),1312

[87] Peterson T.E. *J. Chromatogr.* (1993), *630*, 353

[88] Fanali S. *J. Chromatogr.* (1991), *545*, 437

[89] Heuermann M., Blaschke G. presented at: *4th Int. Symp. on HPCE, Amsterdam 1992*

[90] Lurie I.S., Klein R.F.X., Dal Cason T.A., LeBelle M.J., Brenneisen R., Weinberger R.E. *Anal. Chem.* (1994), *66*, 4019

[91] Soini H., Snopek J., Novotny M.V. *Technical Information Beckman Instruments* (Fullerton, USA), DS-836, 1992

[92] Nishi H., Terabe S. *J. Pharm. Biomed. Anal.* (1993), *11*, 1277

[93] Rogan M.M., Drake C., Goodall D.M., Altria K.D. *Anal. Biochem.* (1993), *208*, 343

[94] Meyer S., Schurig V. *Electrophoresis* (1994), *15*, 835

[95] Fanali S. *J. Chromatogr.* (1989), *474*, 441

[96] Fanali S., Flieger M., Steinerova N., Nardi A. *Electrophoresis* (1992), *13*, 39

[97] Altria K.D. *Chromatographia* (1993), *35*, 177

[98] Soini H., Riekkola M.-L., Novotny M.V. *J. Chromatogr.* (1992), *608*, 265

[99] Wren S.A.C., Rowe R.C. *J. Chromatogr.* (1992), *603*, 235

[100] Wren S.A.C., Rowe R.C. *J. Chromatogr.* (1993), *635*, 113

[101] Gareil P., Gramond J.P., Guyon F. *J. Chromatogr.* (1993), *615*, 317

[102] Pluym A., Van Ael W., De Smet M. *Trends Anal. Chem.* (1992), *11*, 27

[103] Altria K.D., Harden R.C., Hart M., Hevizi J., Hailey P.A., Makwana J.V., Portsmouth M.J. *J. Chromatogr.* (1993), *641*, 147

[104] Heuermann M., Blaschke G. *Pharmazie* (1993), *48*, 74

[105] Ref. 87 taken from [12]

[106] Rawjee Y.Y., Vigh G. *Anal. Chem.* (1994), *66*, 619

[107] Valkó I.E., Billiet H.A.H., Frank J., Luyben K.C.A.M. *J. Chromatogr.A* (1994), *678*, 139

[108] Nardi A., Eliseev A., Bocek P., Fanali S. *J. Chromatogr.* (1993), *638*, 247

[109] Penn S.G., Goodall D.M., Loran J.S. *J. Chromatogr.* (1993), *636*, 149

[110] Mayer S., Schleimer M., Schurig V. *J. Microcol. Separ.* (1994), *6*, 43

[111] Tait RJ., Thompson D.O., Stella V.J., Stobaugh J.F. *Anal. Chem.* (1994), *66*, 4013

[112] Chankvetadze B., Endresz G., Blaschke G. *Electrophoresis* (1994), *15*, 804

[113] Valkó I.E., Billiet H.A.H., Frank J., Luyben K.C.A.M. *Chromatographia* (1994), *38*, 730

[114] Quang C., Khaledi M.G. *J. High Resol. Chromatogr.* (1994), *17*, 99

[115] Belder D., Schomburg G. *J. High Resol. Chromatogr.* (1992), *15*, 686

[116] St.Pierre L.A., Sentell K.B. *J. Chromatogr. B* (1994), *657*, 291

[117] Guttman A., Cooke N. *J. Chromatogr. A* (1994), *680*, 157

[118] Birrell H.C., Camilleri P., Okafo G.N. *J. Chem. Soc. Chem. Commun.* (1994), 43

[119] Yoshinaga M., Tanaka M. *J. Chromatogr. A* (1994), *679*, 359

[120] Belder D., Schomburg G. *J. Chromatogr. A* (1994), *666*, 351

[121] Quang C., Khaledi M.G. *Anal. Chem.* (1993), *65*, 3354

[122] Ruyters H., van der Wal S. *J. Liq. Chromatogr.* (1994), *17*, 1883

[123] Tait R.J., Tan P., Thompson D.O., Stella V.J., Stobaugh J.F. presented at: *5th Int. Symp. on HPCE, Orlando (FL), Jan 25-28, 1993* Poster

[124] Sun P., Barker G.E., Mariano G.J., Hartwick R.A. *Electrophoresis* (1994), *15*, 793

[125] Nishi H., Nakamura K., Nakai H., Sato T. *J. Chromatogr. A* (1994), *678*, 333

[126] Aturki Z., Fanali S. *J. Chromatogr. A* (1994), *680*, 137

[127] D'Hulst A., Verbeke N. *J. Chromatogr.* (1992), *608*, 275

[128] Soini H., Stefansson M., Riekkola M.-L., Novotny M.V. *Anal. Chem.* (1994), *66*, 3477

[129] Stefansson M., Novotny M. presented at: *5th Int. Symp. on HPCE, Orlando (FL), Jan 25-28, 1993* Poster T134

[130] Nishi H., Nakamura K., Nakai H., Sato T., Terabe S. *Electrophoresis* (1994), *15*, 1335

[131] Soini H., Riekkola M.-L., Novotny M.V. *J. Chromatogr. A* (1994), *680*, 623

[132] Stalcup A.M., Agyel N.M. *Anal. Chem.* (1994), *66*, 3054

[133] Kuhn R., Erni R., Bereuter T., Häusler J. *Anal. Chem.* (1992), *64*, 2815

[134] Höhne E., Krauss G.-J.Gübitz G. *J. High Resol. Chromatogr.* (1992), *15*, 698

[135] Kuhn R., Wagner J., Walbroehl Y., Bereuter T. *Electrophoresis* (1994), *15*, 828

[136] Walbroehl Y., Wagner J. *J. Chromatogr. A* (1994), *680*, 253

[137] Kuhn R., Steinmetz C., Bereuter T., Haas P., Erni F. *J. Chromatogr. A* (1994), *666*, 367

[138] Walbroehl Y., Wagner J. *J. Chromatogr. A* (1994), *685*, 321

[139] Kuhn R., Bereuter T., Erni F. in: *Book of Abstracts. 3rd Int. Symp. on Chiral Discrimination, Tübingen 1992* Schurig V. (Ed.) Univ. Tübingen: 1992, Abstr. Nr. 94

[140] Barker G.E., Russo P., Hartwick R.A. *Anal. Chem.* (1992), *64*, 3024

[141] Busch S., Kraak J.C., Poppe H. *J. Chromatogr.* (1993), *635*, 119

[142] Arai T., Ichinose M., Kuroda H., Nimura N., Kinoshita T. *Anal. Biochem.* (1994), *217*, 7

[143] Yang J., Hage D.S. *Anal. Chem.* (1994), *66*, 2719

[144] Ishihama Y., Oda Y., Asakawa N., Yoshida Y., Sato T. *J. Chromatogr. A* (1994), *666*, 193

[145] Valtcheva L., Mohammed J., Pettersson G., Hjertén S. *J. Chromatogr.* (1993), *638*, 263

[146] Chu Y.-H., Whitesides G.M. *J. Org. Chem.* (1992), *57*, 3524

[147] Armstrong D.W., Rundlett K., Reid G.L. *Anal. Chem.* (1994), *66*, 1690

[148] Nishi H., Fukuyama T., Matsuo M. *J. Microcol. Separ.* (1990), 2, 234

[149] Lurie I.S. *J. Chromatogr.* (1992), *605*, 269

[150] Tivesten A., Alatalo A.-M., Holgersson H., Folestad S. in: *15th Int. Symp. on Capillary Chromatography, Riva del Garda 1993*, Sandra P. (Ed.), Hüthig: Heidelberg, 1993, p. 1453

[151] Ref. 82 taken from [15]

[152] Tran A.D., Blanc T. *J. Chromatogr.* (1990), *516*, 241

[153] Kang L., Buck R.H. *Amino Acids* (1992), 2, 103

[154] Dobashi A., Ono T., Hara S. *Anal. Chem.* (1989), *61*, 1984

[155] Dobashi A., Ono T., Hara S., Yamaguchi J. *J. Chromatogr.* (1989), *480*, 413

[156] Otsuka K., Terabe S. *Electrophoresis* (1990), *11*, 982

[157] Otsuka K., Terabe S. in: *12th Int. Symp. on Capillary Chromatography, Kobe, Japan 1990*, Jinno K., Sandra P. (Eds.), Ind. Publ. Consult.: Tokyo, 1990, p. 646

[158] Otsuka K., Kashihara M., Kawaguchi Y., Koike R., Hisamitsu T., Terabe S. *J. Chromatogr. A* (1993), *652*, 253

[159] Otsuka K., Karuhaka K., Higashimori M., Terabe S. *J. Chromatogr. A* (1994), *680*, 317

[160] Mazzeo J.R., Grover E.R., Swartz M.E., Peterson J.S. *J. Chromatogr. A* (1994), *680*, 125

[161] Wang J., Warner I.M. *Anal. Chem.* (1994), *66*, 3773

[162] Terabe S., Shibata M., Miyashita Y. *J. Chromatogr.* (1989), *480*, 403

[163] Cole R.O., Sepaniak J., Hinze W.L. *J. High Resol. Chromatogr.* (1990), *13*, 579

[164] Nishi H., Fukuyama T., Matsuo M. *Anal. Chim. Acta* (1990), *236*, 281

[165] Nishi H., Fukuyama T., Matsuo M., Terabe S. *J. Microcol. Separ.* (1989), *1*, 234

[166] Nishi H., Fukuyama T., Terabe S. *J. Chromatogr.* (1991), *553*, 503

[167] Ueda T., Kitamura F., Mitchell R., Metcalf T., Kuwana T., Nakamoto A. *Anal. Chem.* (1991), *63*, 2979

[168] Terabe S., Miyashita Y., Ishihama Y., Shibata O. *J. Chromatogr.* (1993), *636*, 47

[169] Nishi H., Terabe S. *Electrophoresis* (1990), *11*, 691

[170] Nishi H., Fujimura N., Yamaguchi H., Fukuyama T., Matsuo M., Terabe S. in: *12th Int. Symp. on Capillary Chromatography, Kobe, Japan 1990* Jinno K., Sandra P. (Eds.), Ind. Publ. Consult.: Tokyo, 1990, p. 640

[171] Trisciani A. in: *13th Int. Symp. on Capillary Chromatography, Riva del Garda 1991*, Sandra P. (Ed.), Hüthig: Heidelberg, 1991, p. 1703

[172] Sirén Hx., Jumppanen J.H., Manninen K., Riekkola M.-L. *Electrophoresis* (1994), *15*, 779

[173] Okafo G.N., Bintz C., Clarke S.E., Camilleri P. *J. Chem. Soc. Chem. Commun.* (1992), 1189

[174] Okafo G.N., Camilleri P. *J. Microcol. Separ.* (1993), *5*, 149

[175] Cole R.O., Sepaniak M.J. *LC-GC Int. (US Ed.)* (1992), *10*, 380

[176] Cole R.O., Sepaniak M.J., Hinze W.L. in: *11th Int. Symp. on Capillary Chromatography, Monterey 1990*, Sandra P., Redant G. (Eds.), Hüthig: Heidelberg, 1990, p. 890
[177] Nishi H., Fukuyama T., Matsuo M. *J. Chromatogr.* (1990), *515*, 233
[178] Pruñonosa J., Diez Gascon A., Gouesclou L. *Application Brief Beckman Instruments* (Fullerton, USA), DS-798, 1991
[179] Desiderio C., Fanali S., Küpfer A., Thormann W. *Electrophoresis* (1994), *15*, 87
[180] Herit A.J., Rickards R.W., Thomas R.D., East P.D. *J. Chem. Soc. Chem. Commun.* (1993), 1497
[181] Furuta R., Doi T. *Electrophoresis* (1994), *15*, 1322
[182] Miyashita Y., Terabe S. *Applications Data Beckman Instruments* (Fullerton, USA), DS-767, 1990
[183] Pruñonosa J., Obach R., Diez-Cascón A., Gouesclou L. *J. Chromatogr.* (1992), *574*, 127
[184] Karger A.E., Stoll E., Hänsel W. *Pharmazie* (1994), *49*, 155
[185] Smith J.T., Nashabeh W., El Rassi Z. *Anal. Chem.* (1994), *66*, 1119
[186] Smith J.T., El Rassi Z. *Electrophoresis* (1994), *15*, 1248
[187] Tickle D.C., Okafo G.N., Camilleri P., Jones R.F.D., Kirby A.J. *Anal. Chem.* (1994), *66*, 4124
[188] Otsuka K., Terabe S. *J. Chromatogr.* (1990), *515*, 221
[189] Aiken J.H., Huie C.W. *Chromatographia* (1993), *35*, 448
[190] Ishihama Y., Terabe S. *J. Liq. Chromatogr.* (1993), *16*, 933
[191] Snopek J., Jelínek I., Smolková-Keulemansová E. *J. Chromatogr.* (1988), *438*, 211
[192] Jelínek I., Snopek J., Smolková-Keulemansová E. *J. Chromatogr.* (1991), *557*, 215
[193] Snopek J., Jelínek I., Smolková-Keulemansová E. *J. Chromatogr.* (1989), *472*, 308
[194] Jelínek I., Snopek J., Smolková-Keulemansová E. *J. Chromatogr.* (1988), *439*, 386
[195] Jelínek I., Dohnal J., Snopek J., Smolková-Keulemansová E. *J. Chromatogr.* (1989), *464*, 139
[196] Mayer S., Schurig V. *J. High Resol. Chromatogr.* (1992), *15*, 129
[197] Schurig V. DE 41 36 462 (Cl. C07B57/00), 6 May 1993
[198] Jung M., Mayer S., Schleimer M., Schurig V. *GIT Fachz. Lab. (Chromatographie 1/93)* (1993), 18
[199] Mayer S., Schurig V. *J. Liq. Chromatogr.* (1993), *16*, 915
[200] Schurig V., Jung M., Mayer S., Negura S., Fluck M., Jakubetz H. *Angew. Chem.* (1993), *106*, 2265
[201] Armstrong D.W., Tang Y., Ward T., Nichols M. *Anal. Chem.* (1993), *65*, 1114
[202] Szemán J., Ganzler K. *J. Chromatogr. A* (1994), *668*, 509
[203] Li S., Lloyd K. *Anal. Chem.* (1993), *65*, 3684
[204] Li S., Lloyd D.K. presented at: *5th Int. Symp. on HPCE, Orlando (FL), Jan 25-28, 1993* Poster W208
[205] Li S., Lloyd D.K. *J. Chromatogr. A* (1994), *666*, 321
[206] Sun P., Landman A., Hartwick R.A. *J. Microcol. Separ.* (1994), *6*, 403

3.8 Planar Chromatography

Chromatographic methods can be divided up into two main classes. First, column chromatographic methods with gases or liquids as the mobile phase (GC, gas chromatography; LC, liquid chromatography; EKC, electrokinetic chromatography; cf. Sections 3.5, 3.4, and 3.7, respectively). Secondly, planar chromatographic methods with liquids as the mobile phase (PC, paper chromatography; TLC, thin-layer chromatography). Finally, there are electrophoretic methods applied on layers of paper, which belong to the second main class of chromatographic methods, but they are discussed in Section 3.7.

The following section deals with the planar chromatographic methods PC and TLC. Only information about the stationary phase materials as well as the components of the mobile phases and their concentrations will be given; no details on the separation procedure, e.g., dimensions of chambers or plates, etc. will be provided. In some cases, however, the working temperature or the use of a special technique will be mentioned.

The visualization of the separated compounds (spots) is generally performed by UV/VIS methods (photometric detection; fluorescence detection etc.). One possibility is post-chromatographic reaction with the solution of a derivatization agent (e.g., ninhydrin for amino acids), which can be applied by spraying onto the plate or by immersing the plate into the solution. Another possibility is the pre-chromatographic derivatization of the compounds (e.g., dansylation for amino acids). While for qualitative analyses a visual monitoring under UV or visible light is sufficient, for quantitative analyses more accurate and automated methods such as densitometric scanning with variable or multiple wavelength are preferred. In planar chromatography there are three fundamental techniques for the enantioseparation of chiral compounds:

(i) Separation on achiral stationary phases (layers) via diastereomeric derivatives formed by reaction of the sample components with a chiral derivatizing agent (CDA).

(ii) Separation on achiral stationary phases (layers) through diastereomeric complexes via addition of chiral mobile phase additives (CMA) formed by the sample components with the additive. A special case of this method is the addition of chiral counter-ions (CCI) to the mobile phase.

(iii) Separation on chiral stationary phases (layers) via diastereomeric association complexes formed by the sample components with the stationary phase.

3.8.1 Fundamentals

To facilitate understanding of the terminology to be used later, a brief introduction to the calculation of planar chromatographic data will be given, i.e. important equations for the separation of enantiomers, and some principles (cf. Sections 3.3 and 3.5). The spot migration is shown in Figure 3.8.1 [1].

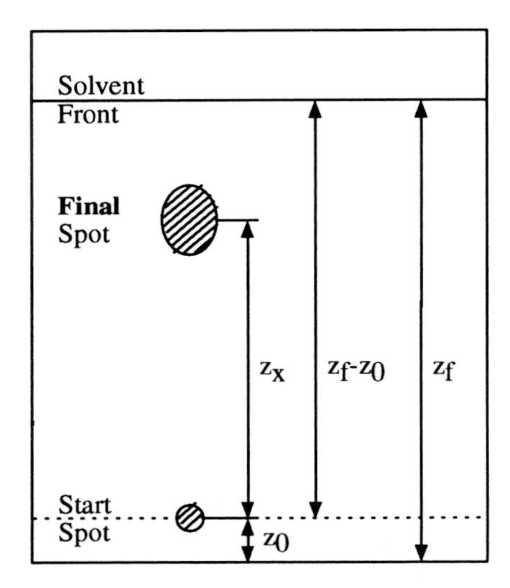

Figure 3.8.1 Spot migration in planar chromatography [1].

The R_f-value (1), the term most used in planar chromatography, is defined as the ratio of the length of the spot migration (= distance between the starting point and most intensive point of a substance spot) and the length from the starting point to the solvent front (all distances are in cm):

(1) $R_f = z_x/(z_f - z_0)$ [dimensionless]

The ΔR_f-value (2) can be used to describe the quality of an enantiomer separation:

(2) $\Delta R_f = R_{f2} - R_{f1}$ [dimensionless]
R_{f1} = R_f-value of the more slowly migrating enantiomer
R_{f2} = R_f-value of the faster migrating enantiomer

The efficiency can be described by the 'height equivalent to a theoretical plate' (HETP) and the total number of plates or the number of plates over the migration distance:

(3a) $H = w_b^2/16z_x$ [cm]
w_b = length of a spot [cm]
(3b) $H = w_{0.5}^2/5.541z_x$ [cm]
$w_{0.5}$ = peak width at halfheight [cm]
(e.g., after densitometric scanning)

(4a) $N = (z_f - z_0)/H$ [dimensionless]
(4b) $N' = (z_f - z_0)N$ [dimensionless]
(4c) $N' = R_f N$ [dimensionless]

The partition coefficient k can be expressed as:

(5) $k = (1-R_f)/R_f$ [dimensionless]

The relative retention α is the ratio of the partition coefficients k_1/k_2 of two neighboring compounds:

(6a) $\alpha = k_2/k_1$ [dimensionless]
(6b) $\alpha = [(1-R_{f2})/R_{f2}]/[(1-R_{f1})/R_{f1}]$ [dimensionless]

Another term, the resolution R_s, is much more important. For the minimal spot separation $R_s = 1$:

(7a) $R_s = 2[(z_{x1}-z_{x2})/(w_{b1}+w_{b2})]$ [dimensionless]
z_{x1} = migration length of the faster migrating enantiomer [cm]
z_{x2} = migration length of the more slowly migrating enantiomer [cm]
$w_{b1}+w_{b2}$ = sum of lengths of the spots of both enantiomers [cm]
(7b) $R_s = 1.18[(z_{x1}-z_{x2})/(w_{h1}+w_{h2})]$ [dimensionless]
$w_{h1}+w_{h2}$ = sum of the peak widths at half height of both enantiomers (e.g. after densitometric scanning) [cm].

Most recently, a new optical topological index for predicting the separation of optical isomers by TLC was introduced. For details see Ref. [2].

3.8.2 *Paper chromatography*

Paper chromatography (PC) is the traditional chromatographic method. It is a particular type of partition chromatography. Normal paper is mainly made of cellulose, a chiral, optically pure compound, which is a linear polysaccharide with 1,4-β-linked D-glucose units. Due to its natural chirality, it promises enantioselectivity for the analysis of optically active compounds. Meanwhile, PC methods are disappearing more and more due to the advantages of TLC-methods (e.g., shorter migration times, higher number of theoretical plates, better separation efficiency, etc.), but viewed historically, PC made TLC possible. Therefore, important highlights of the PC history are represented in this part.

Unsuccessful attempts to separate enantiomers applying PC were reported by several authors. Flood et al. in 1948 [3], for instance, tried to separate DL-arabinose by adding (-)-menthol or (-)-2-pentanol to the mobile phase. Hinman et al. in 1950 [4] were able to separate the diastereomers of the dipeptide valylvaline but not the enantiomers, using an achiral mobile phase. The first separations of enantiomers by

PC were carried out independently by several investigators in 1951 (Table 3.8.1). Dalgliesh postulated in 1952 [14] the 'three-point attachment model' for successful racemate separations, which was adapted from a former model for enzymes and receptors [15,16] (cf. Section 3.4.3). Normally, the separation principle of PC is partition chromatography, but the chiral recognition mechanism of cellulose is also due to adsorption effects (adsorption chromatography) [17].

Table 3.8.1 First published enantiomer separations using paper chromatography.

Compound	Mobile phase	Author	Rf.
2-Naphthylbenzylamine	(+)-Tartaric acid	Bonino/Carassiti	5
Kynurenine (**1**)	[not reported]	Mason/Berg	6
Kynurenine, kynureninesul-phate-der., tryptophan, tyrosine-3-sulphonic acid	Lutidine	Fujisawa	7
Glutamic acid, tyrosine, tyrosine-3-sulphonic acid	(-)-2-(N-Methylamino)-1-phenylpropane (-)-2-(N-Methylamino)-1-phenylpropane; etc.	Sakan et al. and Kotake et al.	8 9
Kynurenine (**1**)	1-Butanol; etc.	Sensi	10
N-α-Acetylkynurenine	1-Butanol-acetic acid	Dalgliesh et al.	11
Ephedrine hydrochloride	1-Butanol-water-acetic acid	Kariyone/Hashimoto	12
*2-Aminophenylacetic acid	Phenol saturated water	Berlingozzi et al.	13

* = Paper was impregnated with (+)-10-camphorsulphonic acid

In Table 3.8.2 an overview of successful enantiomer separations is given. For more detailed reviews see Ref. [17,18]; a comprehensive work on amino acid derivatives was reported by Weichert [19].

Table 3.8.2 Enantiomer separations using paper chromatography.

Compound class	Compound
Alkaloids	Canadine methoiodide [20]; 8-hydroxylaudanosolin [21]; #lysergic acid, #isolysergic acid, isolysergic acid as 2-propylamide [22]; laurifoline iodide, magnoflorine iodide [23]; 1,2,5,6-tetrahydroxyaporphyrine [21]; 2,3,5,6-tetrahydroxyaporphyrine methobromide [24]
Amines	2-Naphthylbenzylamine [5]
Amino acids	**Ala-der. [19]; ##ala.HCl [25]; 3-(3-amino-1-naphthyl)-ala, 3-(4-amino-1-naphthyl)-ala [26]; 2-aminophenylacetic acid [13]; cystathionine [27]; cys [28]; α,ε-diaminopimelic acid [29]; DOPA [19]; glu[8 ;9]; his [30]; **his-der. [19]; kynurenine [6;7;10]; kynurenine-der. [19]; N-α-acetylkynurenine [11]; 5-hydroxykynurenine [31]; kynurenine-sulphate-der. [7]; ##phe.HCl [25]; phe-der. [19];

Compound class	Compound
	2,3-, 2,4-, 2,5-, 2,6-, 3,5-dihydroxy-phe [32]; ser-der. [19]; trp [7]; ##trp [25]; **trp-der. [19]; α-hydroxy-trp [33]; 5-, 6-hydroxy-trp [34]; *6-methyl-trp, *7-methyl-trp [35]; tyr [8;9]; ##tyr [25]; tyr-3-sulphonic acid [8;9]
Flavonoids	Catechin, epicatechin; epigallocatechin, gallocatechin [36;37]
Pharmaceuticals	Adrenaline hydrochloride [38]; ephedrine [12]; isopropyl adrenaline hydrochloride, noradrenaline hydrochloride, pervitin hydrochloride, sympamine sulphate [38]
Pigments	Ommatin D [39]; rhodommatin [40]

* = native cellulose: Whatman No. 3MM paper (Whatman Scientific, Maidstone, Kent, UK); in most of the other cases a Whatman No. 1 paper was used; ** = more than 40 different chiral amino acid derivatives were separated; # = as diastereomeric 2-butylamides after deivatization with *S*-(+)-2-butanol; ## = as menthyl esters after derivatization with 1*R*,3*R*,4*S*-(-)-menthol.

A special type of chiral PC employs ion-exchange paper [41]. This paper was produced by dipping a Whatman No.1 paper into a suspension of diluted silica gel, followed by drying and dipping into a solution of sodium alginate. After ascending development using pyridine/water/1-pentanol/$Na_2HPO_4 \cdot 7H_2O$ (0.5%) (7:7:2:6) as eluent, it was possible to separate several racemic α-amino acids into their enantiomers.

3.8.3 *Thin-layer chromatography*

Since 1958, thin-layer chromatography (TLC) is a commonly used method [42]. Like PC, TLC is a partition chromatographic method, but adsorption effects also influence the separation.

One of the prerequisites was the use of materials with a very fine particle size for the production of thin-layers [42]. An important advantage of TLC was the possibility to apply other materials than cellulose, such as silica gel, aluminium oxide, reversed-phase materials, etc. Its main advantage is the simultaneous analysis of 30 to 60 samples per plate (HPTLC). Simultaneous analysis leads to better reproducibility and a higher precision than subsequent analysis. Furthermore, TLC methods are usually fast, the equipment is much cheaper than that required for HRGC, HPLC or CZE, and some chiral plates can be prepared easily in any laboratory. However, there are also disadvantages, i.e. derivatization is often necessary with almost all TLC-techniques. In addition, with complex multicomponent mixtures only HRGC, HPLC or CZE deliver sufficient results.

For the various TLC techniques different abbreviations exist [43]. Typical abbreviations used in the literature are:

CLC	Centrifugal Layer Chromatography
FFPLC	Forced-Flow Planar Liquid Chromatography
FFTLC	Forced-Flow Thin-Layer Chromatography
HPPC	High Performance Planar Chromatography
HPPLC	High Performance Planar Liquid Chromatography
HPTLC	High Performance Thin-Layer Chromatography
OPLC	Over-Pressured Layer Chromatography
OPPC	Over-Pressured Planar Chromatography
OPTLC	Over-Pressured Thin-Layer Chromatography
PLC	Planar Liquid Chromatography
RPC	Rotation Planar Chromatography
TLC	Thin-Layer Chromatography

In the beginning, PC methods were transferred to TLC. The first enantiomer separation performed by using TLC was by Contractor and Wragg in 1965 [34]. These authors separated tryptophan and 5- and 6-hydroxytryptophan on a cellulose thin-layer. The compounds had already been resolved by PC [7; 34]. In the last eighteen years many different approaches to the separation of racemic mixtures have been developed. Almost all enantiomer separations have been performed by linear chromatography (TLC and HPTLC). One of the main differences between TLC and HPTLC is the particle size (TLC: 10-20 µm; HPTLC: 3-6 µm); in HPTLC the separation efficiency is improved. In only few cases a circular chromatographic method (RPC) has been applied. For reviews see Ref. [18; 44-47]. In the applications mentioned in the following, almost all of the eluents are given in volume per volume (v/v).

3.8.3.1 Chiral derivatization agents (CDA)

Enantiomer separation via diastereomeric derivatives by TLC was first reported by Nitecki et al. in 1967 [48]. These authors separated several diastereomeric 2,5-diketopiperazines which were synthesized from tert-butyl-oxycarbonyl-*S*-dipeptide methyl esters using silica gel plates with diisopropyl ether-$CHCl_3$-AcOH (6:3:1) as eluent [48]. This technique can be used for the determination of the enantiomeric purity of dipeptides formed from a racemic amino acid and a non-racemic amino acid.

Generally, most of the chiral derivatization agents (CDA) already described for GC or HPLC (for details see Sections 3.5 and 3.4, respectively) can also be used in TLC. In Table 3.8.3 several CDAs used for thin-layer chromatography and various applications are summarized.

Table 3.8.3 Chiral derivatization agents (CDA) used in thin-layer chromatographic analysis.

CDA	Deriv.	Compound	Plate	Eluent
Alkanols				
S-(-)-1-Phenylethanol [49]* [CDA-1]	Esters	Cyclophosphamide [49]	SiO_2	$CHCl_3$-EtOH (9:1)
Amines				
S-(-)-Phenylethyl amine (PEA) [50] [CDA-2]*	Amides	2-(6-Chloro-2-carbazolyl)propanoic acid [=carprofen) [50]	?	?
		2-(2-naphthyl)propionic acid, 2-(4-biphenyl)propionic acid 2-(4-isobutylphenyl)propionic acid (= ibuprofen), 2-(2-fluorenyl)-propanoic acid (= cicloprofen) [51]	SiO_2	$CHCl_3$-cyclohexane-THF (271:224:4)
R-(+)-dto.		6-Chloro-5-cyclohexyl-1-indancarboxylic acid (= clidanac) [52]	SiO_2	Benzene-acetonitrile (100:13)
		2-(3-benzoylphenyl)propanoic acid (= ketoprofen), 2-(4-thiophenoylphenyl)propanoic acid (= suprofen), 2-[4-(2-isoindolinyl-1-one)phenyl] propanoic acid (= indoprofen) [53]	SiO_2	$CHCl_3$-ethyl acetate (15:1)
S-1,1'-Naphthylethylamine (NEA) [54] [CDA-3]*	Amides	2-[3-Chloro-4-(3-pyrrolin-1-yl) phenyl]propionic acid (= pirprofen [54]	?	MeOH-CH_2Cl_2
Aminoalkanols				
(+)-2-Amino-1-butanol* [CDA-4] [55]	Amides	Iso-/lysergic acid [55]	SiO_2	DEE-EtOH-0.75 M imidazol (18:1:1)
		Iso-/lysergic acid [55]	Al_2O_3	$CHCl_3$-acetone-1M imidazol (10:9:1)
1*R*,2*R*-(-)-2-amino-1-(4-nitrophenyl)-1,3-propanediol [56] [CDA-5] *	Carboxamides	Lactic acid, 3-bromo-2-methylpropionic acid, mandelic acid, Z-permethrinic acid, N-acetyl-phe, 2-methoxy-3-benzoylpropionic acid, fenoprofen, naproxen [56]	SiO_2	$CHCl_3$-EtOH-Acetic acid (18:2:1)
1*S*,2*S*-(+)-dto. [56]		(for same compounds [56])		
Acids				
S-(+)-Leucine [57] [CDA-6] *	Amides	2-(2-Fluorenyl)propanoic acid (=cicloprofen) [57]	SiO_2	Benzene-DEE-MeOH (8:3:1)
S-(+)-2-Methoxy-2-phenylacetic acid [58] [CDA-7] *	Ester	Ethyl 4-(dimethylamino)-3-hydroxybutanoate [58]	SiO_2	Diethyl ether
1-O-Benzyl-4,6-O-benzylidene-N-acetyl muramic acid [59] [CDA-8]	Amides	Ala, leu, phe (methyl esters) [59]	SiO_2	Benzene-acetone (3:1)

CDA	Deriv.	Compound	Plate	Eluent
1,2,3,4-O-Diisopropylidene-galacturonyl-glycine [59] [CDA-9]	Amides	Phe methyl ester [59]	SiO_2	Benzene-acetone (3:1)
Acid amides				
S-1-Fluoro-2,4-dinitrophenyl-5-alanine amide (FDAA) [60] (= Marfey's reagent) [CDA-10] *	Peptides	Proteinogenic amino acids, ethionine, nleu, nval [60]	SiO_2	Acetic acid-tBME (various mixtures)
		Proteinogenic amino acids, cystine, ethionine, nleu, nval [60]	RP-18	MeOH-0.3 M NaAc (pH 4.0)
	Amines	2-Aminoalkanes, norephedrine, 2-amino-1-alkanols, 2-amino-2-phenylethanol and rel. compounds [61]	RP-18	MeOH-H_2O (various mixtures)
Acid anhydrides				
S-Leucine N-carboxyanhydride [62] [CDA-11]	Peptides	3,4-Dihydroxy-phe (DOPA) [62]	SiO_2	Et Ac-formic acid-H_2O (15:1:7)
		Proteinogenic amino acids [63]	SiO_2	Different eluents
		His [63]	MCC	1-BuOH-AcOH-H_2O (4:1:1)
Acid chlorides				
N-TFA-*S*-(-)-prolyl chloride (TPC) [64] [CDA-12] *	Amides	Amphetamine [64]	SiO_2	$CHCl_3$-MeOH (197:3)
N-Benzyloxycarbonyl-S-prolyl chloride (ZPC) [64] [CDA-13]	Amides	Amphetamine [64]	SiO_2	Hexane-$CHCl_3$-MeOH (10:9:1)
		Methamphetamine [64]	SiO_2	Diisopropyl ether-Et Ac-hexane-AcCN (1:2:2:2)
S-(+)-Benoxaprofen chloride [65] [CDA-14]	Amides	Amphetamine, methamphetamine, 1-PEA, tranylcypromine [65]	SiO_2	Toluene-CH_2Cl_2-THF (5:1:1)
		Metoprolol, oxprenolol, propranolol [66]	SiO_2	Toluene-acetone (10:1)
R-(-)-2-Methoxy-2-trifluoromethyl-2-phenylacetyl chloride (MTPA) [67] [CDA-15] *	Esters	2-Hydroxyhexadecanoic acid [67]	SiO_2	Heptane-methyl formate-DEE-AcOH (25:20:5:1)
S-(+)-Flunoxaprofen chloride [68] [CDA-16]	Amides	1-PEA, tranylcypromine [68]	SiO_2	Toluene-CH_2Cl_2-THF (7:1:1)
S-(+)-Naproxen chloride [69] * [CDA-17]	Amides	Proteinogenic amino acids (as methyl esters) [69]	SiO_2	Toluene-CH_2Cl_2-THF (5:1:2) (NH_3-atmosphere)

CDA	Deriv.	Compound	Plate	Eluent
S-(-)-N-1-(2-Naphthylsulphonyl)-2-pyrrolidinecarbonylchloride [70] [CDA-18]	Amides	Proteinogenic amino acids [70]	SiO_2	$CHCl_3$-H_2O (49:1)
Isocyanates				
R-(+)-1-Phenylethylisocyanate (PEIC) [71] * [CDA-19]	Carbamates	Aliphatic alcohols (e.g. 2-alkanols, 1-octen-3-ol [71]	SiO_2	Benzene-DEE (19:1)
S-(-)-dto*.[71]		(for same compounds [71])		
R-(-)-1,1'-Naphthylethylisocyanate(NEIC) [72] [CDA-20]*	Carbamates	Pharmaceuticals (e.g., propranolol, alprenolol, metoprolol, bunitrolol, oxprenolol, pindolol) [72]	SiO_2	Benzene-DEE-acetone (88:10:5)
		Flecainide [73]	SiO_2	Toluene-CH_2Cl_2-THF (5:4:1)
		meta-O-Dealkylated flecainide, cibenzoline [73]		Toluene-CH_2Cl_2-THF-MeOH (50:37:10:3)

*=commercially available (suppliers: Aldrich, Fluka, Sigma); 1=performance with centrifugal thin-layer chromatography; 2 = plate coating: silica gel treated with phenylmethylvinyl chlorosilane

3.8.3.2 Chiral stationary phases (CSP)

Cellulose

The use of cellulose (CSP-1) as the stationary phase for TLC can be considered as a continuation of paper chromatography. Therefore, the first investigations were performed on cellulose plates transferring the already-developed PC methods to TLC. Reported enantioseparations by means of cellulose layers and its derivatives are summarized in Table 3.8.4. In the main amino acids and peptides have been investigated by several groups. Standard cellulose showed only a weak enantioselectivity for underivatized amino acids, except those containing aromatic groups (e.g., tyrosine) and threonine with ethanol-pyridine-water (2:3:1) as eluent [74]. Cellulose exhibits different structures under hydrophilic or hydrophobic conditions. The hydrophobic condition of native cellulose has been discussed by Yuasa et al. [75] as the reason for the successful enantioseparations of underivatized amino acids.

For the separation mechanism it is assumed that several helical chains of cellulose form secondary structures containing chiral cavities [76]. The separation of enantiomers is effected by their different fit into the lamellar chiral layer structure of the cellulose. Working at a reduced temperatures, e.g., at 0°C, may improve the enantioselectivity due to the increased hydrophobicity, but with a drastic increase in the developing times [75]. The enantioselectivity is also dependent on the eluent system. Various methyltryptophans with water as eluent were only separated into enantiomers at a minimum temperature of 40°C. With other eluents, like 0.5 mol lithium chloride or 0.76 mol ammonium sulphate, the separation was achieved at room temperature and also at -5° C [77].

The influence of different types of cellulose on the enantioselectivity was first studied by Munier et al. [78] and later by Lederer [35]. Comparison of several commercial cellulose plates revealed the advantage of microcrystalline cellulose. For instance, for methyltryptophans or kynurenine the resolution was much better with microcrystalline cellulose than with normal cellulose plates.

Only a few studies have been performed using triacetylcellulose (TAC; CSP-2). The first successful attempt was carried out by Hesse and Hagel in 1973 [80], resolving Tröger's base (see Table 3.8.4). Later, new microcrystalline triacetylcellulose plates were reported by Faupel [79]. These plates are commercially available as 'OPTI-TAC' from Antec, Switzerland.

Table 3.8.4 Cellulose and derivatives as chiral stationary phases (CSP) for TLC.

Compound Class	Example	Eluent
A) Cellulose (CSP-1) [34] [1 microcrystalline cellulose = MCC; 2 native cellulose] (ready-to-use cellulose plates are commercially available from many manufactures)		
Amino acids	3-/5-Hydroxykynurenine, 3-methoxykynurenine, trp, 5-/6-hydroxy-trp [34]	1-BuOH-pyr.-H_2O (1:1:1)
	Diaminodicarboxylic acids [81]	MeOH-H_2O-AcOH (20:5:1)
	o-/m-/p-tyr and der., DOPA,	MeOH-H_2O (various mixtures)
	Z-β-(3,4-dihydroxyphenyl)serine [82]	
	14 proteinogenic amino acids[75]	EtOH-pyr.-H_2O (var. mix.)
	Arg, lys [83]	AcCN-pyr.-1 mM NaOH (2:2:1)
	Asp, glu [83]	AcCN-pyr.-1 mM HCl (2:2:1)
	Asn, gln [83]	AcCN-pyr.-H_2O (2:2:1)
	Cys [83]	EtOH-H_2O (6:1)
	1-/5-/6-/7-Methyl-trp [77]	H_2O (40° C)
	1-/5-/6-/7-Methyl-trp [77]	0.5 M LiCl (20° C)
	4-/5-/6-Fluoro-trp [35;84]	[1] 1 M NaCl
	4-Methyl-trp, 5-methoxy-trp [85]	[1] 1 M NaCl
Dipeptides	Ala-ala, asp-ala, lys-ala, phe-ala, phe-phe, trp-trp, tyr-tyr [86]	[1] Pyr.-H_2O (2:1)
	Gly-trp [85]	[1] 1 M NaCl
Pharmaceuticals	Isoxepac and its propanoic acid analogue [87]	MeOH-H_2O (various mixtures)

Compound Class	Example	Eluent
B) Triacetylcellulose (TAC) [80] (CSP-2) ('OPTI-TAC' from Antec, Switzerland)		
Ketones	2-Phenylcyclohexanone [88]	EtOH-H_2O (4:1)
Pharmaceuticals	5-Benzoyl-2,3-dihydro-6-hydroxy-1H-indene 1-carboxylic acid (= oxindanac benzyl ester) [79]	EtOH-H_2O (17:3)
Miscellaneous	Tröger's base [80]	EtOH
	2,2,2-Trifluoro-1-(9-anthryl)ethanol [88]	EtOH-H_2O (4:1)

Ligand-exchange chromatography (LEC)

In 1981 Busker et al. [89] synthesized new optically active proline derivatives designed for metal chelates to be used in liquid chromatography. Later, Günther et al. [90] introduced copper(II) complexes of these proline derivatives, especially of 2*S*,4*R*,2'*RS*-N-(2'-hydroxydodecyl)-4-hydroxyproline, as the chiral stationary phase for TLC (CSP-3). Today, this CSP is commercially available from Macherey & Nagel, Düren, Germany ('CHIRALPLATE') and from Merck, Darmstadt, Germany ('CHIR'). The wide range of applications, in particular for the resolution of amino acids and their derivatives, can be seen in Table 3.8.5. The synthesis [90,91] of 'CHIRALPLATE' (CSP-3) starts with the impregnation of an RP-18 silica gel plate with a mixture of methanol and a solution of 0.25 % copper(II) acetate (1:9). After drying at 110°C a further impregnation with a methanolic solution of (2*S*,4*R*,2'*RS*)-N-(2'-hydroxydodecyl)-4-hydroxyproline is performed followed by drying at room temperature.

Mack et al. [92] used 'CHIR' with concentration zones: a small band, a so-called concentration zone, is placed in front of the actual separation layer. This layer section is made of an inert silica gel with extremely wide pores and a very small specific surface area. These properties almost completely preclude interaction between the sample molecules and the concentration zone. During development the applied sample spots travel with the front of the mobile phase. On reaching the borderline of the two layer sections the circular spots are concentrated into narrow bands. Hence, an improved starting position and a higher efficiency of the subsequent separation process is obtained, especially in the case of large sample volumes.

The basic separation principle of this CSP is ligand exchange chromatography (LEC). The name 'ligand exchange' was introduced by Helferich in 1961 [93] for a column liquid chromatographic technique. The first separation of optical isomers by LEC was performed by Rogoszin and Davankov in 1968 [94]. Later, this method was transferred to reversed-phase HPLC [95] (cf. also Section 3.4). The prerequisites for a successful separation of optical isomers are the following: The compounds to be separated must contain at least two polar functional groups in the right order, which can be active at the same time as ligands for the copper ion [45]; they have to be in the anionic form for complex formation [95]. A model describing the solute-sorbent interactions in such an LEC system should account for the following facts:

(i) Formation of the mixed ligand sorption complex at the surface of the reversed phase support. This is considered to be the only process accounting for the chiral recognition of the amino acid enantiomers, and its extent of formation governs the retention of the enantiomers. The formation of the sorption complex is dependent on the Cu(II) concentration and the pH-value of the eluent.

(ii) Hydrophobic interactions have been shown to contribute significantly to the retention of solutes and to affect enantioselectivity. The interactions can be regulated by the type and concentration of organic solvent in the mobile phase.

(iii) For a given pair of enantiomers the *R*-enantiomer is always more strongly retained than the *S*-enantiomer, with the exception of cysteine and histidine [95].

Taking into account the phenomena described above, a description of a model for enantiomeric resolution can be envisaged; the N-alkyl chain penetrates the reversed-phase interface layer and is oriented parallel to the octadecyl chains. Such an arrangement creates a highly stable structure and enables the selector to have the best interaction with the desired enantiomer. As a result the alkyl radical interacts with the hydrophobic support from the opposite direction of the main coordination plane of the Cu(II) chelate. When an *R*-enantiomer of an amino acid interacts with the adsorbed complex, the α-chain of the amino acid (due to the steric arrangement) is in close proximity to the hydrocarbon chain of the reversed-phase stationary phase. On the other hand, the *S*-enantiomer lacks this property. This leads to a stronger interaction of the *R*-enantiomer than of the *S*-enantiomer [96]. In Figure 3.8.2 the mechanism of chiral ligand exchange chromatography is demonstrated by complex formation of copper with the chiral selector [92]. The chiral selector is racemic at the 2'-position (dodecyl side-chain). The use of a selector with a defined chirality at this position did not improve the enantioselectivity, which shows that this hydroxyl group is not involved in the ternary complex (cf. Figure 3.8.2) [97].

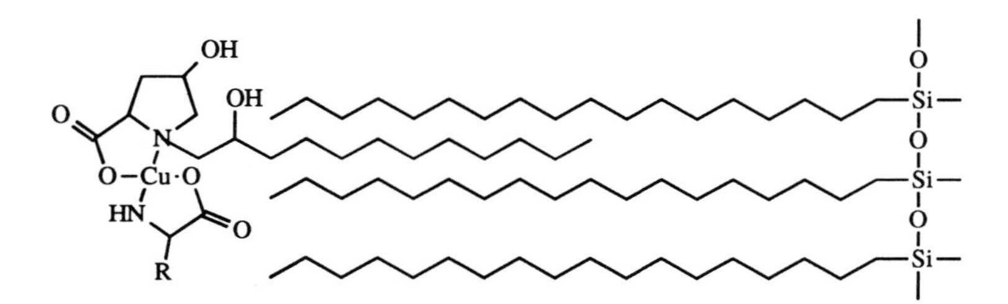

Figure 3.8.2 Mechanism of chiral ligand exchange chromatography on CSP-3 as a chiral selector [92]; R = residue of an α-amino acid.

Nyiredy et al. [98] showed the applicability of forced-flow planar chromatography for the separation of enantiomers on 'CHIRALPLATE'. Circular or linear overpressured layer chromatography (OPLC), microchamber rotation planar chromatography (M-RPC) and ultra-microchamber rotation planar chromatography (U-RPC) with the following advantages are described: Very rapid analysis times (approximately 10 min for qualitative and between 45 and 60 min for quantitative analyses); depending on the separation to be performed and the method employed, between 50 and 80 samples may be applied on a single 20 x 20 cm 'CHIRALPLATE' [98].

Brinkman and Kamminga [99] optimized the separation conditions for normal pressure TLC (shorter plates of 5 instead of 13 cm, no heat treatment, unsaturated chamber, changed eluent: 80 % acetonitrile) reducing the analyzing times dramatically. With modern HPTLC methods quantitative analysis of optical isomers is also possible. Using commercially available plates an ee-value of 99.8 % is detectable [100].

Table 3.8.5 TLC-Separation of enantiomers with CSP-3 [88]; (CHIRALPLATE/ = Macherey-Nagel, FRG; CHIR/ = Merck, FRG).

Compounds	Eluent	Ref.
Amino acids, dipeptides, and derivatives		
alanine	F	101
ala-ala	A	86
ala-phe	A	102
α-butyl-ala	A	103
3-chloro-ala	A	104
3-(3-cyclohexenyl)-ala	A	105
3-cyclohexyl-ala	A	105
3-cyclopentyl-ala	A	90
α-ethyl-ala	A	103
α-heptyl-ala	A	103
α-hexyl-ala	A	103
(1-naphthyl)-ala	A	106
(2-naphthyl)-ala	A	106
α-nonyl-ala	D	103
α-octyl-ala	D	103
α-pentyl-ala	A	103
α-phenyl-β-ala	A	105
α-propyl-ala	A	103
α-sulpho-β-ala	A	105
2-(3-thienyl)-ala	A	105
2-(2-thienyl)-ala	A	105
2-aminobutanoic acid	A	104
2-amino-2-butyl butanoic acid	A	103
2-amino-2-hexyl butanoic acid	A	103
2-amino-2-methyl butanoic acid	A	107
2-amino-2-pentyl butanoic acid	A	103
2-amino-2-propyl butanoic acid	A	103
2-aminopimelic acid	A	105
2-aminotetraline 2-carboxylic acid	A	108
2-amino-6-hydroxytetraline 2-carboxylic acid	A	108

Compounds	Eluent	Ref.
asp-ala	A	86
Z-3-methyl-asp	A	105
α-methyl-asp	A	107
N-methyl-asp	A	109
PTH-asp	B	110
2-azabicyclo[3.3.0]octane 3-carboxylic acid	A	111
*cysteine**	A	90
S-(2-chlorobenzyl)-cys	A	90
penicillamine* (= 2-amino-3-thio-3-methylbutanoic acid)	A	90
S-[2-(2-pyridyl)ethyl]-cys	A	90
S-(3-thiabutyl)-cys	A	90
S-(2-thiapropyl)-cys	A	90
***DOPA** (= 3,4-dihydroxyphenylalanine)*	B	112
α-methyl-DOPA	B	113
glutamic acid	A	106
4-methylene-glu	A	105
α-methyl-glu	A	107
PTH-glu	B	110
glutamine	A	90
*glycine***		
allyl-gly	A	105
(2-cyclohexenyl)-gly	A	105
(3-cyclohexenyl)-gly	A	105
cyclohexyl-gly	A	105
cyclopentyl-gly	A	106
gly-ile	B	114
gly-leu	B	114
gly-phe	B	104
gly-trp	B	114
gly-val	B	114
(1-methylcyclopropyl)-gly	A	104
α-methyl-phenyl-gly	A	107
phenyl-gly	A	104
isoleucine	A	90
allo-ile	A	104
leucine	C	106
N-formyl-tert-leu	A	104
leu-leu	B	114
α-methyl-leu	A	107
N-methyl-leu	A	102
α-methyl-nor-leu	A	107
nor-leu	A	90
tert-leu	A	106

Compounds	Eluent	Ref.
lysine		
lys-ala	A	86
methionine	A	104
ethionine (= 2-amino-4-ethylthio-butanoic acid)	A	106
ethionine-sulphone	A	106
homo-met	***	115
isopropionine (= 2-amino-4-isopropylthio-butanoic acid)	A	105
met-met	B	114
met-sulphone	A	106
α-methyl-met	A	106
seleno-met	A	106
phenylalanine	A	89
α-allyl-phe	A	107
4-amino-phe	A	100
3-bromo-phe	A	90
4-bromo-phe	A	90
4-chloro-phe	A	90
α-difluoromethyl-phe	A	107
2,6-dimethyl-phe	A	108
E-/*Z*-2,b-dimethyl-phe	A	108
N,N-dimethyl-phe	B	106
2-fluoro-phe	A	106
homo-phe	A	104
E-β-hydroxy-phe	A	108
4-iodo-phe	A	89
3-methoxy-α-methyl-phe	A	116
E-/*Z*-β-methyl-4-nitro-phe	A	108
2-methyl-phe	A	108
4-methyl-phe	A	105
α-methyl-phe	A	106
E-/*Z*-β-methyl-phe	A	105
N-methyl-phe	A	104
α-(2-methylthio)ethyl-phe	A	107
4-nitro-phe	A	90
phe-ala	A	86
phe-phe	A	86
pipecolic acid	F	101
proline	B	89
Z-4-hydroxy-pro	A	104
serine	F	101
O-acetyl-ser	A	105
O-benzyl-ser	A	90
α-methyl-ser	B	106
O-methyl-ser	A	105
PTH-ser	B	110
tetrahydroisoquinoline 3-carboxylic acid	A	108

Compounds	Eluent	Ref.
Z-tetrahydro-4-methylisoquinoline 3-carboxylic acid	A	108
tetrahydro-5-methylisoquinoline 3-carboxylic acid	A	108
thyroxine	A	106
tryptophan	A	89
N-carbamyl-trp	H	117
5-bromo-trp	A	104
5-hydroxy-trp	A	105
5-methoxy-trp	A	90
4-methyl-trp	A	104
5-methyl-trp	A	90
6-methyl-trp	A	89
7-methyl-trp	A	90
α-methyl-trp	A	107
trp-trp	A	86
tyrosine	B	89
m-tyr	A	105
o-tyr	A	105
O-benzyl-tyr	A	90
2,5-dimethyl-tyr	A	108
3-fluoro-tyr	A	104
2-methyl-tyr	A	108
α-methyl-tyr	A	106
E-/*Z*-β-methyl-tyr	A	108
N-methyl-m-tyr	B	106
O-methyl-tyr	A	90
2,5,O-trimethyl-tyr	A	108
tyr-tyr	A	86
valine	A	104
α-methyl-val	A	107
N-methyl-val	B	106
nor-val	A	99
α-methyl-nor-val	A	116
Hydroxycarboxylic acids		
4-bromomandelic acid	J	92#
4-chloromandelic acid	J	92#
3,4-dihydroxymandelic acid	G	101
2-hydroxybutanoic acid (sodium salt)	G	18
2-hydroxydocosanoic acid	G	18
2-hydroxyhexadecanoic acid	G	18
2-hydroxyhexanoic acid	L	2
3-hydroxymandelic acid	G	101
4-hydroxymandelic acid	G	101
4-hydroxy-3-methoxymandelic acid	G	101
2-hydroxy-3-methylbutanoic acid	G	101
Z-2-hydroxy-3-methylpentanoic acid	G	101
2-hydroxy-4-methylpentanoic acid (sodium salt)	G	101
2-hydroxy-4-methylthio-butanoic acid (sodium salt)	G	101
2-hydroxyoctanoic acid	G	18

Compounds	Eluent	Ref.
2-hydroxypentanoic acid	G	18
2-hydroxy-2-phenylpropanoic acid	G	18
2-hydroxy-3-phenylpropanoic acid	G	101
2-hydroxypropanoic acid (= lactic acid)	G	18
2-hydroxytetradecanoic acid	G	18
mandelic acid	G	101
O-methylmandelic acid	J	46
Miscellaneous		
3-amino-3,5-dimethyl-4-pentanolide HCl	A	104
1-tert-butyl-3-phenyl-3-azetidinone	E	118
1-adamantyl-3-phenyl-3-azetidinone	E	118
6-benzoylamino-2-chlorohexanoic acid	I	92#
noradrenaline (as Schiff base of salicylaldehyde)	K	76#
Thiorphan	***	119
Thiazolidine-der. formed by penicillamine and aldehydes	A	120

[only the first citation is mentioned; in some cases more eluents were reported; PTH = N-phenylthiohydantoin-derivative]
* = as thiazolidine-der.; ** = achiral; *** = eluent not given; # = CHIR (Merck); all other separations 'CHIRALPLATE' (Macherey-Nagel); A = methanol-water-acetonitrile (1:1:4) [90]; B = methanol-water-acetonitrile (5:5:3) [90]; C = methanol-water (1:8) [100]; D = 1-propanol-water-acetonitrile (2:1:2) [103]; E = hexane-ethyl acetate (10:1) [118]; F = acetone-methanol-water (5:1:1) [101]; G = dichloromethane-methanol (9:1) [101]; H = 1 mM copper(II) acetate in 5 % aqueous methanol (pH 5.8) [117]; I = methanol-water-acetonitrile (5:2:5) [92]; J = dto. + 0.05 M/l potassium dihydrogenphosphate [92]; K = chloroform-methanol (9:1) (80 % saturated with water) [76]; L = water-acetonitrile (2:3) [2].

Besides the above-mentioned 'CHIRALPLATE' and 'CHIR', further CSPs for LEC have been developed. Günther et al. [90] synthesized a series of similar proline derivatives to optimize the enantioselectivity. The structures of some of these derivatives are shown in Figure 3.8.3.

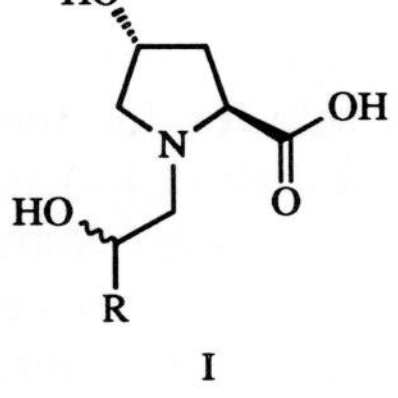

I: CSP-3: $R = C_{10}H_{21}$
II: CSP-4: $R = C_{10}H_{21}$
I: CSP-5: $R = C_{12}H_{25}$
I: CSP-6: $R = C_{14}H_{29}$

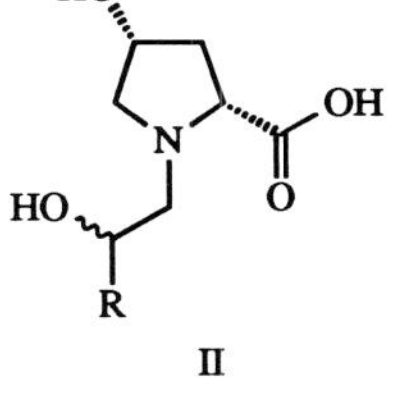

Figure 3.8.3 Structures of proline derivatives used as CSPs for LEC.

In a similar manner, Weinstein [121] used a Cu(II)-complex of *S*-N,N-dipropylalanine (CSP-7) on RP-18 silica gel plates for LEC. With this stationary phase all dansylated proteinogenic amino acids were separated into their enantiomers, except proline, using an acetic acid buffer as eluent. Nearly the same results have been obtained with a copper complex of *S*-N,N-didecyl-alanine as the stationary phase (CSP-8) [96].

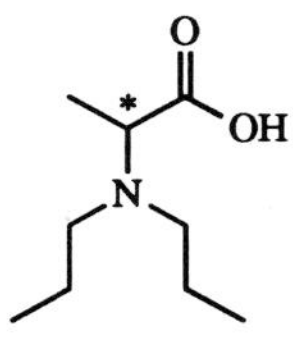

(*S*)-N,N-dipropyl-alanine
(CSP-7)

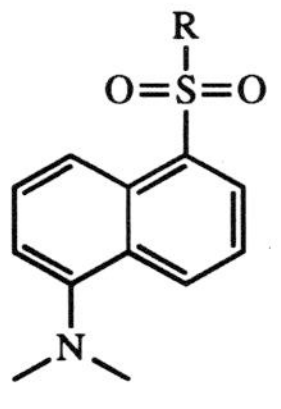

5-Dimethylamino-1-naphthalene
sulphonyl = Dansyl (R= amino acid)

Marchelli et al. [122] synthesized several types of alkylene-bis(amino acid amides) such as 1,2-ethylene-bis(phenylalanine-amide) (CSP-9). Using water-acetonitrile as eluent at different pH values, it was possible to separate dansylated amino acids (e.g., asp, glu, leu, nleu, phe). With a water-acetonitrile-2 mmol diphenylalanine complex as eluent other dansylated amino acids have been resolved (e.g., ser, thr, met, nval). Surprisingly, all amino acids showed a higher R_f-value for the *R*-enantiomer, except leu, nleu and nval were separated in a reversed order of elution [122].

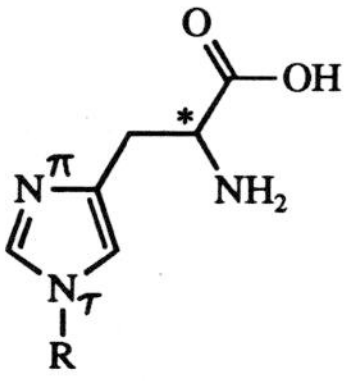

R = $CH_2(CH_2)_8CH_3$
N-Decylhistidine

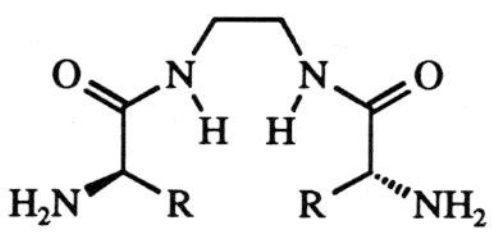

R = C_6H_5-CH_2-
(= Phenylalanine)

Remelli et al. [123] used *S*-N(t)-decylhistidine (CSP-10) as chiral selector for the separation of phenylalanine and tyrosine with methanol-water-acetonitrile (1:1:4) as the eluent. In a direct comparison the results for these amino acids were better than that obtained with 'CHIRALPLATE' or 'CHIR' [123]. Further chiral compounds have been separated, e.g., trp, 5-methyl-trp, 6-methyl-trp, α-methyl-phe, α-methyl-tyr, and DOPA, with methanol-acetonitrile-tetrahydrofuran-water (7.3:9:33.9:52.9) as eluent. Small but significant stereoselective effects were found in the formation of ternary Cu(II) complexes of *R*- or *S*-histidine with amino acids, where the homochiral complex formation (containing ligands of the same chirality) is favoured. Thermodynamic and spectroscopic studies have shown that these ternary complexes are nearly octahedral; histidine behaves as a tridentate ligand, binding the Cu(II) ion with its amino and pyridino nitrogens in equatorial positions and with a carboxylate oxygen in a distorted axial position. This behaviour has been discussed as the basis for the stereoselectivity found [123].

Sinibaldi et al. [124] used, in contrast to the already-mentioned complexes, a chiral polymer *S*-phenylalanine for LEC ('PPAA', CSP-11; cf. Figure 3.8.4). PPAA was synthesized by reaction of ethylene glycol diglycidyl ether with *S*-phenylalanine amide dissolved in methanol. Then RP-18 plates were immersed in a water-acetonitrile solution (7:3) containing 3 mmol copper(II) acetate and PPAA. This plates can be used for the enantiomer separation of dansylated amino acids with various mixtures water-acetonitrile as eluent [124].

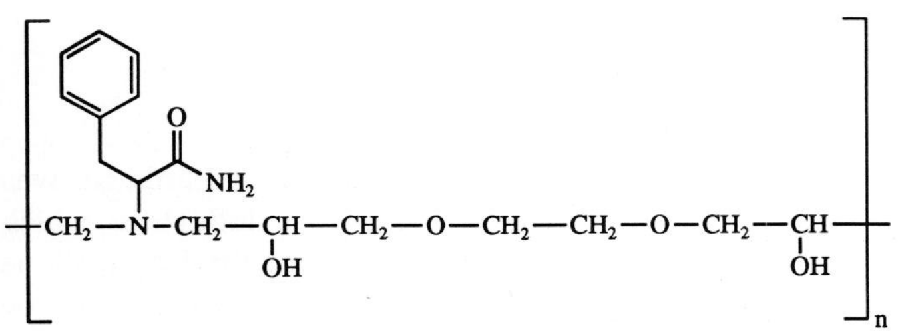

O

NH_2

$-CH_2-N-CH_2-CH-CH_2-O-CH_2-CH_2-O-CH_2-CH-$

OH OH

n

Figure 3.8.4 Copolymer of ethylene glycol diglycidyl ether with *S*-phenylalanine amide [124].

Charge-transfer complexation (Pirkle-type CSPs)

In 1981 Pirkle et al. [125] introduced a new type of chiral stationary phase for HPLC exhibiting π-donor or π-acceptor properties (for details see HPLC Section 3.4). Wainer et al. [126] transferred this separation principle of the so-called 'brush-type' CSP to TLC with *R*-N-(3,5-dinitrobenzoyl)phenylglycine which was ionically bonded to aminopropyl silanized silica gel-coated plates (CSP-12) for the separation of 2,2,2-trifluoro-1-(9-anthryl)ethanol with hexane-isopropanol (19:2) as eluent.

A similar approach has been taken by Wall [127] with *R*-N-(3,5-dinitrobenzoyl)leucine (CSP-13) and the same working conditions as Wainer et al. [126]. Wall also reported on the improvement by covalent bonding of CSP-12 and CSP-13 on amine-modified silica gel plates (CSP-14 and CSP-15, respectively) using N-ethoxycarbonyl-2-ethoxy-1,2-dihydroquinoline (EEDQ) as a catalyst (note: both CSPs destroy the fluorescence of the indicator). This plates show improved stability. The same compound with the same eluent was studied.

Further investigations by Wall [128] using all four CSPs revealed the capability of these phases for separating pharmaceutical compounds (e.g., hexobarbital, oxazepam, lorazepam, propranolol, atenolol, metoprolol) and 1,1'-bis(2-naphthol) with mixtures of hexane-isopropanol or hexane-isopropanol-acetonitrile as eluents.

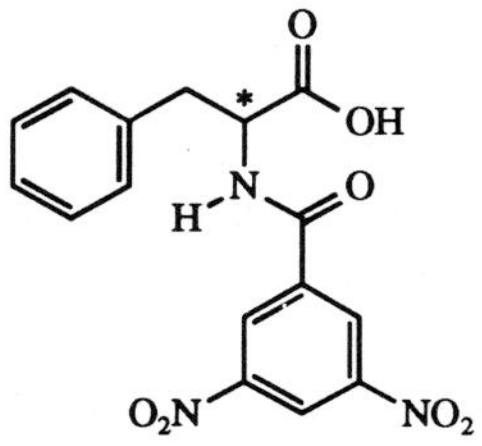

N-(3,5-Dinitrobenzoyl)-
phenylglycine

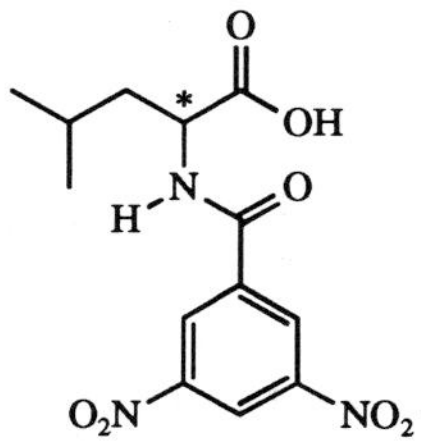

N-(3,5-Dinitrobenzoyl)-
leucine

Cyclodextrin stationary phases

Cyclodextrin which is covalently bonded to silica gel without nitrogen linkages was introduced by Armstrong [129]. This author used α-, β- and γ-cyclodextrins which were bonded via alkyl chlorosilane derivatives (e.g., 7-octenyldimethylchlorosilane or 3-glycidoxypropyldimethylethoxysilane) to silica gel. This material can be applied for HPLC and TLC. Using a silica-bonded β-cyclodextrin (CSP-16) several racemic compounds such as dansylated amino acids (e.g., ala leu, met, val), amino acid 2-naphthylamides (e.g., ala, met), as well as ferrocene derivatives (e.g., 1-ferrocenyl-1-methoxyethane, 1-ferrocenyl-2-methylpropanol, (1-ferrocenylethyl)thioglycolic acid) have been separated using mixtures of methanol and 1 % aqueous triethyl ammonium acetate (pH 4.1) as eluents [130]. Mechanistic considerations as well as the structures of cyclodextrins can be found in Sections 3.5 (GC) and 3.4 (HPLC).

Alak and Armstrong [130] described another important factor, i.e. the binder, for the production of cyclodextrin stationary phases. The binder is necessary to ensure that the stationary phase adheres to the support and has sufficient mechanical strength to withstand spotting, development and general handling procedures. However, the binder can also alter the efficiency, selectivity, development time and detection unless it is properly utilized. In the case of β-cyclodextrin-bonded phase plates one has to be careful that the binder does not form a strong inclusion complex with β-cyclodextrin rendering it ineffective for further separations. To prevent complex formation polymeric binders were used, because they are too large to fit into the cyclodextrin cavity [130].

Wilson [87] prepared silica gel plates with non-covalent bonded β-cyclodextrin (CSP-17) by dipping silica plates into a mixture of 1 % β-cyclodextrin in ethanol-dimethyl sulphoxide (4:1). After drying, this procedure was repeated. Using these plates the pure mandelic acid enantiomers gave different R_f values with methanol-water as eluent. In varying the percentage of methanol from 0 to 100 % no change in the R_f values of either *R*- or *S*-mandelic acid was observed. An attempt to separate mixtures of the mandelic acid enantiomers was without success, yielding a single spot with an R_f between those of the pure enantiomers [87].

Organic acids as CSPs

Bhushan and Ali [131] produced silica gel plates containing (+)-tartaric acid (CSP-18). With these plates and chloroform-ethylacetate-water (28:1:1) as eluent phenyl-thiohydantoin-derivatives (PTH) of proteinogenic amino acids could be ressolved. The forming of PTH derivatives works without racemization. The separation is a result of (+)(-)- and (+)(+)-diastereomeric salts of the PTH amino acids with tartaric acid [131]. The same authors also used a mixture of (+)-tartaric acid and (+)-ascorbic acid (CSP-19) to impregnate silica gel plates. With butyl acetate-chloroform (1:5) as eluent several PTH amino acids were separated [131].

Silica gel impregnated with *S*-aspartic acid (CSP-20) for the alkaloids hyoscyamine [(±)-hyoscyamine = atropine) and colchicine with 1-butanol-chloroform-acetic acid-water (3:6:1:1) as the eluent has been reported by Bhushan and Ali [132]. Aspartic acid is anionic above its pI of 3.0. Since the pH of the solvent system is also greater than 3.0 it is thought that ionic interaction between anionic aspartic acid and the components of racemic mixtures of alkaloids existing as large cations produces diastereomeric salts leading to enantiomeric separation [132]. The resolution only took place at or around 0°C but not at room temperature.

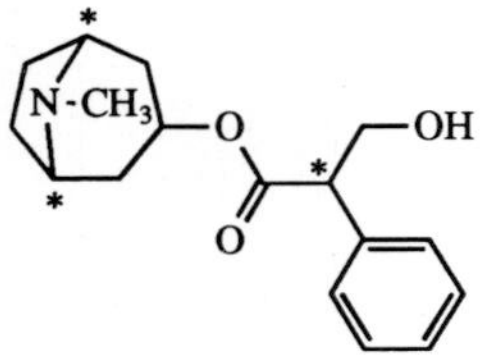

Hyoscyamine

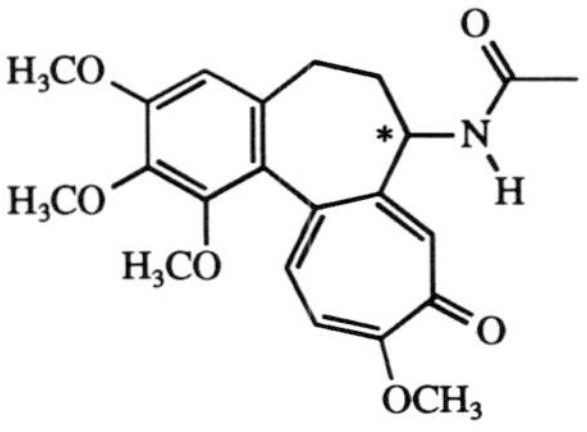

Colchicine

Alkaloids as CSPs

The use of (-)-brucine (CSP-21) as stationary phase for TLC was reported by Bhushan and Ali [133]. These authors impregnated silics gel with (-)-brucine (adjusted at pH 7.1 to 7.2) for the separation of underivatized amino acids using 1-butanol-acetic acid-chloroform (3:1:4) as eluent. As the possible separation mechanism it has been suggested that amino acids, which have isoelectric points < 7 and therefore exit as anions at pH 7.1 to 7.2, interact with the basic brucine to give diastereomeric salts. With a different eluent further amino acids were separated [134].

Silica gel impregnated with 0.2 % natural berberine (achiral) [135], containing (-)-hydrastine (3:2) (CSP-22) [136] has been used for the separation of underivatized proteinogenic amino acids and DOPA with an eluent containing ethanol-acetic acid-chloroform (3:1:6).

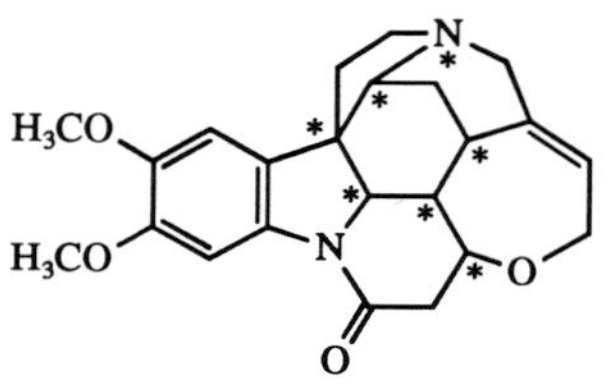

Brucine

The separation principle is similar to that for brucine, i.e. the amino acids form diastereomeric salts (-)(-) and (-)(+) with hydrastine [136].

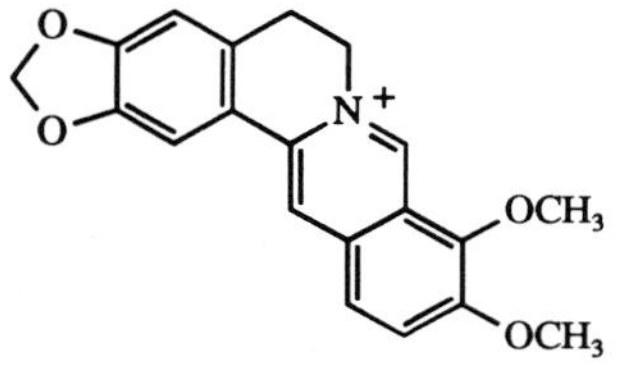

Berberine

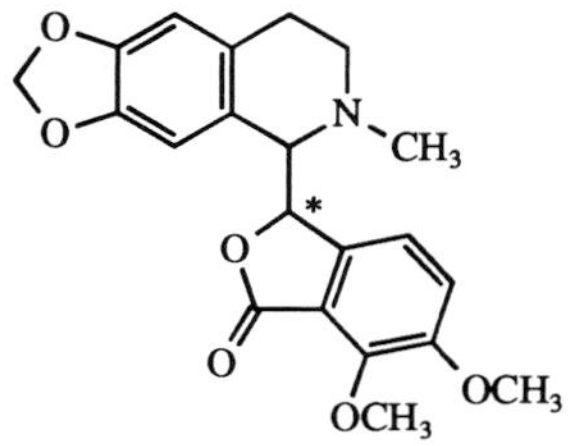

Hydrastine

Other CSPs for TLC

N-*R*,*R*-1,3-*E*-Chrysanthemoyl-*S*-valine chemical bonded on 3-aminopropyl silanized silica gel (CSP-23) was used for the separation of N-3,5-dinitrobenzoyl-valine methyl ester with hexane-1,2-dichloroethane-ethanol (50:20:1) as the eluent [137].

Commercially available aminopropyl-silanized silica plates were derivatized with *R*-(-)-1-(1-naphthyl)ethyl isocyanate (NEIC) (CSP-24) [138]. With an eluent containing hexane-isopropanol-acetonitrile (20:8:1) separations of pharmaceuticals such as 2-(4-isobutylphenyl)propionic acid (ibuprofen), naproxen, fenoprofen, flurbiprofen, benoxaprofen as 3,5-dinitroanilyl-derivative and 1-phenylethylamine, tocainide as 3,5-dinitrobenzoyl-derivative (DNB) were obtained. One of the key aspects in the chiral recognition mechanism is a π-π interaction between a π-acidic moiety on the solute and the π-basic naphthyl moiety on the stationary phase [138].

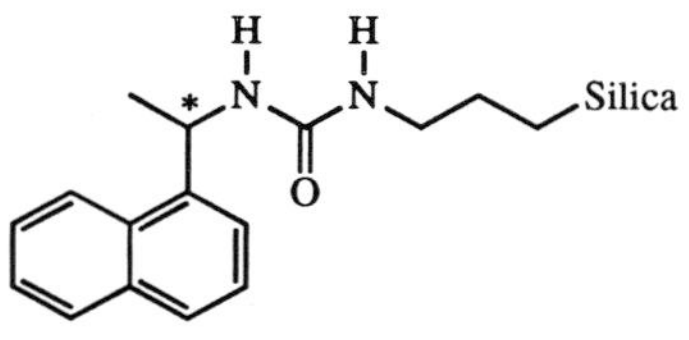

1,1'-Naphthylethylurea CSP

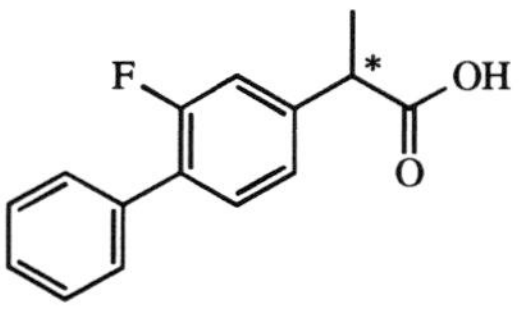

Flurbiprofen

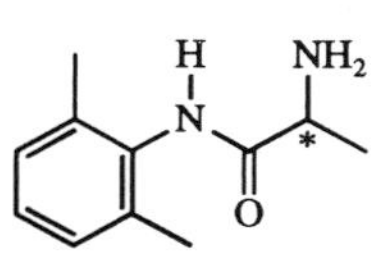

Tocainide

2-Aza-N-2-hydroxyalkylbicyclo[3.3.0]octane-3-carboxylic acid has also been applied successfully as a stationary phase (CSP-25) [139].

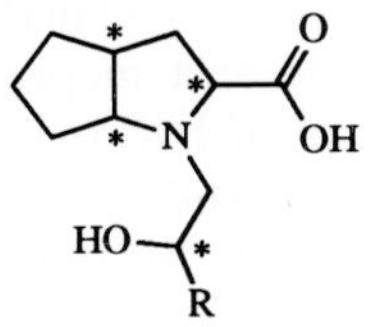

New CSP (R = alkyl)

The biopolymer chitin (CSP-26), a polysaccharide made of N-acetyl-D-glucosamine with β-1,4-glycosidic bondings has also been used as chiral stationary phase [140]. About 90 % of all amino groups are acetylated. Transition metal ions like copper(II) can be strongly bound by the free amino groups. CSP-26 was made by mixing chitin with 0.5 M of an aqueous solution of a copper(II) salt at room temperature. Then this chitin-derivative was washed with water and dried at 80°C (3 h). Using methanol-water (1:1) and other eluents for the separation of amino acids, generally very high α-values were obtained. However, only six different amino acids have been studied.

3.8.3.3 Chiral mobile phase additives (CMA)

Organic acids (and derivatives) as CMA

As early as the 1960's attempts were made to add organic acids like D-galacturonic acid (CMA-1) to the eluent. The separation of ephedrine with 2-propanol-1mol aqueous D-galacturonic acid (94:3) as the eluent was obtained on silica gel plates and aluminium oxide plates as well as on silica gel plates impregnated with CMA-1 [141, 142]. Bauer et al. [143] used cyclohexane saturated with (+)-diethyl tartrate (CMA-2) as an eluent on (+)-diethyl tartrate impregnated silica gel plates for the separation of the chiral 1,2-(2-oxobutylene)ferrocene. Yoneda and Baba [144] used a 1:1-mixture of 0.3 mol (+)-sodium tartrate-0.2 mol aluminium chloride (CMA-3) on silica gel plates for the separation of the chiral tris(ethylenediamine)cobalt(III) complex.

Cyclodextrins and derivatives as CMA

Cyclodextrins as chiral mobile phase additives were first used for the separation of achiral positional isomers [145], but underivatized cyclodextrins usually exhibit poor efficiencies [44]. With greater success, in 1988, Armstrong [146] applied 0.3 mol hydroxypropyl-β-cyclodextrin (CMA-4) in a mixture of acetonitrile-water (7:13) as the mobile phase to separate racemic mixtures of methionine-2-naphthylamide, benzyl-2-oxazolidinone and 5-(4-methylphenyl)-5-phenylhydantoin on RP-18 plates.

One advantage of hydroxypropyl-β-cyclodextrins is their higher solubility in water than that of native β-cyclodextrin. In the same year, Armstrong also introduced the analogous hydroxyethyl-β-cyclodextrin derivative [147].

The following compounds were separated into their enantiomers: dansylated amino acids (e.g., leu, met, nleu, phe, thr, val), ala-2-naphthylamide and mephenytoin using mixtures of acetonitrile-water containing 0.3 or 0.4 mol of hydroxypropyl-β-cyclodextrin (CMA-4). Diastereomers were also separated [147]. The average molar substitution with hydroxypropyl groups was 0.6 ± 0.1. In 1991, Duncan and Armstrong [148] pointed out that higher degrees of substitution of hydroxypropyl-β-cyclodextrin lead to lower enantioselectivity. The changes in chiral selectivity between the native and the derivatized cyclodextrin may result from the loss of hydrogen-bonding groups and/or the gain of additional interaction sites around the mouth and the base of the derivatized cyclodextrin cavity. Armstrong et al. [149] improved the solubility of native β-cyclodextrin in water by adding urea. With an eluent of 0.1 mol β-cyclodextrin (CMA-5) in a mixture of acetonitrile-water (3:7), which was saturated with urea, several enantiomeric pairs such as dansylated amino acids (2-aminobutyric acid, asp, glu, leu, met, nleu, nval, phe, ser, thr, trp, val) were separated. With methanol-water containing β-cyclodextrin, urea and sodium chloride the enantiomers of mephenytoin, alanine-2-naphthylamide hydrochloride, and 2,2'-binaphthyldiyl-N-benzyl-monoaza-16-crown-5 were resolved. N'-benzylnornicotine and N'-(2-naphthylmethyl)nornicotine were separated with methanol-aqueous triethyl ammonium acetate (1 %; pH 7.1) containing β-cyclodextrin, urea and sodium chloride. Finally, with a mixture of acetonitrile-water containing β-cyclodextrin, urea and sodium chloride several more racemic mixtures were separated such as (1-ferrocenyl-2-methylpropyl)thioethanol, (1-ferrocenylethyl)thiophenol, 2-chloro-2-phenylacetyl chloride, and menthyl-4-toluenesulphinate [149].

Adopting the same technique Lepri et al. [150] used RP-18 plates for the separation of several chiral compounds, e.g., dinitropyridylated and dinitrophenylated amino acids like leu, phe, trp. Later, the same group [151] successfully separated methylthiohydantoin-derivatives of several amino acids (e.g., phe, tyr), as well as met-2-naphthylamide, leu-2-naphthylamide, leu-4-nitroanilide, and ala-4-nitroanilide [152]. Recently, LeFevre [153] described the separation of dansylated amino acids on RP-18 plates using a similar eluent. The author used a mixture of methanol or acetonitrile with an aqueous solution of 0.2 mol β-cyclodextrin saturated with ammonia and containing 0.6 mol sodium chloride to separate the following dansylated (dns) amino acids: ala, arg, asn, cit, cystine (di-dns), gln, his (mono- and di-dns), ile, allo-ile, lys (di-dns), N-methyl-val, orn, pipecolic acid, pro, tyr (di-dns).

Maltosyl-β-cyclodextrin as the chiral mobile phase additive (CMA-6) was introduced by Duncan and Armstrong [154]. Using different RP-plates, the authors performed enantiomer separation of amino acid derivatives such as alanine- and methionine-2-naphthylamide, dansylated leu and val, underivatized val, as well as N'-(2-naphthylmethyl)nornicotine with the eluent acetonitrile-water (3:7) containing 0.4 mol maltosyl-β-cyclodextrin and 0.6 mol sodium chloride.

Albumins as CMA

Bovine serum albumin (BSA) as the chiral mobile phase additive (CMA-7) was introduced by Lepri et al. [155]. These authors compared the separation quality of different commercially available RP-18 plates such as $RP_{18}W/UV_{254}$ and Sil C_{18}-50 UV_{254} (Macherey-Nagel, Düren, FRG) as well as HPTLC $RP_{18}W/F_{254s}$ (Merck, Darmstadt, FRG) using different contents of BSA, buffers and isopropanol (cf. Table 3.8.6) for several amino acid derivatives and various other compounds [85; 155-159].

BSA is a highly enantioselective additive which yields high α-values and very good resolutions R_S (e.g., for DNP-nval $\alpha = 12.19$ and $R_S = 5.0$). BSA concentrations higher than 5 % must be avoided, since the development time increases substantially with increasing BSA concentration (e.g., 40 min without BSA becomes 120 min when the concentration of BSA is 6 %). R_f-values also increase considerably with increasing concentrations of BSA.

The effect of different pH values has also been studied. At pH 6.9 the development time was slightly longer than that observed at pH 4.7 (pI 4.7 for BSA); the solvent front was more irregular. Elution at pH 9.2 resulted in low retention of all the compounds, which made it impossible to separate the optical isomers [155].

Using different solutions of sodium tetraborate containing 6 % BSA and 12 % isopropanol (pH 9.3-9.9) [85] for trp and its derivatives (cf. Table 3.8.6), the eluent at pH 9.8 showed the best results. Later, other amino acid derivatives such as 9-fluorenylmethoxycarbonyl (Fmoc) [156], 4-nitroanilides [157], methylthiohydantoin (MTH) [157] or phenylthiohydantoin (PTH) [157] were also studied.

The PTH derivatives of *S*-met, *S*-trp and *S*-val showed racemization during the analysis (cf. Table 3.8.6) [157]. This racemization did not occur under slightly different conditions (8 % BSA at pH 4.1).

Additionally, the enantiomers of PTH-phe were resolved, which was not possible under the former conditions (7 % BSA at pH 3.5). Only one of the two different RP-18 plates used is applicable for the separation of kynurenine and 3-(1-naphthyl)-ala (only $RP_{18}W/UV_{254}$ but not Sil C_{18}-50 UV_{254}; for further conditions see Table 3.8.6) [157]. BSA shows enantioselectivity for racemates with structures completely different from those of amino acids, their derivatives, and similar compounds such as hydroxy acids, even uncharged molecules like 1,1'-bis(2-naphthol).

Variation in eluent pH can lead to a reverse of the elution order due to the effect of pH both on the conformation of the protein in solution and the charge of the solute [158]. The ionic strength of the eluent also had a notable influence on the chromatographic behaviour. The presence of 1 % sodium chloride in the mobile phase produced more compact spots and increased the retention, probably because of the reduction in the ionic interactions (electrostatic attraction) between BSA and the optical isomers, as well as the increase of the hydrophobic interactions between the amino acid derivatives and both the stationary phase and BSA [158; 159].

The stereoselectivity was also noticeably influenced by the temperature. The selectivity increased when the temperature was reduced, e.g., from 20°C to 10°C [158]. More polar amino acids (e.g., asp, ser) could not be resolved, regardless of the experimental conditions used [158]. This latter problem was solved using dansylated derivatives.

Finally, asp, glu, ser and thr were also separated with BSA as the chiral mobile phase additive [159]. Sil C_{18}-50 UV_{254} plates were found unsuitable for the separation of dansylated amino acids with BSA in the mobile phase [159]. Comparison of BSA and ß-cyclodextrin as CMAs revealed higher α-values with BSA than with β-cyclodextrin [159].

Table 3.8.6 Separation of enantiomers with bovine serum albumin (BSA) as the chiral mobile phase additive (CMA) (RP-18 plates were used) [85; 155-159].

Compounds	Der.	BSA [%]	2-Pr.[%]	Buffer/Solution	pH	Ref.
Amino acids, dipeptides and derivatives						
alanine	DNPy	4	2##	0.50 M acetic acid#	**	158
	Fmoc	6	23	0.10 M acetate	4.9*	156
	4-NA	8	3	0.10 M acetate	**	157
3-cyclohexyl-ala	Fmoc	5	23	0.10 M acetate	4.9*	156
3-(1-naphthyl)-ala	-	6	6	0.05 M tetraborate	9.3	157
2-aminobutanoic acid	DNP	4	2##	0.10 M acetate	**	158
	Dns	5	2	0.10 M acetate	4.7	159
arginine						
N-α-benzoylarginine-7-amido-4-methylcoumarin	-	6	2	0.05 M carbonate	9.8	157
aspartic acid	Dns	5	2	0.10 M acetate	4.7	159
glutamic acid	Dns	5	2	0.10 M acetate	4.7	159
glycine [chiral der.]						
phenyl-gly	DNB	5	2	0.10 M acetate	4.7	155
gly-trp	-	6	12	0.05 M tetraborate	9.8	85
N-phthalyl-gly-trp	-	3	2	0.10 M acetate	**	158
isoleucine	PTH	7	2	0.50 M acetate	3.5	157
leucine	DNB	5	2	0.10 M acetate	4.7	155
DNP		4	2	0.05 M phosphate	6.9	155
	DNPy	5	2	0.10 M acetate	4.7	155
	Dns	5	2	0.50 M acetic acid	3.4#	159
	Fmoc	5	36	0.10 M acetate	4.9*	156
	4-NA	8	20	0.10 M acetate	**	157
nleu	DNP	4	2	0.05 M phosphate	6.9	155
	DNPy	5	2	0.10 M acetate	4.7	155
	Dns	6	2	0.10 M acetate	4.7	159
	Fmoc	5	23	0.10 M acetate	4.9*	156
	2-NPS	6	6	0.10 M acetate	**	158
methionine	DNP	3	2	0.05 M phosphate	6.9	155
	DNPy	2	2	0.10 M acetate	4.7	155
	Dns	6	2	0.10 M acetate	4.7	159

Compounds	Der.	BSA [%]	2-Pr.[%]	Buffer/Solution	pH	Ref.
	Fmoc	6	12	0.10 M acetate	4.9*	156
	PTH	7	2	0.50 M acetate	3.5	157
ethionine-sulphone	DNP	4	2##	0.50 M acetic acid#	**	158
met-sulphone	DNP	5	2	0.10 M acetate	4.7	155
met-sulphoxide	DNP	5	2	0.10 M acetate	4.7	155
phenylalanine	DNPy	2	2	0.10 M acetate	4.7	155
	Dns	6	2	0.10 M acetate	4.7	159
	Fmoc	5	36	0.10 M acetate	4.9*	156
	PTH	8	2	0.50 M acetate	4.1	157
4-nitro-phe	tBoc	6	6	0.05 M carbonate	**	158
pipecolic acid	DNP	4	2##	0.10 M acetate	**	158
proline	Fmoc	5	23	0.10 M acetate	4.9*	156
	MTH	9	2	0.10 M acetate	4.9	157
serine	Dns	7	2	0.50 M acetic acid	3.4	160
threonine	Dns	7	2	0.50 M acetic acid	3.4	159
tryptophan	-	6	12	0.05 M tetraborate	9.8	85
	tBoc	8	2	0.50 M acetic acid	**	158
	CBZ	5	2	0.50 M acetic acid	**	158
	Dns	5	2	0.10 M acetate	4.7	159
	Fmoc	6	12	0.10 M acetate	4.9*	156
	PTH	7	2	0.50 M acetate	3.5	157
trp-amide	-	6	12	0.05 M tetraborate	9.8	85
N-acetyl-5-methyl-trp	-	3	2	0.05 M phosphate	**	158
4-fluoro-trp	-	6	2	0.05 M carbonate	**	158
5-fluoro-trp	-	6	2	0.05 M carbonate	**	158
6-fluoro-trp	-	6	2	0.05 M carbonate	**	158
4-methyl-trp	-	6	12	0.05 M tetraborate	9.8	85
5-methyl-trp	-	6	12	0.05 M tetraborate	9.8	85
6-methyl-trp	-	6	12	0.05 M tetraborate	9.8	85
7-methyl-trp	-	6	12	0.05 M tetraborate	9.8	85
5-methoxy-trp	-	6	12	0.05 M tetraborate	9.8	85
tyrosine	PTH	7	2	0.50 M acetate	3.5	157
valine	Fmoc	5	36	0.10 M acetate	4.9*	156
	Dns	5	2	0.10 M acetate	4.7	159
	PTH	7	2	0.50 M acetate	3.5	157
nval	DNP	4	2	0.10 M acetate	4.7	155
	DNPy	4	2##	0.50 M acetic acid#	**	158
	Dns	5	2	0.10 M acetate	4.7	159
	Fmoc	5	23	0.10 M acetate	4.9*	156
	2-NPS	6	6	0.10 M acetate	**	158
Miscellaneous						
1,1'-bis(2-naphthol)	-	5	11##	0.05 M phosphate	**	158
binaphthyl-2,2'-diyl hydrogen phosphate	-	5	2	0.50 M acetic acid#	**	158

Compounds	Der.	BSA [%]	2-Pr.[%]	Buffer/Solution	pH	Ref.
3,5-dinitro-N-(1-phenylethyl)benzamide	-	6	20	0.10 M carbonate	**	158
O-hippuryl-3-phenyllactic acid	-	6	2	0.05 M carbonate	9.8	157
β-hydrastine	-	5	10##	0.05 M carbonate	**	158
kynurenine	-	6	6	0.05 M tetraborate	9.3	157
4-nitrophenyl-β-thiofucopyranoside	-	5	11##	0.05 M phosphate	**	158
2,2,2-trifluoro-1-(9-anthryl)ethanol	-	6	20	0.05 M tetraborate	9.8	157

* = pH before addition of 2-propanol; ** = pH-value not given; # = contains 1 % of sodium chloride; ## = at 10°C (all other separations at room temperature); tBoc = N-tert-Butyloxycarbonyl; BSA = Bovine serum albumine; CBZ = N-Benzyloxycarbonyl; DNB = N-3,5-Dinitrobenzoyl; DNP = N-2,4-Dinitrophenyl; DNPy = N-Dinitropyridyl; Dns = N-Dansyl (=N-5-dimethylamino-1-naphthalenesulphonyl); Fmoc = N-9-Fluorenylmethoxycarbonyl; MTH = Methylthiohydantoin; 4-NA = N-4--Nitroanilide; 2-NPS = N-2-Nitrophenylsulphenyl; 2-Pr. = 2-Propanol; PTH = Phenylthiohydantoin

Other CMA for TLC

Tivert and Backman [161] reported the separation of propranolol, metoprolol and alprenolol with dichloromethane as eluent containing 5 mM *S*-N-benzyloxycarbonyl glycyl-proline (CMA-8) as chiral counter-ion (CCI) and 0.4 mM ethanolamine using a diol-modified silica gel plate. Duncan et al. [162] separated several pharmaceuticals (e.g., 2-amino-1-phenyl-1-propanol, pindolol, propranolol, 2-amino-1-(3-hydroxyphenyl)ethanol [norphenylephrine], isoproterenol, timolol, methoxamine) using silica gel plates or diol-modified silica gel plates with different eluent systems.

Furthermore, Duncan et al. [162] employed 1*R*-(-)-ammonium-10-camphorsulphonate (CMA-9) as an additive as well as N-benzyloxycarbonyl-derivatives (BC) of amino acids and dipeptides (e.g., *S*-N-BC-isoleucyl-proline CMA-10) for pharmaceuticals like 2-amino-1-(3-hydroxyphenyl)ethanol (norphenylephrine), 2-amino-1-(4-hydroxyphenyl)ethanol (octopamine), isoproterenol, metoprolol, propranolol and timolol with silica gel plates of diol-modified silica gel plates, using dichloromethane-methanol or dichloromethane-isopropanol with different mixtures as eluents. All these CMAs form diastereomeric ion-pairs [162].

References

[1] Geiss F. *Fundamentals of Thin Layer Chromatography (Planar Chromatography),* Hüthig: Heidelberg , (1987)

[2] Pyka A. *J. Planar Chromatogr.* (1993),*6*, 282

[3] Flood A.E., Hirst E.L., Jones J.K.N. *J. Chem. Soc.* (1948), 1679

[4] Hinman J.W., Caron L.E., Christensen H.N. *J. Am. Chem. Soc.* (1950), *72*, 1620

[5] Bonino G.B., Carassiti V. *Nature* (1951),*167*, 569
[6] Mason M., Berg C.P. *Federation Proc.* (1951),*10*, 221
[7] Fujisawa Y. Osaka Shiritsu Daigaku Igaku Zasshi (*J. Osaka City Med.Center*) (1951), 1, (C.A. 1954, 48: 13550d)
[8] Sakan T., Nakamura N., Senoh S. *J. Chem. Soc. Japan* (1951),*72*, 745
[9] Kotake M., Sakan T., Nakamura N., Senoh S. *J. Am. Chem. Soc.* (1951), *73*, 2973
[10] Sensi P. *Acta Vitaminol.* (1951), *5*, 10
[11] Dalgliesh C.E., Knox W.E., Neuberger A. *Nature* (1951), *168*, 20
[12] Kariyone T., Hashimoto Y. *Nature* (1951), *168*, 739
[13] Berlingozzi S., Serchi G., Adembri G. *Sperim. Sez. Chim. Biol.* (1951), 2, 89
[14] Dalgliesh C.E. *J. Chem. Soc.* (1952), 3940
[15] Easson L.H., Stedman E. *Biochem. J.* (1933), *27*, 1257
[16] Ogston A.G. *Nature* (1948), *162*, 963
[17] Buss D.R., Vermeulen T. *Ind. Eng. Chem.* (1968), *60*, 12
[18] Günther K. in: *Handbook of Thin-Layer Chromatography*, Sherma J., Fried B. (Eds.) Dekker: New York, (1991), p. 541
[19] Weichert R. *Ark. Kemi* (1969), *31*, 517
[20] Tomita M., Sugamoto M. Yakugaku Zasshi *J. Pharm. Soc. Jap.* (1962), *82*, 1141
[21] Franck B., Blaschke G. *Liebig's Ann. Chem.* (1966), *695*, 144
[22] Macek K., Vanétek S. *Coll. Czech. Chem. Comm.* (1959), *24*, 315
[23] Kikkawa I. Yakugaku Zasshi *J. Pharm. Soc. Jap.* (1961), *81*, 732
[24] Franck B., Schlinghoff G. *Liebig's Ann. Chem.* (1962), *659*, 123
[25] Akiyama I., Onaya M., Hayakawa A., Tsuzuki Y. *Chem. Lett.* (1972), 269
[26] Zaltzman-Nierenberg, Daly J., Guroff G., Udenfriend S. *Anal. Biochem.* (1966), *15*, 517
[27] Blaschko H., Hope D.B. *Biochem. J.* (1956), *63*, 7P
[28] De Ligny C.L., Nieboer H., De Vijlder J.J.M., Van Willigen J.H.H.G. *Recueil Trav. Chim.* (1963), *82*, 213
[29] Rhuland L.E., Work E., Denman R.F., Hoare D.S. *J. Am. Chem. Soc.* (1955), *77*, 4844
[30] Weichert R. *Acta Chem. Scand.* (1955), *9*, 547
[31] Makino K., Takahashi H. *Science* (1953), *118*, 699
[32] Lambooy J.P. *J. Am. Chem. Soc.* (1954), *76*, 133
[33] Miwa T., Ohsuka A., Sakan T. *J. Chem. Soc .Jap.* (1953), *74*, 113
[34] Contractor S.F., Wragg J. *Nature* (1965), *208*, 71
[35] Lederer M. *J. Chromatogr.* (1992), *604*, 55
[36] Roberts E.A.H., Wood D.J. *Biochem. J.* (1953), *53*, 332
[37] Mayer W., Merger F. *Liebig's Ann. Chem.* (1961), *644*, 65
[38] Alessandro A., Caldarera C.M. *Boll. Chim. Farm.* (1954), *93*, 404
[39] Butenandt A., Biekert E., Koga N., Traub P. *Hoppe Seyler's Z. Physiol. Chem.* (1960), *321*, 258
[40] Butenandt A., Biekert E., Kübler H., Linzen B., Traub P. *Hoppe Seyler's Z. Physiol. Chem.* (1963), *334*, 71
[41] El Din Awad A.M., El Din Awad O.M. *J. Chromatogr.* (1974), *93*, 393
[42] Stahl E. *Angew.Chem.* (1983), *95*, 515
[43] Sherma J., Fried B. (Eds.) *Handbook of Thin-Layer Chromatography*, Dekker: New York, (1991)

[44] Ward T.J., Armstrong D.W. in: *Chromatographic Chiral Separations*, Zief M., Crane L.J. (Eds.) Dekker: New York, (1988), p. 131

[45] Martens J., Bhushan R. *Chem. Ztg.* (1988), *112*, 367

[46] Martens J., Bhushan R. *J. Pharm. Biomed. Anal.* (1990), *8*, 259

[47] Wilson I.D., Spurwaz T.D., Witherow L., Ruane R.J., Longden K. *Recent Advances in Chiral Separations*, Stevenson D., Wainer I.D. (Eds.), Plenum: New York, (1991), p. 159

[48] Nitecki D.E., Halpern B., Westley J.W. *J. Org. Chem.* (1968), *33*, 864

[49] Jarman M., Stec. W.J. *J. Chromatogr.* (1979), *176*, 440

[50] Kemmerer J., Koechlin B., Rubio F. *Federation Proc.* (1978), *37*, 605

[51] Maître J.-M., Boss G., Testa B., Hostettmann K. *J. Chromatogr.* (1986), *356*, 341

[52] Tamura S., Kuzuna S., Kawai K., Kishimoto S. *J. Pharm. Pharmacol.* (1981), *33*, 701

[53] Rossetti V., Lombard A., Buffa M. *J. Pharm. Biomed. Anal.* (1986), *4*, 673

[54] Günther K., Rausch R. in: *Proc. 4th Int. Symp. on Instrumental High Performance Thin-Layer Chromatography, Selvino 1987*, Traitler H., Studer A., Kaiser R.E. (Ed.), Inst. Chromatogr.,: Bad Dürkheim , (1987), p. 155

[55] Sarsúnová M., Semonsky M., Cerny A. *J. Chromatogr.* (1970), *50*, 442

[56] Slégel P., Vereczkey-donáth B., Ladányi L., Tóth-Lauritz M. *J Pharm. Biomed. Anal.* (1987), *5*, 665

[57] Lan S.J., Kripalani K.J., Dean A.V., Egli P., Difazio L.T., Schreiber E.C. *Drug Metab. Dispos.* (1976), *4*, 330

[58] Comber R.N., Brouillette W.J. *J. Org. Chem.* (1987), 52, 2311

[59] Ko-odziejczyk A.M., Arendt A. *Polish J. Chem.* (1979), *53*, 1017

[60] Ruterbories K.J., Nurok D. *Anal. Chem.* (1987), *59*, 2735

[61] Heuser D., Meads P. *J. Planar Chromatogr.* (1993), *6*, 624

[62] Barooshian A.V., Lautenschleger M.J., Harris W.G. *Anal. Biochem.* (1972), *49*, 569

[63] Barooshian A.V., Lautenschleger M.J., Greenwood J.M., Harris W.G. *Anal. Biochem.* (1972), *49*, 602

[64] Eskes D. *J. Chromatogr.* (1976), *117*, 442

[65] Weber H., Spahn J., Mutschler E., Möhrke W. *J. Chromatogr.* (1984), *307*, 145

[66] Pflugmann G., Spahn H., Mutschler E. *J. Chromatogr.* (1987), *416*, 331

[67] Beneytout J.L., Tixier M. *J. Chromatogr.* (1986), *351*, 363

[68] Spahn H. *J. Chromatogr.* (1988), *427*, 131

[69] Büyüktimkin N., Buschauer A. *J. Chromatogr.* (1988), *450*, 281

[70] Nishi H., Ishii K., Taku K., Shimizu R., Tsumagari N. *Chromatographia* (1989), *27*, 301

[71] Freytag W., Ney K.H. *J. Chromatogr.* (1969), *41*, 473

[72] Gübitz G., Mihellyes S. *J. Chromatogr.* (1984), *314*, 462

[73] Martin E., Spahn H., Mutschler E. in: *Chiral Separations*, Stevenson D., Wilson I.D. (Eds.) Plenum: New York, (1988), p. 185

[74] Yuasa S., Shimada A. *Science Rep. College Gen. Educ.*, Osaka Univ. 1982, 13

[75] Yuasa S., Shimada A., Isoyama M., Fukuhara T., Itoh M. *Chromatographia* (1986), 79

[76] Mack M., Hauck H.E. *Chromatographia* (1988), *26*, 197

[77] Kuhn A.O., Lederer M., Sinibaldi M. *J. Chromatogr.* (1989), *469*, 253

[78] Munier R.L., Drapier A.-M., Gervais C. *C. R. Acad Sci. Paris* D (1976), *282*, 1761

[79] Faupel M. in: *Proc. 4th Int. Symp. on Instrumental High Perforformance Thin-Layer Chromatography, Selvino 1987,* Traitler H., Studer A., Kaiser R.E. (Ed.), Inst.Chromatogr.: Bad Dürkheim, (1987), p. 147

[80] Hesse G., Hagel R. *Chromatographia* (1973), *6*, 277

[81] Chimiak A., Potofski T. *J. Chromatogr.* (1975), *115*, 635

[82] Bach K., Haas H.J. *J. Chromatogr.* (1977), *136*, 186

[83] Fukuhara T., Isoyama M., Shimada A., Itoh M., Yuasa S. *J. Chromatogr.* (1987), *387*, 562

[84] Kieu H.T., Lederer M. *J. Chromatogr.* (1993) ,*635*, 346

[85] Lepri L., Coas V., Desideri P.G., Zocchi A. *J. Planar Chromatogr.* (1992), 234

[86] Wang K.T., Chen S.T., Lo L.C. *Fresenius Z. Anal. Chem.* (1986), *324*, 338

[87] Wilson I.D. in: *Bioactive Analytes, Including CNS Drugs, Peptides, and Enantiomer,* Reid E., Scales B., Wilson I.D. (Eds.), (1986), p. 277

[88] Ref. 89 taken from [18]

[89] Busker E., Martens J., Weigel H. DE 31 43 726 (Cl.C07D207/16), 11 May 1983

[90] Günther K., Martens J., Schickedanz M. DE 33 28 348 (Cl.C07C99/12), 14 Feb 1985

[91] Günther K., Martens J., Schickedanz M. *Angew. Chem.* (1984), *96*, 514

[92] Mack M., Hauck H.E., Herbert H. *J. Planar Chromatogr.* (1988), *1*, 304

[93] Helferich F. *Nature* (1961), *189*, 1001

[94] Rogozhin S.V., Davankov V.A. DE 19 32 190 (Cl.B01d), 08 Jan 1970 (C.A. 1970, 72: 90875c)

[95] Davankov V.A., Bochkov A.S., Kurganov A.A. *Chromatographia* (1980), *13*, 677

[96] Grinberg N., Kalász H., Han S.M., Armstrong D.W. in: *Modern Thin-Layer Chromatography,* Grinberg N. (Ed.) Dekker: New York, (1990), p. 313

[97] Ref. 31 taken from [45]

[98] Nyiredy S., Dallenbach-Toelke K., Stichler O. *J. Chromatogr.* (1988), *450*, 241

[99] Brinkman U.A.T., Kamminga D. *J. Chromatogr.* (1985), *330*, 375

[100] Rausch R. in: *Recent Advances in Thin-Layer Chromatography,* Dallas F.A.A., Read H., Ruane R.J., Wilson I.D. (Eds.), Plenum: New York, (1988), p. 151

[101] Günther K. *J. Chromatogr.* (1988), *448*, 11

[102] Günther L., Rausch R. in: *Proc. 3rd Int. Symp. on Instrumental High Performance Thin-Layer Chromatography, Würzburg 1985,* Kaiser R.E. (Ed.), Inst. Chromatogr.: Bad Dürkheim, (1985), p. 469

[103] Brückner H. *Chromatographia* (1987), *24* ,725

[104] Günther K., Schickedanz M. *Naturwissenschaften.* (1985), *72*, 149

[105] Neuzil E., Lacoste A.M., Denois F. *Biochem. Soc. Trans.* (1986), *14*, 642

[106] Günther K. *GIT Fachz. Lab. Suppl.* (1986), 6

[107] Günther K., Schickedanz M., Drauz K., Martens J. *Fresenius Z. .Anal. Chem.* (1986), *325*, 298

[108] Toth G., Lebl M., Hruby V.J. *J. Chromatogr.* (1990), *504*, 450

[109] Euerby M.R. *J. Chromatogr.* (1990), *502*, 226

[110] Mack M., Hauck H.E. in: *Proc. Int. Symp. on Instrumental Thin-Layer Chromatography / Planar Chromatography, Brighton 1989,* Kaiser R.E. (Ed.), Inst. Chromatogr.: Bad Dürkheim, (1989), p. 141

[111] Martens J., Lübben S. *Tetrahedron* (1990), *46*, 1231

[112] Günther K., Martens J., Schickedanz M. *Fresenius Z. Anal. Chem.* (1985), *322*, 513

[113] Martens J., Günther K., Schickedanz M. *Arch. Pharm.* (1986), *319*, 572

[114] Günther K., Martens J., Schickedanz M. *Angew. Chem.* (1986), *98*, 284
[115] Vriesema B.K., ten Hoeve W., Wynberg H., Kellogg R.M., Boesten W.H.J., Meijer E.M., Schoemaker H.E. *Tetrahedron Lett.* (1986), *26*, 2045
[116] Brückner H., Bosch I., Graser T., Fürst P. *J. Chromatogr.* (1987), *395*, 567
[117] Gont L.K., Neuendorf S.K. *J. Chromatogr.* (1987), *393*, 343
[118] Calet S., Urso F., Alper H. *J. Am. Chem. Soc.* (1989), *111*, 931
[119] Ref. 129 taken from [18]
[120] Kovács-Hadady K., Kiss I.T. *Chromatographia* (1987), *24*, 677
[121] Weinstein S. *Tetrahedron Lett.* (1984), *25*, 985
[122] Marchelli R., Virgili R., Armani E., Dossena A. *J.Chromatogr.* (1986), *355*, 354
[123] Remelli M., Piazza R., Pulidori F. *Chromatographia* (1991), *32*, 278
[124] Sinibaldi M., Messina A., Girelli A.M. *Analyst* (1988),*113*, 1245
[125] Pirkle W.H., Finn J.M., Schreine J.L., Hamper B.C. *J. Am. Chem. Soc.* (1981), *103*, 3964
[126] Wainer I.W., Brunner C.A., Doyle T.D. *J. Chromatogr.* (1983), *264*, 154
[127] Wall P.E. in: *Proc. Int. Symp. on Instrumental Thin-Layer Chromatography / Planar Chromatography, Brighton 1989,* Kaiser R.E. (Ed.), Inst. Chromatogr.: Bad Dürkheim, (1989), p. 237
[128] Wall P.E. *J. Planar Chromatogr.*(1989), *2*, 228
[129] Armstrong D.W. US 4,539,399 (Cl.B01D15/08), 03 Sep 1985
[130] Alak A., Armstrong D.W. *Anal. Chem.* (1986), *58*, 582
[131] Bhushan R., Ali I. *J. Chromatogr.* (1987),*392*, 460
[132] Bhushan R., Ali I. *Chromatographia* (1993), *35*, 679
[133] Bhushan R., Ali I. *Chromatographia* (1987), *23*, 141
[134] Bhushan R. *J. Liq. Chromatogr.* (1988), *11*, 3049
[135] Bhushan R., Ali I. *Fresenius Z. Anal. Chem.* (1988), *329*, 793
[136] Bhushan R. *Fresenius Z. Anal. Chem.* (1989), *333*, 144
[137] Ti N. Farumashia 1985, 21, 747 (C.A. 1985 ,103, 226700k) taken from [18]
[138] Brunner C.A., Wainer I. *J. Chromatogr.* (1989), *472*, 277
[139] Martens J., Bhushan R., Lübben S. in: *Proc. Int. Symp. on Instrumental Thin-Layer Chromatography / Planar Chromatography, Brighton 1989,* Kaiser R.E. (Ed.), Inst. Chromatogr.: Bad Dürkheim, (1989), p. 155
[140] Ró`y-o J.K., Malinowska I. *J. Planar Chromatogr.* (1993), *6*, 34
[141] Paris R.R., Sarsúnová M., Semonsky M. *Ann. Pharm. Fr.* (1967), *25*, 177
[142] Sarsúnová M., Semonsky M. CZECH 128,028 (Cl.C07d), 30 Nov 1966
[143] Bauer K., Falk H., Schlögl K. *Monatsh. Chem.* (1968), *99*, 2186
[144] Yoneda H., Baba T. *J. Chromatogr.* (1970), *53*, 610
[145] Armstrong D.W. *J. Liq. Chromatogr.* (1980), *3*, 895
[146] Armstrong D.W. in: *Proc. 4th Int. Symp. on Cyclodextrins, München 1988,* Huber O., Szejtli J. (Eds.), Kluwer: Dordrecht, (1988), p. 437
[147] Armstrong D.W., Faulkner Jr. J.R., Han S.M. *J.Chromatogr.* (1988), *452*, 323
[148] Duncan J.D., Armstrong D.W. *J. Planar Chromatogr.* (1991), *4*, 204
[149] Armstrong D.W., He F.-Y., Han S.M. *J. Chromatogr.* (1988), *448*, 345
[150] Lepri L., Coas V., Desideri P.G., Checchini L. *J. Planar Chromatogr.* (1990), 311
[151] Lepri L., Coas V., Desideri P.G. *J. Planar Chromatogr.* (1990), 533

[152] Lepri L., Coas V., Desideri P.G. *J. Planar Chromatogr.* (1991), 338

[153] LeFevre J.W. *J.Chromatogr.* (1993), *653*, 293

[154] Duncan J.D., Armstrong D.W. *J. Planar Chromatogr.* (1990) ,*3*, 65

[155] Lepri L., Coas V., Desideri P.G. *J. Planar Chromatogr.* (1992), 175

[156] Lepri L., Coas V., Desideri P.G. *J.Planar Chromatogr.* (1992), 294

[157] Lepri, L, Coas V., Desideri P.G, Pettini L. *J. Planar Chromatogr.* (1992), 364

[158] Lepri L., Coas V., Desideri P.G., Pettini L. *J. Planar Chromatogr.* (1993), 100

[159] Lepri L., Coas V., Desideri P.G., Santianni D. *Chromatographia* (1993), 297

[160] Remelli M., Fornasari P., Dondi F., Pulidori F. in: *3rd Int. Symp. on Chiral Discrimination, Tübingen 1992,* Schurig V. (Ed.), Univ. Tübingen, (1992)

[161] Tivert A.-M., Backman A. in: *Proc. Int. Symp. on Instrumental Thin-Layer Chromatography / Planar Chromatography, Brighton 1989,* Kaiser R.E. (Ed.), Inst. Chromatogr.: Bad Dürkheim, (1989), 225

[162] Duncan J.D., Armstrong D.W., Stalcup A.M. *J. Liq. Chromatogr.* (1990), *13*, 1091

3.9 Other methods

3.9.1 Counter-Current Chromatography

Few articles dealing with enantiomer separation by means of counter-current chromatography (CCC) can be found in the literature. Two main techniques of CCC have been applied for this purpose; (i) droplet counter-current chromatography (DCCC or DCC); and (ii) rotation locular counter-current chromatography (RLCCC or RLCC).

The separation principle of DCCC is based on the partition (partition chromatography) of the sample between a current of droplets of the mobile phase and a liquid stationary phase filled into a column. DCCC grew out of the observation that a light phase with a low wall-surface affinity formed discrete droplets that rose through the heavy phase with visible evidence of very active interfacial motion. Under ideal conditions each droplet could become a 'plate', if kept more or less discrete throughout the system. A DCCC instrument consists basically of 200 to 600 long vertical columns (length = 20-60 cm) of narrow-bore glass silanized tubing (1.5-2 mm i.d.) interconnected in series by capillary PTFE tubes. After filling the whole system with stationary phase a sample can be injected into a sample chamber. The mobile phase is then pumped through the sample chamber and inserted with the capillary tube into the bottom of the first glass column of wider bore. A steady stream of ascending droplets is formed. When a droplet reaches the top of the column, it is transferred to the bottom of the next column through the PTFE tubing, thus generating new droplets. Under suitable conditions, the capillary tubing allows only the mobile phase to flow. A small amount of stationary phase may also enter the PTFE tubing initially, but the effect is negligible. As the mobile phase moves through the column, turbulance within the droplet promotes efficient partitioning of the solute between the two phases. Depending on the separation problem, the mobile phase may be either heavier or lighter than the stationary phase. When lighter, the mobile phase is transferred to the bottom of the column (ascending mode) and, when heavier, the mobile phase is transferred to the top of the column (descending mode). Every solvent mixture that yields two immiscible layer phases can, in principle, be used for DCCC, but the formed droplets have to be smaller than the inner diameter of the column in order to pearl through the stationary liquid phase [1].

The other technique, RLCCC, is performed with an apparatus containing glass columns that are segmented into loculi by perforated PTFE disks. The columns are interconnected with capillary PTFE tubing. The whole system is inclined from the horizontal at an angle of 25-40°. The columns are filled with the specific heavier phase (= stationary phase). The sample is introduced with the specific lighter phase (= mobile phase) at the bottom of the first column. The light phase displaces the heavy phase up to the height of the capillary opening in the disk, then the light phase flows to the next loculus. This process is repeated along the whole column. Every loculus is kept half full with the heavy phase, while the light mobile phase is pumped continuously through the whole system. Due to the separation problem the phases can be reversed, i.e. the lighter phase is used as the stationary phase and the heavier phase as the eluant [2].

Preliminary investigations by Bowman et al. [3] in 1968 showed that partial resolution of organic racemates, e.g. camphoric acid, could be obtained by partitioning the racemates between an aqueous phase and an optically active ester of (-)-tartaric acid (e.g., diisopentyl tartrate). Further developments were described by Prelog et al. [4] in 1982. Salts of enantiomeric α-amino alcohols like norephedrine with lipophilic anions like hexafluorophosphate could be separated by partition between an aqueous and a lipophilic phase containing esters of tartaric acid.

The first enantiomer separation using CCC was reported by Domon et al. [5] in 1982. The authors were able to separate a racemic mixture of norephedrine with an RLCCC instrument equipped with 16 columns (45 cm x 11 mm i.d.) divided by centrally perforated PTFE disks into 37 loculi each. The separation was carried out at 2-3°C in the descending mode with a flow-rate of 17-20 ml/h (final eluate volume, 1650 ml) and a rotation speed of 60-70 rpm (40° slope). The stationary phase was an aqueous solution of 0.5 mol sodium hexafluorophosphate (pH 4). A solution of 0.3 mol *R,R*-di-5-nonyl tartrate in 1,2-dichloroethane was used as mobile phase. Although no baseline separation was achieved, practically pure enantiomers were obtained by the RLCC technique. The authors expected that a complete resolution could be achieved using an apparatus with more loculi [5].

The resolution of isoleucine and other amino acid enantiomers by means of DCCC was reported by Takeuchi et al. [6] in 1984. The authors used a DCCC instrument equipped with a column made by interconnecting 400 pieces of PTFE tubing of two sizes, 40 cm x 4 mm and 50 cm x 1 mm i.d., alternatively in series with a flow of 1.1 ml/min (descending mode; final eluate volume, 4400 ml). The stationary and mobile phases used were 1-butanol containing 2 mmol N-dodecyl-*S*-proline (**1**), and acetate buffer (pH 5.5) containing 1 mmol copper(II) acetate, respectively.

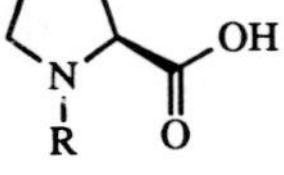

(R = dodecyl)

1 **2** 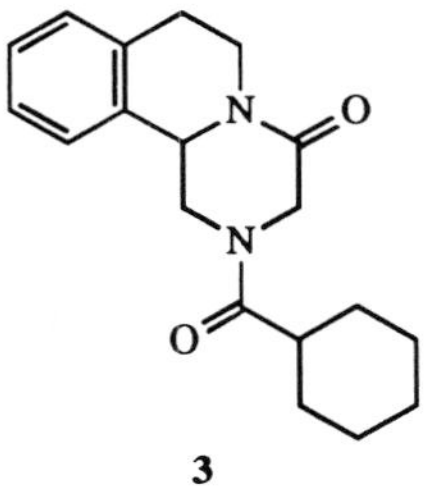 **3**

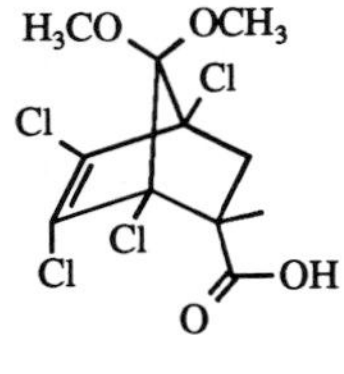

The DCCC baseline separation of a halocarboxylic acid (**2**) was reported by Oka and Snyder [7] in 1986. The instrument was equipped with 225 glass columns (40 cm x 1.9 mm i.d., 9 racks of columns with 25 columns per rack), interconnected in series by PTFE tubing (0.5 mm i.d.). The flow-rate was 10 ml/h (final eluate volume, 400 ml) using the descending mode. The stationary phase was a mixture of chloroform-methanol-water (7:13:8). The mobile phase was a 0.01 mol phosphate buffer (pH 7) containing 0.1 mol *R*-(-)-2-amino-1-butanol.

In a particular case of CCC the technique is applied continuously. While Ching et al. [8] described the preparative resolution of praziquantel (**3**) enantiomers using a liquid-solid system with microcrystalline cellulose triacetate as the stationary phase [8], Watabe [9] reported on a gas-liquid system for the enantiomer separation of organic compounds.

Sato et al. [10] applied this technique for N-trifluoroacetyl-α-amino acid isopropyl esters. N-hexadecanoyl-*S*-valine-tert-butylamide or N-hexadecanoyl-*S*-leucine-tert-butylamide were used as chiral liquid phases which were diluted in an achiral hydrocarbon phase (see also GC, Section 3.5). For technical details on continuous counter-current gas-liquid chromatography (CCGLC) see Ref. [11-13].

Recently, industrial technologists have developed an enantioselective counter-current extraction technique capable of scaling up to ton quantities [14]: In one experiment, two heptane solutions in a cell flow parallel to one another, in opposite directions, separated by a water-saturated cellulose. One heptane solution is 10% in dihexyl D-tartrate and the other is 10% in dihexyl L-tartrate. Racemic norephedrine is added to one of the solutions. Heptane solvent and dihexyl tartrate are hydrophobic and cannot cross the water-saturated membrane. The fairly polar norephedrine crosses the membrane back and forth. As liquid flow continues, each enantiomer of norephedrine is concentrated on the side of the membrane containing the tartrate enantiomer with which it has a greater affinity.

References

[1] Hostettmann K. *Planta Medica* (1980), *39*, 1

[2] Hostettmann K., Hostettmann M. *GIT Fachz.Lab.* (1981) *(Suppl.Chromatogr.)*, 22

[3] Bowman N.S., McCloud G.T., Schweitzer G.K. *J. Am. Chem. Soc.* (1968), *90*, 3848

[4] Prelog V., Stojanac Z.., Kovacevic K. *Helv.Chim.Acta* (1982), *65*, 377

[5] Domon B., Hostettmann K., Kovacevic K., Prelog V. *J. Chromatogr.* (1982), *250*, 149

[6] Takeuchi T., Horikawa R., Tanimura T. *J. Chromatogr.* (1984), *284*, 285

[7] Oka S., Snyder J.K. *J. Chromatogr.* (1986), *370*, 333

[8] Ching C.B., Lim B.G., Lee E.J.D., Ng C. *J.Chromatogr.* (1993), *634*, 215

[9] Watabe K. JP 01 09,941 [89 09,941] (Cl.C07B57/00), 13 Jan 1989 (C.A. 1989,111: 56601z

[10] Sato K., Watabe K., Ihara T., Hobo T. in: *3rd Int. Symp. Chiral Discrimination, Tübingen 1992,* Schurig V. (Ed.), Tübingen, 1992

[11] Watabe K., Kanda H., Sato K., Hobo T. *J.Chromatogr.* (1992), *590*, 289

[12] Sato K., Watabe K., Ihara T., Hobo T. *J.Chromatogr.* (1993), *629*, 291

[13] Sato K., Motokawa O., Watabe K., Ihara T., Hobo T. *Separ .Sci. Technol.* (1993), *28*, 1409

[14] Stinson S. C., *Chem. Eng. News* (1994), *40*, 38

3.9.2 *'Pseudo'-Racemates*

In addition to the already-described methods with chiral auxiliaries, using either chiral derivatization agents (CDA) or chiral stationary phases (CSP), another gas chromatographic method can be used for the resolution of enantiomers, i.e. the preparation of 'pseudo'-racemates with subsequent achiral GC-MS analysis. This technique is mainly used for studies of metabolic pathways of drugs and pharmaceuticals.

The method was described for the first time by Gal et al. [1] in 1975. The authors synthesized the single enantiomers of 2-amino-1-(2,5-dimethoxy-4-methylphenyl)-propane as well as the enantiomers of the hexadeutero analogue (perdeuteriated in the methoxy groups). A 'pseudo'-racemic mixture of the *R*-amine and the *S*-$^{2}H_{6}$-amine was applied to rats. After metabolization in the rat liver microsomes, the amines were N-hydroxylated. These metabolites were transformed into N,O-bis(pentafluoropropionyl) derivatives and then analyzed by gas chromatography-chemical ionization mass spectrometry (GC-CIMS) using selected ion monitoring (SIM). The enantiomeric excess (ee) can be determined, for instance, using the height ratio of both molecular peaks (m/e = 417 and 423 for the non-deuterated and the deuterated amine, respectively).

Subsequently, several further applications using this deuterium labelling technique were reported; they are outlined in Table 3.9.1. Nowadays, this technique is being more and more replaced by direct methods, because there is a larger choice of chiral stationary phases for GC and HPLC.

Table 3.9.1 Applications of deuterium labelling.

Compound	No. of ^{2}H-atoms	Author	Year	Ref.
Propranolol	2	Ehrsson	1976	2
Propoxyphene	2	McMahon/Sullivan	1976	3
Amphetamine	3	Gal/Wright/Cho	1976	4
Methadone	3, 5 or 8	Nakamura et al.	1982	5
4'-Methylthiopropranolol	6	Easterling et al.	1982	6
Propranolol	2 or 5	Walle et al.	1983	7
Prenylamine	2	Schmidt et al.	1992	8

References

[1] Gal J., Gruenke L.D., Castagnoli Jr. N. *J. Med. Chem.* (1975), *18*, 683

[2] Ehrsson H. *J. Pharm. Pharmac.* (1976), *28*, 662

[3] McMahon R.E., Sullivan H.R. *Res. Commun. Chem. Pathol. Pharmacol.* (1976), *14*, 631

[4] Gal J., Wright J., Cho A.K. *Res. Commun. Chem. Pathol. Pharmacol.* (1976), *15*, 525

[5] Nakamura K., Hachey D.L., Kreek M.J., Irving C.S., Klein P.D. *J. Pharm. Sci.* (1982), *71*, 40

[6] Easterling D.E., Bai S.A., Walle U.K., McCarthy J.R., Walle T. *Fed. Proc. (Abstr.)* (1982), *41*, Abstr. 6227, 1336

[7] Walle T., Wilson M.J., Walle U.K., Bai S.A. *Drug Metab. Dispos.* (1983), *11*, 544
[8] Schmidt E.K., von Unruh G.E., Paar W.D., Dengler H.J. *Biol. Mass Spectrom.* (1992), *21*, 103

3.9.3 *Immunological methods*

An area of the analysis of chiral compounds that has received relatively little attention until recently is that of stereospecific immunoassay [1], although the pioneering studies of Landsteiner and van der Scheer showed that antibodies could distinguish between the single enantiomers of 2-[N-(4-aminobenzoyl)amino]-2-phenylacetic acid [2] or among the three possible tartaric acid stereoisomers [3]. For more detailed information about immunological methods the special literature [4] is recommended.

Nearly fifty years later, in 1976, Kawashima et al. [5] developed a radio-immunoassay (RIA) technique that was able to distinguish between the propranolol enantiomers. Cook et al. [6-9] introduced other stereoselective RIAs against several barbituric acid derivatives [6,7], warfarin derivatives [8] and antimalarial drugs [9]. Cross-reactivity of the chiral antibodies with the 'wrong' enantiomer in all these studies was found to be less than 3 %, provided that the stereochemical purity of the drug used to prepare the antigen was sufficiently high. Small impurities in the antigen may result in disproportionately large errors in the assay, as has been shown with warfarin. Cross-reactivity with metabolites may also result in problems when the site of metabolism is far removed from the chiral centre in the drug, as is the case, e.g., with 4-hydroxywarfarin [8].

An advantage of chiral immunoassays is that antisera developed for the analysis of the stereoisomers of one drug may be used for the determination of the stereoisomers of a related compound. Thus, antisera developed for the RIA of *S*-pentobarbital [7] showed a high cross reactivity (> 80 %) with *S*-secobarbital [6], but only a 2 % cross-reactivity with its *R*-antipode.

It has to be taken into consideration that racemic haptens may yield two enantioselective antibodies which cannot be assumed to be present in equal proportions. The use of such antisera could thus produce data of limited value [9]. Further stereoselective RIAs were, for instance, applied by Findlay et al. for pseudoephedrine [10] and by Midha et al. for ephedrine [11]. Sahui-Gnassi et al. [12] used a selective antibody to *S*-propranolol for an enzyme-linked immunosorbent assay (ELISA). More applications are outlined in an extensive review [4].

Besides the above-mentioned immunoassays, a stereospecific radioreceptor assay was applied by Nahorski et al. [13] for propranolol enantiomers. Radioreceptor analysis is dependent on the ability of a drug and any active metabolites to compete with a radio-labelled ligand for preparations of its biological receptors, which act as chiral surfaces. Although such assays are problematical in terms of the receptor systems employed, they do allow chiral discrimination with minimal sample preparation [1].

References

[1] Hutt A.J., Caldwell J. in: *Metabolism of Xenobiotics*, Gorrod J.W., Oelschlager H, Caldwell J. (Eds.), Taylor & Francis: London, 1988

[2] Landsteiner K., van der Scheer J. *J. Exper. Med.* (1928), *48*, 315

[3] Landsteiner K., van der Scheer J. *J. Exper. Med.* (1929), *49*, 407

[4] Cook C.E. in: *Drug Stereochemistry - Analytical Methods and Pharmacology*, Wainer I.W., Drayer D.E. (Eds.), Dekker: New York, 1988, p. 45

[5] Kawashima K., Levy A., Spector S. *J. Pharmacol. Exper. Therap.* (1976), *196*, 517

[6] Cook C.E., Myers M.A., Tallent C.R., Seltzman T., Jeffcoat A.R. *Fed. Proc. (Abstr.)* (1979), *38*, Abstr. 2713, 742

[7] Cook C.E., Seltzman T.P., Tallent C.R., Lorenzo B., Drayer D.E. *J. Pharmacol. Exper. Therap.* (1987), *241*, 779

[8] Cook C.E., Ballentine N.H., Seltzman T.B., Tallent C.R. *J. Pharmacol. Exper. Therap.* (1979), *210*, 391

[9] Cook C.E., Seltzman T.P., Tallent C.R., Wooten J.D. *J. Pharmacol. Exper. Therap.* (1982), *220*, 568

[10] Findlay J.W.A., Warren J.T., Hill J.A., Welch R.M. *J. Pharm.Sci.* (1981), *70*, 624

[11] Midha K.K., Hubbard J.W., Cooper J.K., Mackonka C. *J. Pharm.Sci.* (1983), *72*, 736

[12] Sahui-Gnassi A., Pham-Huy C., Galons H., Warnet J.-M., Claude J.-R., Duc H.-T. *Chirality* (1993), *5*, 448

[13] Nahorski S.R., Batta M.I., Barnett D.B. *Eur. J. Pharmacol.* (1978), *52*, 393

3.9.4 *Electrodes, membranes and sensors*

A new field of stereospecific analysis is the use of enantioselective electrodes, membranes and sensors (optodes). In 1987, Shinbo et al. [1] prepared polymeric membrane electrodes based on the chiral crown ether 2,3:4,5-bis [1,2-(3-phenyl-naphtho)]-1,6,9,12,15,18-hexaoxacycloeicosa-2,4-diene. The electrodes showed good

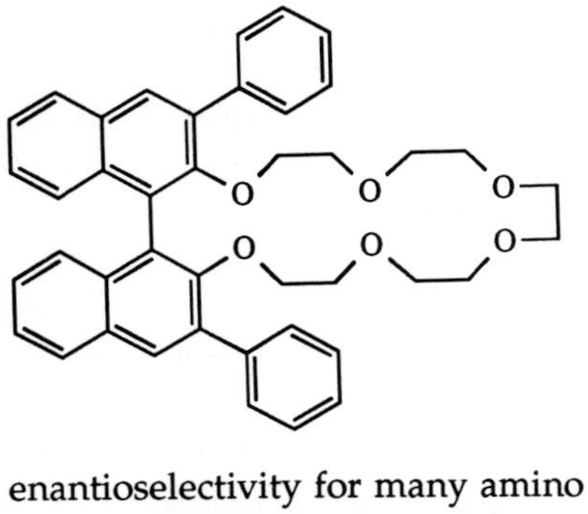

enantioselectivity for many amino acid methyl esters; they responded preferably to the enantiomer that formed the more stable complex with the crown ether.

He et al. [2] presented a scheme for sensing optical isomers of biogenic amines like propranolol. Recognition of one of the propranolol enantiomers is accomplished by specific interaction of the amine with an optically active substrate like dibutyl tartrate in a PVC membrane. Propranolol, which is present in the protonated ammonium form at physiological pH value, is carried into the membrane and a proton is simultaneously released from the proton carrier (a lipophilic phenolic xanthene dye which undergoes protolytic dissociation in the membrane), which thereby suffers a colour change. The sensor responded to propranolol, but also to other biogenic amines, e.g., 1-phenylethylamine or norephedrine in the 20 μM to 10 mM range, but it has a pH-dependent response.

Bates and co-workers [3] used a peroctylated α-cyclodextrin as a potentiometric ion-selective electrode to measure enantiomeric purity of ephedrine in the presence of serum cations. The optically active sensor was prepared by incorporating the peroctylated α-cyclodextrin into solvent polymeric membranes. These electrodes exhibited high enantioselectivity in binding ephedrine and also pseudoephedrine [4].

A new principle for a calorimetric biosensor for the determination of enantiomeric excess (ee) was reported by Hundeck et al. [5] in 1993. Two enzymes, immobilized on polymer resins, were used in the system. One of them hydrolysed only one enantiomer, while the other converted both. It was thus possible to determine the concentration of the analyte (e.g. phenylalanine methyl ester), as well as the ratio of the enantiomers. The sensor can be used in aqueous and organic phases. Other enantioselective enzyme electrodes, e.g., for lactic or malic acid, have also been described [6].

References

[1] Shinbo T., Yamaguchi T., Nishimura K., Kikkawa M., Sugiura M. *Anal. Chim. Acta* (1987), *193*, 367

[2] He H., Uray G., Wolfbeis O.S. *Proc. SPIE Int. Soc. Opt. Eng.* (1990), 1368 (Chem., Biochem., Environ. Fiber Sens. II), 175

[3] Bates P.S., Kataky R., Parker D. *J. Chem. Soc. Chem. Commun.* (1992), 153

[4] Kataky R., Bates P.S., Parker D. *Analyst* (1992), *117*, 1313

[5] Hundeck H.G., Weiß M., Scheper T., Schubert F. *Biosensors & Bioelectronics* (1993), *8*, 205

[6] Mizutani F., Yabuki S., Katsura T. *Anal. Sci. Suppl.* (1991), *7*, 871

ANNEX

List of chiral substanes analyzed by the treated techniques

Chiroptical methods

Nuclear magnetic resonance

Liquid chromatography

Gas chromatography

Supercritical fluid chromatography

Electrophoresis

Planar chromatography

Counter-current chromatography

'Pseudo racemates'

Immunological methods

Electrodes, membranes, sensors

INDEX